The Ring-Net Fishermen

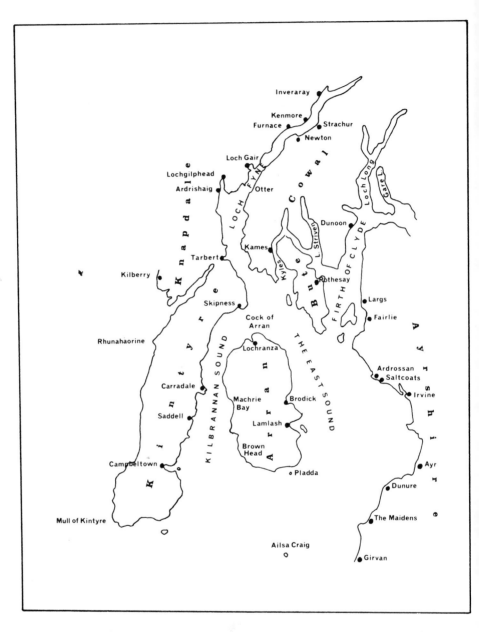

Frontispiece. Map of Lochfyne, Kilbrannan Sound, Firth of Clyde and Ayrshire Coast, showing principal fishing communities

The Ring-Net Fishermen

ANGUS MARTIN

With Drawings by Will Maclean

JOHN DONALD PUBLISHERS LTD.
EDINBURGH

This book is dedicated, with love, to the non-human creatures of Earth, especially to those trees whose regrettable deaths have made possible its production.

ISBN 0 85976 064 2

The publishers acknowledge the financial assistance of the Scottish Arts Council in the publication of this volume.

Printed by Bell & Bain Ltd., Glasgow
Phototypesetting by Wright Printers Ltd., Dundee

Preface

THIS book had its origin in a programme of tape-recording in the fishing communities of Tarbert, Campbeltown, and Carradale, begun in April, 1974. The initial stimulus to activity was provided by the artist Will Maclean, who was by then already engaged in researching the history of ring-netting. His concern was primarily with visual documentation, and the result was an exhibition of photographs, plans, and drawings presented in Glasgow at the beginning of 1978, and now the property of the Scottish National Gallery of Modern Art, Edinburgh.

My own fascination with ring-netting was fostered in early childhood within a family which had engaged in the method for four generations. When I finished with school in 1967, it was to ring-net fishing I went, thus representing the fifth − and probably the final − generation of my family active in the fishery.

This book could have been better had I worked longer at it, but my involvement has been perhaps too passionate, and, as with most passionate involvements, had finally to be retreated from.

I have tried to render accurately in this book the speech forms of Kintyre and Ayrshire as I heard − and recorded − them in the homes of the fishermen. If their appearance in cold print should offend any persons, particularly among those to whom quotes are attributed, I can plead only that by transforming them into 'standard' English an absurd falsification would have been committed, damaging to the essential character of the book. Gaelic and Scots terms have likewise been incorporated without reservation.

My greatest debt is to the retired fishermen of Kintyre and Ayrshire, many of whom are now gone. Without exception, they answered my many questions with a patience and courtesy which transcended the − as must often have seemed to them − naivety of my interest. I shall forget none of them, and not solely because they supplied the needs of this book.

The grand old man of Tarbert, Hugh McFarlane, who died on 3 June, 1979, at the age of 94 years, alone I am compelled to mention. His importance to this work could not be over-stressed. His knowledge, dignity, warmth, and irrepressible cheerfulness impressed me always. Many a fine *ceilidh* I had with him, and in the final reckoning of human values, perhaps no better words could be said of any man.

My gratitude is extended to: John Tuckwell, of John Donald Publishers Ltd., for his patient direction and encouragement, without which this work would probably not have been completed; Will Maclean, fondly remembering the passage we worked together; Cathy Wilson, Islay and London, and Lachie Paterson, Carradale, for their varied support, trustingly given; Andy and Margie Alexander, The Maidens, Ayrshire, and Donald and Cathy Macleod, Castleton Farm, Lochfyne-side, for their hospitality and assistance while collecting in these areas; George Campbell Hay, Edinburgh, for having made available to me his knowledge of the traditions and languages of his native Tarbert; Professor Derick Thomson and Kenneth MacDonald, Department of Celtic, University of Glasgow, for assistance in identifying and explaining difficult Gaelic terms − to the latter I am particularly indebted for his efforts in the transcription of tape-recorded Gaelic; Eric R. Cregeen, School of Scottish Studies, University of Edinburgh, for his advice and encouragement; Ian A. Fraser, School of Scottish Studies, and Rev. Donald Mackenzie, Auchterarder, for advice on the Gaelic elements of Kintyre place-names.

1. Pre-eminent among Kintyre folk-historians of the fishing industry, Tarbert's Hugh McFarlane, pictured six years before his death in 1979. Photograph by Alex Coupar, Spanphoto, Dundee

For checking technical chapters: Andrew Alexander, Matthew Sloan, and John T. McCrindle, all of The Maidens, Ayrshire (ring-netting); John Ritchie, Fraserburgh, and Robert Bruce, Boddam, Peterhead (line-fishing); William Whyte and Fred Stephen, both of Fraserburgh (drift-netting); Captain Hugh McShannon, Campbeltown, Duncan McSporran, Dalintober, Roger Finch, Ipswich, and − in particular − Robert Smith, Museum of Country Life, Auchindrain, Argyll (sailing and ballasting).

Collectively, to the staffs of Campbeltown and Ayr fishery offices; the Scottish Record Office, West Register House, Edinburgh; the School of Scottish Studies, University of Edinburgh; Campbeltown Public Library; the Mitchell Library, Glasgow; the National Library of Scotland; the Herring Industry Board; the Marine Laboratory, Department of Agriculture and Fisheries, Aberdeen; the Record Office, House of Lords; and the *Courier* Office, Campbeltown.

For varied assistance, to J. M. Steven, Edinburgh; Andrew Noble, Fraserburgh; W. R. P. Bourne, Department of Zoology, University of Aberdeen; Dr. G. A. Boxshall, British Museum (Natural History); J. K. Bates, Scottish Record Office; Commander R. C. Burton, R.N., Naval Historical Branch, Ministry of Defence; Neil McKellar, White Fish Authority, Edinburgh; Murdo Macdonald, archivist, Argyll and Bute District Council, Lochgilphead; W. P. Miller, James N. Miller & Sons, boat-builders, St. Monans; T. Reid, Thomas Reid & Sons, engineers, Paisley; J. Stuart, J. & W. Stuart, net-makers, Musselburgh; A. Campbell, Kelvin Diesels Ltd., Glasgow; R. J. Raggett, GEC-Marconi Electronics Ltd; Charles Stewart, Campbeltown; William McDougall and Ian Y. Macintyre, both of Tarbert, Argyll; Mungo Munro, Dunure; Gilbert Clark, Port Charlotte, Islay; Duncan Colville, Machrihanish, Kintyre; and Rosemary Bigwood, Edinburgh.

Illustrations: Figs. 21, 30, 32, 41, 42, 44, 45, 52, 53, 58, 60, 65, and 66 were drawn by Will Maclean for his documentary exhibition, *The Ring-Net,* first shown at the Third Eye Centre, Glasgow, in 1978. These appear by courtesy of the Scottish National Gallery of Modern Art, the property of which the material now is. The valued co-operation of Douglas Hall and Margaret Mackay of the Gallery in making available copies of these drawings is gratefully acknowledged. Acknowledgement is due also to Tom Scott of Edinburgh who produced the copies.

Figs. 22, 35, 36, 37, 38, and 43 were drawn by Will Maclean specifically for this book.

The majority of the photographs contained in this book were also included in *The Ring-Net* exhibition, but are credited to their original owners, to whom the gratitude of the author is extended. Prints of many of these photographs were made especially for publication by James Blyth, Falkland, to whom thanks are due.

Figs. 2, 3, 33, 34, and 49 appear by permission of the Controller of Her Majesty's Stationery Office. Drawn by James Stewart of Edinburgh, they were originally published in B. F. Primrose's *Paper* of 1852 (refer to References and Notes, p. x).

Contents

References and Notes

Abbreviations of written sources

A.D.   Records of the Lord Advocate's Department, Scottish Record Office.

A.F.   Agriculture and Fisheries, Scottish Record Office.

E.R.C.   Evidence to the Royal Commission, 1862, 1864, and 1877 (subsequently published with reports – see R.R.C.).

F.B.E.   Scottish Fishery Board Enquiry, Edinburgh evidence, June-July, 1856. Bound volume of hand-written transcripts. A.F. 7/9.

F.B.R.   Fishery Board for Scotland annual report.

N.S.A.   New (Second) Statistical Account, Argyllshire.

O.S.A.   Old (First) Statistical Account.

Paper   Paper by the Secretary of the Board of Fisheries on drift-net and trawl-net fishing for herring – B. F. Primrose. Published 1852. A.F. 37/142.

R.F.B.   Register of Fishing Boats.

R.R.C.   Report of the Royal Commission –

1. Report of the Royal Commission on the operation of the Acts relating to trawling for herring on the coasts of Scotland, 1863 (T. H. Huxley, Lyon Playfair, and C. F. Maxwell).

2. Report of the Commissioners appointed to enquire into the sea fisheries of the United Kingdom, 1865 (T. H. Huxley, James Caird, and George Shaw Lefevre).

3. Report on the herring fisheries of Scotland, 1878 (Frank Buckland, Spencer Walpole, and Archibald Young).

Oral sources

1. The tape-recordings made by the author during the period 1974-6, in the fishing communities of Kintyre, Lochfyneside, and Ayrshire, are lodged, together with transcripts, in the archive of the School of Scottish Studies, University of Edinburgh. The transcripts are arranged in accordance with the chapter organisation of this book.

2. The tape-recordings made by the author in the farming and post-crofting-fishing communities of Kintyre and mid-Argyll in 1977 are likewise, with transcripts, preserved in the archive of the School of Scottish Studies. Material from that collection is indicated 'S.S.S.' in the body of the text. The material support of the University of Edinburgh, in that project, the author here gratefully acknowledges.

3. The series of tape-recordings begun on 10 December, 1976, and continued February 1978 to January 1979, was made for the *Historical Dictionary of Scottish Gaelic* project, which is being managed from the Department of Celtic, University of Glasgow, where tapes and transcripts together are lodged. Material from that collection is indicated 'H.D.S.G.' in the text, which does not, however, signify that the term or terms will ultimately have a place in the published dictionary. Scots and English terms, incidentally recorded but nevertheless belonging to that collection, are indicated as above. The material support of the University of Glasgow, in that project, is here gratefully acknowledged.

The dates accompanying informants' names, in the References and Notes sections appended to each chapter, apply, unless otherwise indicated, to the days on which recordings were made.

List of Informants

Tarbert: Hugh McFarlane, 1884*; David McFarlane, 1895*; John Weir, 1895*; H.D.S.G. only – Dugald McFarlane, 1899; Angus McFarlane, 1913; William McCaffer, 1913; Robert Ross, 1926.

Campbeltown: John McWhirter, 1886*; David McLean*, 1888; Dugald Blair, 1889; Archibald Stewart, 1889*; Robert Morans, 1891*; James MacMillan, 1901*; Robert McGown, 1902*; Duncan Blair, 1902*; George Newlands, 1902; John McIntyre, 1911.

Carradale: John Conley, 1886*; Donald McIntosh, 1893; Robert Conley, 1896; H.D.S.G. only – Duncan 'Denis' McIntosh, 1899*; Calum Buchanan, 1902; Robert Paterson, 1902; Dugald Campbell, 1903; Matthew MacDougall, 1909.

Dalintober: Duncan McSporran, 1888; Henry Martin, 1891; Angus Martin, 1895.

Lochfyneside: Samuel Turner, 1888*; Robert MacNab, 1900* (Minard); Donald MacVicar, 1898 (Low Kames); Jessie Campbell MacBrayne, 1892, Margaret Campbell, 1899 (sisters, Silvercraigs); Robert McLachlan, 1902 (Ardrishaig).

Ayrshire: John 'Jake' McCrindle, 1888; Thomas Sloan, 1892*; Thomas McCrindle, 1897*; John Turner McCrindle, 1902; Matthew Sloan, 1917; (Thomas) Andrew Alexander, 1928 (all of Maidens); James McCreath, 1897 (Girvan).

N.B. Years of birth accompany names. Those informants who, at the completion of this work, are dead, have been indicated by an asterisk.

Introduction

THE evolution of the Scottish ring-net from a crude assemblage of drift-nets hauled on to the sandy beaches of lower Lochfyne in the mid-1830s, to a sophisticated deep-water method, requiring powerful motor boats and an array of electronic equipment, took 120 years to complete. Ironically, with the attainment of maximum development, a rapid decline began which, within the past decade, signifies the unmistakable, and probably irreversible, end of that method of herring fishing.

The ring-net represented, from the very day of its origin, a significant technological advance. Its catching capacity was limited only by the weight of fish which it could bear. As, in the present century, the purse seine-net and mid-water trawl have rendered the ring-net virtually obsolete, so, in the nineteenth century, it threatened and finally destroyed the traditional drift-net fishery on Lochfyne.

Acceptance of the method was, however, not achieved without cost to the ring-net fishermen. On the fishing grounds to which it was successively introduced, angry resistance was encountered. That resistance was manifest first on Lochfyne, and the cost which was paid there is calculable in terms of imprisonment, starvation, and blood. That first contest between tradition and innovation was the most protracted and most bitter, but would, on lesser scales, be repeated wherever the ring-net was taken. This book is substantially concerned with these disputes.

The ring-net in its earliest forms can be defined only as a beach seine-net. The term ring-net is applicable to the method only in the final stage of its evolution, when its operation ceased to be dependent on contact with the shore. The terminological confusion is increased by the fishermen's names for the net throughout the nineteenth century, viz. trawl, scringe or screenge, and draught. The first term preponderated, and invaded legislative and other documents. It, therefore, shall be used in this book to the exclusion of the other terms.

There is a fundamental distinction between drift-nets and seine-nets. The former functions passively − as a drifting wall of netting into which shoals might or might not swim and become enmeshed − whereas the latter functions actively, by being hauled on to land or into a boat.

In its original form the trawl-net cannot be considered to have possessed any distinctive features in either its design or in the customary ways of using it. Beach seines are referred to, and in some cases depicted, in the records of the oldest civilisations,[1] and were also used by so-called primitive peoples. When Captain James Cook visited the South Pacific island of Tongataboo in 1777 with the *Resolution*, he observed the native fishermen surround a shoal of fish on an offshore bank with a net which he described as 'a long net, like a seine'. The fishermen removed the catch from within the surround netting using lap-nets extended between two poles.[2]

1

The principle of first enclosing a shoal within walls of netting, and then removing it with smaller nets was employed, albeit on a larger scale, on the Norwegian coast. A net was extended across the entrance to a fjord, and the fish hauled ashore within the fjord using small seine-nets.[3]

The pilchard fishermen of Cornwall and the south and south-west coasts of Ireland worked in a similar way. A description of pilchard-seining at St Ives in 1864 detailed an operation of greater sophistication, involving two and sometimes three separate nets joined to enclose a shoal located close inshore. The fish were thus contained within a 300- or 400-yard (274.32 or 365.76 m.) circumference of netting.

The nets, which were 60 yards (54.86 m.) deep, were then hauled inshore by warp and capstan until the foot-rope along the bottom of the nets touched ground. With the shoal enclosed within the diminished circle, the net was secured. A small 'tuck-seine', of about 140 yards (128.02 m.) length, was then put to use. It was set like an ordinary seine inside the ring of the other nets, and hauled into a boat until only the bunt – a bag of slack netting – remained, containing a part of the catch, which was then discharged into the boat in large baskets. A great catch might be removed from the net over as prolonged a period as two weeks, thus supplying the curing stations with convenient quantities.[4]

But the operational limitations imposed upon the fishermen of Cornwall and Ireland by the shore-based character of the method precluded the development of an offshore fishery. As was commented in 1865: 'It was . . . only when the shoals of pilchards came into the bays, where alone these nets can be used, that any fish were caught.'[5]

That disadvantage was exposed when, about the middle of the nineteenth century, a drift-net fishery for pilchards began in Cornish waters. The seine-fishermen complained that the setting of drift-nets outside the bays broke up the shoals and prevented them from entering shallow water, and that the drift-nets, when worked in the bays, interfered with the operation of the seines.[6]

These disputes between the traditional seine-fishermen of Cornwall and the progressive drift-net fishermen, who followed the shoals, are distinctly parallel with the disputes which locked two classes of fishermen on Lochfyneside in a malevolent 20-year-long struggle. The irony is that the roles of tradition-bearer and tradition-challenger were reversed.

(When, in the early years of the present century, the Campbeltown trawl-fishermen ventured to Donegal Bay – where there seems to have been no tradition of seining – they were surprised to observe local fishermen using drift-nets to surround herring in shoal waters. The resemblance to trawling ended at that, however, because their object remained to get the fish to mesh, and that they achieved by going into the circle in their small boats and splashing the water with oars to scatter the fish into the netting.[7] An identical device was used by Minch drift-net fishermen. Eight nets, sewn into one, sometimes yielded 16 and 18 crans of herring.[8])

The third and final coastal area of the British Isles where a considerable fishery with a form of beach seine began, in advance of that on Lochfyne, was the Firth of Forth. There a fishery on sprats, or *garvies* – the Scots name – opened annually in

November, and continued until the end of February. The main fishing grounds lay in the shallow estuary west of Queensferry, but at times the shoals were pursued as far in as Kincardine and Alloa.[9]

Trawl-nets were said to have been introduced to Queensferry – the second most important trawl-fishing community on the Forth, after Newhaven[10] – in 1829.[11] The first official reference to 'trauls' on the Firth of Forth appeared in 1834, in connection with complaints about the destruction of herring fry.[12]

The seine-nets used in garvie fishing were, like those for pilchards, designed exclusively for inshore work, and were generally hauled to the land. No deep-water fishery developed there; the very nature of the Forth estuary precluded such a development. Ring-netting would evolve only in the deep, clear waters of Lochfyne and the Kilbrannan Sound, and by 1930 rank, with deep-water purse-seining, its close relation, as the most efficient method of herring fishing in the world.

REFERENCES AND NOTES

1. R. Balls, *Fish Capture*, London, 1961, 35.
2. J. Cook, *Voyages Round the World, 1768-1780*, Edinburgh, 1890, 136.
3. J. R. Crewe, *R.R.C.* (1863), App. F, 36.
4. *R.R.C.* (1865), App. III, Sean or Seine Nets, 4.
5. *Ib.*, liii.
6. *Ib.*
7. A. Stewart, 1 May, 1974, and D. Blair, 5 June, 1974.
8. K. MacRae, Portree, Skye, 19 February, 1974, recorded by W. Maclean.
9. Dr. L. Playfair and Vice-Admiral H. Dundas, Report on the Sprat Fishery of the Firth of Forth, pub. with Draft Resolution by the Board of Fisheries and Statement upon the Herring Fisheries Act, 1860, 41. 1861. A.F. 37/143.
10. *Ib.*
11. J. Methuen, fish-curer, letter to Fish. Bd., 12 December, 1859. A.F. 37/16.
12. *F.B.R.* (1834), 2.

1
The Violent Birth: An Historical Introduction

THE geography of the area in which the principal events of this history occurred may be conceived as two stretches of sea, closely contained by land. The first, Lochfyne, angles narrowly in a south-westerly direction from inland Argyll, and, 22 miles along its course, turns southerly to spread towards the expanse of the Firth of Clyde. Opening south-westerly off the widening lower loch, the Kilbrannan Sound extends, formed in the west by the long, undulating arm of Kintyre, and in the east by the great, hard head of Arran. These waters were good for herring, and the main fishing towns on their shores faced eastwards across them: Inveraray at the northern end of Lochfyne; Lochgilphead and Ardrishaig below the neck of the narrow upper loch; Tarbert, 10 miles south by sea; and, at the southern end of the Kilbrannan Sound, Campbeltown.

The Lochfyne herring fishery was probably of ancient origin, being essential to the subsistence economy of the populations along the lochsides. A native fishery on a large scale evidently began in the seventeenth century, [1] but the loch had been attracting fishermen from beyond Argyll since the previous century, and probably before. In 1555 it was reported to the Scottish parliament that the fishermen of the western burghs 'sic as Irwin, Air, Dumbertane, Glasgow' had yearly 'in all tymes bygane . . . resortit to the fisching of Loch Fyne . . . for making of hering and uthers fischeis'. [2]

The existence of a valuable fishery at Campbeltown in the seventeenth century is indicated by the presence there of a tacksman responsible for collecting the Assize of Herring on behalf of the Earl of Argyll. The Assize was a tax on all herring landed in Scotland, imposed by government in the reign of King James IV. [3] By the eighteenth century, if not before, the Campbeltown fleet was operating about the Western Isles. Some fishermen, in 1742 and 1743, had avoided payment of the Assize in these waters, [4] claiming exemption on the ground that the tacksman of the Assize had failed to protect them against illegal tolls levied by Mackinnon in Skye and other chiefs in the west, in 1732 and 1733, and at other times. [5] The fishing boats were mainly, if not exclusively, undecked, and were owned by the fishermen themselves. Each man had a share in the vessel and took a corresponding share of the profits. [6] A transition in both boat design and the system of ownership was, however, imminent.

That transition began about 1760 with the establishment of fleets of bigger, decked vessels at ports around the Scottish coast, but principally at Greenock, Rothesay, and Campbeltown. Lochfyneside investment in these craft, known as *busses,* was evidently on a very slight scale. [7] The busses were fitted out to capitalise on a government scheme for the encouragement of the British fisheries. The scheme – instituted in 1750 with the formation of the short-lived and unsuccessful Society of the Free British Fishery[8] – originated in jealousy of the wealth which the Dutch nation was accumulating by its well-established buss fishery off the British coasts,[9] and also by a need to train seamen for the navy.[10]

The investors in the buss fishery were known as 'adventurers'. In Campbeltown, the number of shareholders to each vessel was generally from two to six, principally merchants, though not a few ship-masters also held shares.[11] The crews – from six to 18 men on each buss, depending on its size[12] – were hired on agreed wages, varied according to age and experience, and took no share of profits.[13]

Government inducement was originally in the form of bounty payments calculated according to the tonnage of decked vessels built for the fishery after 1750. That allowance was restricted to vessels of less than 80 tons, in length a maximum of about 55 feet (16.76 m.).[14] A reduction in the tonnage bounty, from 30/- to 20/- in 1785, was accompanied by the introduction of a bounty of 4/- per barrel of cured herring.[15] This answered the objections of such opponents of the tonnage bounty as Adam Smith, who criticised the buss-owners for fitting out their vessels with the 'sole purpose of catching, not the Fish, but the Bounty'.[16] Government assistance to open-boat fishermen was offered for the first time, 1/- per barrel, increased to 2/- in 1795.[17] Thus stimulated, the small-boat fishery expanded and proved more productive, yet with less reliance on government subsidy. That factor, combined with a diminishing need to train seamen for the navy, decided the government, in 1800, to phase out the tonnage bounty.[18]

The 'adventurers', who readily acknowledged their dependence on government finance, began, after 1800, to withdraw their capital from the fishery.[19] But the reversion to an open-boat fishery was gradual, and may not have been completed.

The Campbeltown fishing industry, dependent again upon the unaided resources of the fishermen, was slow to recover. As the buss fleet was disposed of, open boats were probably reverted to. About 1840, however, the fishermen of Campbeltown began to adopt a 'superior description of boat',[20] and the rebuilding of the distant-water fleet began. In that year, some of the new boats at Tarbert and Lochgilphead were part-decked and of a larger build than the old vessels. At a cost of £120 – before equipping with masts, sails, nets, anchors, etc. – their value was double that of the smaller boats.[21] By 1843, 'half-deckers' – 22 feet (6.71 m.) long by the keel and 9 feet (2.74 m.) of beam – were in use in upper Lochfyne.[22]

The Campbeltown smacks, as these craft were known, were probably of similar design. By 1852, both the Lochfyneside and Campbeltown smacks were sailing to the Isle of Man and Lewis herring fisheries, with crews of four, five, and six men. The advantage to the navy of such vessels was again a noteworthy consideration. 'These boats,' one commentator remarked, 'often make voyages of considerable length, whereby the crews become acquainted with the use of compass, are inured to the sea, and gain the rudiments of seamanship, making this class of boat a valuable nursery for seamen.'[23]

Some of these vessels were probably also involved in the herring fishery in the Sounds of Scalpay and Raasay, off the eastern shore of Skye. The fishing operation there (but not necessarily elsewhere) corresponded with the traditional British buss method. Two or three crews, operating in small boats and with perhaps 10 drift-nets each, supplied an associated smack, on the decks of which the herring were usually cured.[24] The preference for curing on deck, rather than on shore, denotes, in that part of the operation, a shift to the Dutch method. The process was described in 1864 by a Greenock curer, John McLean: 'We put them on the vessel's deck, throw salt on them, gut them, and put them in tubs and (then) into barrels: we let them stand till next day and head them up.'[25]

In 1809, with the foundation of the Board of the British White Herring Fishery (hereafter referred to simply as the Fishery Board), the real expansion of the Scottish herring fisheries had begun. Supervision of the buying, curing, and exportation of herring, maintenance of order on the fishing grounds, the building of fishery piers and harbours, and the compilation of statistics relating to the herring fisheries, were the principal duties of the Board.[26] It originally comprised seven members, but by subsequent legislation was increased to 15 members, including, *ex officio*, the Lord Advocate and the Solicitor-general for Scotland.[27] The practical organisation of the herring fisheries was, however, the responsibility of a team of paid officials, including, by the mid-nineteenth century, 39 fishery officers.[28] The Board's administrative office, in Edinburgh, was staffed by a secretary and three clerks. In 1849, Bouvarie Francis Primrose was appointed secretary on an annual salary of £524.[29] His term of office would span the violent controversies of two decades.

The Beginnings of Ring-Netting

The Scottish ring-net, or trawl-net as it was improperly called, originated from irregular use of traditional drift-nets in the 1830s. The year cannot be established with certainty, but the available evidence indicates either 1833 or 1835.

Ninian Ballantyne, who began fish-curing at Tarbert in 1826, claimed to have purchased the first great catch of herring taken in the new way. He remembered that in December, 1833, two men had stretched drift-nets across the mouth of a Lochfyne bay and enclosed 6,700 barrels of herring, a massive catch even by modern standards. He paid out £400 or £500 for his share of the catch, the rest of which went to farmers and other fish-curers in the neighbourhood.[30] The date 1835 is given in a less detailed report which appeared in the *Edinburgh Observer* two years later.[31]

Within a few years nets were being designed at Tarbert specifically for trawling. These nets first attracted the attention of the fishery officer at Inveraray, Alexander Sutherland, in December, 1836, when he reported that six 'trals' were being worked at Otter Spit, on the eastern shore of Lochfyne. He inspected the nets, but found none with the meshes less than the legal size of one inch from knot to knot,[32] established by legislation in 1808.[33]

Suspicions of the use of small-meshed trawls in the herring fishery increased, but Sutherland's warnings to the Tarbert fishermen involved were disregarded, and he complained to the Fishery Board that he could not possibly suppress the 'evil' single-handed.[34]

The first seizure of a trawl-net was accomplished in 1842. The owner, Peter Black of Tarbert, admitted that it had been used for herring fishing, and Sutherland advised that it should be condemned and destroyed, warning that 'unless examples are made there will be no end to this illegal mode of fishing'.[35]

The outcome of the seizure is not on record, but the net was most probably returned to its owner. The periodic seizures of nets by Sutherland during the period 1844 to 1847 were frustrated by non-prosecution of the owners. The adverse factors included the expense of employing lawyers,[36] and 'a conflicting opinion amongst several of the Justices of the Peace in Argyllshire as to the propriety of putting a stop to trawling for herrings'.[37]

By 1849, Lochfyneside lairds and others not directly involved in the herring fishery were entering the trawling debate. These interventions would exercise a decisive influence upon the issue.

From Inveraray, at the upper end of Lochfyne, where resistance to trawling would later develop with severity, James Robertson, chamberlain to the Eighth Duke of Argyll, reported opposition to trawling among drift-net fishermen, who considered the method 'hurtful to the Herring Fishing in its proper season'.[38]

One landowner, Mungo Campbell of Strachur, on the north-eastern shore of Lochfyne, complained in November, 1848, of 'what is considered encroachments on the legitimate mode of taking herrings'. He referred to 'a class of boats, not owned by the regular fishers of herrings, (which) have of late years had recourse to trawling for fish . . . and drag them to shore on spots favourable for the operation, such as the Sand reef of Otter, and this they do *at all Seasons and at all hours* of the day or night'. His express concern was that trawling could not be 'put an end to . . . unless made illegal by an act of Parliament'.[39]

This was advised against by John Campbell of Stonefield Estate, on which most of the Tarbert fishermen lived. His was the sole influential voice raised in support of the trawl-fishermen. In December, 1848, he forwarded to the Fishery Board a petition in favour of trawling, bearing 216 signatures of fishermen and local curers.[40] He claimed, in an accompanying letter, that the fishery for small herring had been abandoned two years previously, and that not only was trawling productive of greater quantities of herring, but the quality of the catches was superior: 'At the commencement of the season large takes of herring were made . . . of a size and quality superior to what was ever seen in this quarter.'[41]

By the end of 1849 the issue had blown wide open. Two petitions condemning trawling and bearing the signatures of 394 Lochfyneside fishermen were submitted to the Board, augmented by a petition signed by 11 Glasgow fish-curers, a numerically small group, but of immense influence.[42]

The Fishery Board's general inspector, John Miller, had been sent to Lochfyneside in July, 1849, to investigate the issues,[43] and in May, 1850, he presented a report based on the opposing arguments set out in the petitions, and those elicited from the fishermen during his private enquiries. The division between the drift- and the trawl-fishermen − the one group holding to traditional practices and ideals, and the other rejecting them − was already pronounced.

2. The alleged interference with drift-nets by trawl-fishermen, who are here seen shooting a net among fleets of drift-nets set inshore. On the beach, a second crew of trawl-fishermen wait, smoking and talking, for the other end of the net to be taken to land and hauling begun

The main objections to trawling were:

1. The destruction of herring spawn and fry.
2. The dumping of dead fish after heavy catches, when the boats had been filled to capacity, and the repellent effect on the herring shoals caused by resultant putrefaction.
3. Interference by trawl-fishermen, who plundered the drift-net catches and hauled their trawls under the trains of nets to the shore.
4. 'Unless trawling is entirely prevented, the Lochfyne fishing which has for centuries been famed for its herrings will be annihilated and its industrious fishermen ruined.'

3. The alleged plundering of drift-net catches by trawl-fishermen: 'These crews are constantly guilty of lifting the drift-nets and stealing the fish out of them.'

The arguments in favour of trawling were:

1. 'It is free to everyone.'
2. Trawling was the cheaper of the two methods, the cost of a trawl-skiff being about £10, and a complete net from £10 to £12, compared with £35 to £40 for a small drift-net boat, and from £30 to £35 for a train of nets.
3. The trawl had a greater earning capacity, a crew of trawl-fishermen sometimes gaining from £60 to £100 at a haul. The earnings of the trawl-boats in 1848 ranged from £20 to £200, whereas the drift-net boats' earnings did not exceed £140.
4. 'So successful has the trawl-net fishery been at Tarbert within the last three years, that between £500 and £600 has been paid to one carpenter in Tarbert for trawl-boats, and fully that amount expended for nets.'

Miller's sympathy lay with the drift-net fishermen.[44] He conceded in his report that trawling was the more efficient method, but acknowledged 'grounds to suspect that the new mode of catching leads to extermination'. He also offered the hostile analogy: 'The results of the trawl fishery is (*sic*) not dissimilar to what the State Lottery formerly was to its general supporters; dispensing a few immense prizes, the news of which keeps up excitement, and induces many to embark in the same line, who in the end it may be apprehended will share in loss and disappointment.' Miller was not convinced, however, that prohibition of trawling by law could yet be justified, and urged that the enquiry be continued.[45] His advice was not taken.

Condemnation by Law

In April, 1851, the Fishery Board began, in collaboration with the Lord Advocate J. A. Moncrieff, Sir William Gibson Craig of the Treasury, and the Government solicitors, the drafting of a bill to amend the existing fisheries legislation.[46] The bill, though largely a consolidation of salting and branding regulations, relevant to the curing trade, included a clause calculated to secure the prohibition of trawling around the entire Scottish coast.[47] There can be no doubt that the clause originated with the Fishery Board. Its secretary, B. F. Primrose, was particularly anxious that it should be enacted.[48] When the significance of the parliamentary proceedings had fully registered, the Tarbert fishermen organised a dramatic bid to halt the bill. On 26 June a deputation from the village, headed by Neil McCalman, a merchant there with a heavy investment in trawl-boats and gear,[49] waited on the Lord Advocate at the Fishery Board office in Edinburgh. The deputation urged upon him that trawling, instead of catching small fish, was the best method of taking the very largest fish. McCalman, to illustrate the case, pulled two large herring from his pockets and compared them with smaller specimens, which he had also brought with him, and which he stated had been caught in drift-nets.

The claim that trawl-nets destroyed small herring was central to the case against the method, and Moncrieff was suddenly anxious to have the contradiction resolved. That afternoon, John Miller was sent to Campbeltown with instructions from Primrose to buy at random some herring from the drift-net boats as they came in from fishing. He returned to Edinburgh with 'as large if not larger fish than those taken by the trawl'. These fish were shown to Moncrieff.[50] On 24 July, 1851, the bill, which had been introduced by Moncrieff and Craig, became an Act of Parliament.[51]

The sixth clause of Act 14th and 15th Victoria, chapter 26, rendered illegal for herring fishing all nets other than 'the usual Drift Net'. Offenders could be imprisoned for a maximum period of 20 days, failing payment of the imposed fine.[52]

Late in December, with the implementation of the Act imminent, the Fishery Board's assistant inspector, Laurence Lamb, arrived in Tarbert to assess the mood of the fishermen. None of them, he reported, yet showed 'symptoms of violence'. Their mood was rather one of 'sullen indifference'. Some continued to trawl, although the fishing was slack, and trawl-nets were in evidence 'hanging round the harbour as usual'. Some fishermen assured him that to stop trawling would 'take some trouble'.[53]

In May, 1852, as the fishermen prepared for the opening of the herring season, a fishery officer advised the Board: 'I have reason to believe that trawling will commence for this cause, that the most abandoned of the trawlers are unable to furnish themselves with drift-nets.'[54]

The Defiance of Law

The Act, in its initial year of enforcement, 1852, was not accepted by the trawl-fishermen, and the Fishery Board was petitioned during the season with complaints of impoverishment.[55] The stationing on Lochfyne of two ships, the Board's own cutter the *Princess Royal* and the more formidable H.M.S. *Porcupine*, loaned by the Admiralty, considerably curtailed their activities.

The trawl-fishermen's complaints were put to the Lord Advocate, Moncrieff, by Neil McCalman, the Tarbert merchant and boat-owner. He wrote: 'Ever since the failure of the potato crop became general the position of fishermen on this coast is very much changed, with little prospect of improvement. Finding that the new Act would come into operation in January last, they with much difficulty succeeded in procuring drift nets with which hitherto they have been so unsuccessful that the greater portion of them with their families are in absolute starvation.'[56]

The opponents of trawling were no less active in pressing their interests. Four petitions, collectively bearing more than 1,000 signatures, reached the Board in December, 1852, from the Lochfyneside communities. The repeated protest was against the dumping of massive quantities of herring by the trawlers: '(They) surround the fish in shallow water, take a certain quantity of them and make off in order to escape detection, and leave the rest to spoil in the water.'[57]

Legal proceedings against the few trawl-fishermen whose nets were seized during the fishing season failed almost entirely. B. F. Primrose later explained: 'Many men in the Highlands bear exactly the same name and live in the same place. Even with the utmost care we served some (warrants) on the wrong men, and others never appeared . . . We soon found ourselves regularly in the hands of the lawyers. These law proceedings swelled the law bills and threatened to become most expensive, for it appeared there would be no end to it.'[58]

The Shooting of Colin McKeich

If the Fishery Board expected greater success in suppressing trawling in 1853, its expectation was disappointed. H.M.S. *Porcupine* was a mere week stationed on Lochfyne when, during an evening encounter with patrolling naval crews, 28-year-old Colin McKeich of Tarbert was shot and wounded. The *Porcupine's* commander, George Jackson, surprisingly did not learn of the incident until two days afterwards, on 19 June, when the procurator-fiscal of Campbeltown, Charles Mactaggart, appeared on board his ship flourishing an arrest warrant.[59]

Lieut. Jackson took the *Porcupine* to Campbeltown immediately so that the investigation could be conducted in the presence of the sheriff and at a distance from Tarbert. The outcome was that gunner Philip Turner, a 34-year-old Cornishman, and Peter Rennie, a 20-year-old marine from Howgate of Hawick, were implicated in the case.[60]

The Tarbert community was in a 'state of great excitement', a local laird, John Campbell of Kilberry, reported to the Lord Advocate on 19 June. 'As you may fancy,' he wrote, 'the sight of human blood has excited the poor fellows greatly . . . I have used my endeavours to preserve order among the poor fellows who I hear believe they all may be shot like as many mad dogs.'[61]

Colin McKeich had been at the tiller of the skiff *Annan* on the night that he was wounded. The boat was one of a company of four pulling south from Tarbert. Half-a-mile east of Skipness jetty, McKeich saw right ahead a four-oared boat. He recounted: 'Some person from the boat hailed us and told us to lower our sail. I then saw that she was one of the boats belonging to the *Porcupine*.

'We lowered our sail and this boat came alongside of us. There was an officer and four men in her. They searched our boat but found no trawl-net in it and they then let us go. They then went alongside of the boat we were towing and as I was aware that there was a trawl-net on board another of the boats in our company, belonging to Archibald Campbell, I called out in Gaelic to Campbell to look out for himself as the cutter's boat was here.

'I had scarcely said this when a shot was fired not from the boat that had boarded us, but from another boat belonging to the steamer, outside of her, and a ball or slug struck and wounded me on the right shoulder.'[62]

Immediately after McKeich had been shot, a fellow-crewman, John Johnstone, called out, 'Take care and not kill people.'[63] Three more shots were fired.[64] Johnstone felt McKeich's arm and 'found a hole in it, and something warm, which I took for blood flowing'.[65]

The evidence of the accused, Turner and Rennie, was contradictory,[66] as Lord Cowan, before whom the case was tried at Inveraray on 22 September, acknowledged. The jury returned a verdict finding both men guilty of reckless and culpable discharging of firearms. They were sentenced to three months' imprisonment,[67] but petitions organised in Inveraray and Lochgilphead resulted in their being recommended for a Royal pardon, which was granted. On 21 October, almost a month after beginning their sentence, Turner and Rennie were released, and returned to the *Porcupine*.[68] The *Glasgow Herald*, reporting their release, commented that the verdict which had convicted them 'certainly reflected little credit on the brains of the jurymen'.[69]

As for the unfortunate Colin McKeich, he carried the injurious ball in his shoulder until the end of his life. The doctors who examined him had decided that, as it had lodged in a muscle, pain and inflammation would have resulted from any attempt to remove it.[70] He returned to fishing but was, as he himself admitted, 'by no means able to do a man's work'.[71]

The *Porcupine* remained on Lochfyne until the end of October, but achieved little real success. Jackson had admitted on 2 July, not quite a month after his arrival, that trawling could not be 'effectually stopped' by the force at his command. The trawl-fishermen's 'determination, cunning, and ingenuity' had been developed to such a degree that his role was restricted to preventing fishing in the coastal areas which his boats were able to patrol, and the fishermen were beginning to avoid even that risk, by separating and going to 'opposite and different quarters'. Some crews were even arranging to be towed by sailing-boats to a safe distance from where his ship was lying, and there they would have a vessel 'to take their fish from them the moment they are caught'.[72]

The outbreak of war against Russia in 1854, and the consequent demands for warships and seamen placed upon the Admiralty, meant that the Fishery Board was compelled to rely solely upon its own small sailing-vessel, the *Princess Royal*, for supervisory duties. Trawling did not, however, increase. During the season, cholera broke out violently on Lochfyneside, particularly about Tarbert, and its ravages ended fishing.[73] The villagers were said to have considered the epidemic 'a judgement of Providence on trawling'.[74]

Trawling revived in 1855, and in 1856 received a 'great impetus' by the publication of a Treasury Commission report which recommended the repeal of the Act of 1851, 'which has no other result than to keep a considerable population in the habitual and successful violation of law'.[75] The Fishery Board, believing that the report would influence the government to repeal the Act, relaxed its supervision of the Lochfyne fisheries.[76]

B. F. Primrose had, by then, long wished for a repeal.[77] His personal disillusionment with the legislation began in the very first year of its enforcement, during a visit to Tarbert. Realising that the enforcement of the law was having a counter-productive effect on the fishery, he had suffered a 'considerable qualm'.[78] He admitted to having feared the stringency of the Act, but, aware of the feelings of the drift-net fishermen and curers, his intention had been to operate the legislation in a 'modifying spirit'.[79] That disillusionment would stay with him and lay him − and the Fishery Board − open to repeated attacks by the opponents of trawling.

The numbers of trawl-boats fishing on Lochfyne had continued to increase until, by 1858, there were 119 at work, compared with 40 in 1850.[80] In 1858, however, the charge was set that would, a year later, explode the Fishery Board's hope of a repeal, and clear the way for a fresh and aggressive campaign against trawling.

The Years of Anger

In 1858 the Tarbert fishermen ventured for the first time north of Otter Spit and into the narrow upper loch, which until then had been considered unsuitable for trawling.[81] Complaints reached the Fishery Board towards the end of October of disturbances at Inveraray between the visitors from Tarbert and the local drift-net fishermen. During the excitement 'some Inveraray men pushed some Tarbert men over the quay'.[82] One of the trawlers present described the incident as merely a 'small squabble' caused by local 'aggressors'. At the height of the dispute a Tarbert fisherman had been chased down the quay with a hay-fork.[83]

The violence at Inveraray was, however, considered serious enough to warrant the presence of assistant inspector Laurence Lamb with a crew of hired men. On their arrival in Lochfyne they rowed among the fleet of boats between Inveraray and the head of the loch, Lamb warning all the fishermen against 'attempts at violence, aggression or disorder'.[84]

At the beginning of November he conducted an enquiry with the drift-net fishermen who had suffered damage to their gear by the cutting of nets and theft of buoys, but he judged that none of them could offer reliable evidence that the damage had been done by trawlers. The fishermen admitted to him that mischief had been going on before any trawlers arrived.[85] Lamb remained at Inveraray until 17 December.[86]

With the ending of the fishing season, the drift-net fishermen turned their energies to campaigning for the prohibition of trawling. The organiser was Inveraray fisherman Martin Munro, and petitions signed by almost 1,800 fishermen were presented in the House of Commons and in the House of Lords. These petitions, which enumerated the 'evils of trawling', also included a calculated threat. If trawling were not prohibited, 'the petitioners will not be answerable for the riots and bloodshed that are liable, to result from collisions between drift-net fishermen and trawlers'.[87] In February a deputation from Lochfyneside, led by Alexander Finlay, M.P. for Argyll, met B. F. Primrose in Edinburgh to press the drift-net fishermen's case.[88]

The tension fizzled and sparked and finally subsided, and in the meantime a decision on the legal issue was deferred. The threat of renewed hostilities remained, however, and towards the end of July, 1859, Laurence Lamb, back in Inveraray, reported that the trawl-fishermen were declaring openly their intention of forcing their way up the loch should the fishing fail at Tarbert, while the drift-net fishermen, for their part, were speaking freely of their determination to oppose any such invasion.[89]

The Tarbert men did venture into the upper reach of the loch, and fulfilment of the warning of 'riots and bloodshed' which Martin Munro had written into the petitions earlier that year seemed imminent. For Munro, early in August, was plotting just such a crisis.

On the night of 5 August 'the Loch was full of Trawlers hear (*sic*) and got a terrable (*sic*) fishing', wrote Furnace fisherman William McFarlane to his commander-in-chief at Inveraray, Munro. The time for action had arrived.

Next day, a Friday, a mass meeting was held in the little village of Minard, 10 miles south of Inveraray, and fishermen from both sides of the loch assembled 'for to be at them and take their trawls from them'. But after the force had mustered it was annoyingly discovered that most of the Tarbert men had gone home for the week-end. The confrontation was consequently postponed until Monday, when, at three o'clock, the fishermen would re-assemble. This McFarlane communicated to Munro and his Inveraray associates – 'They hoop (*sic*) you will be all as one man on Monday.'[90] Little did the conscientious McFarlane realise, as he drafted with difficulty his crude messages in an alien language, that on Monday Munro would betray him.

On Monday morning, the sheriff-substitute of Argyll, J. Maclaurin, and the procurator-fiscal of Inveraray, Duncan Maclullich, set off for Minard, having been alerted to the plot by Martin Munro, who handed over McFarlane's letter. He and some of the other ringleaders had 'become alarmed at the state to which they had brought matters'.[91] At Minard, Maclaurin and Maclullich distributed among the gathered fishermen copies of a proclamation which warned of 'serious breach of the peace and probably bloodshed' if the planned attack were carried out.[92]

After they had started on their return journey to Inveraray, an estimated 300 fishermen crossed the loch in defiance of the warning, and surrounded four Tarbert smacks and 18 skiffs at anchor off Otter Spit. Threats and angry words were exchanged, but without leadership the planned attack was not attempted, and the fleet of drift-net boats was withdrawn to a triumphant discharge of guns by some of the 80 Tarbert fishermen. (Their possession of guns, they explained, was for the shooting of sea-fowl for food, a custom when away from home.)[93]

That day Lamb had interviews with Maclaurin, Maclullich, and James Robertson, chamberlain to the Duke of Argyll. They were all in agreement that the danger was over for the present. Maclaurin, however, stated ominously that he 'would not be responsible for the preservation of the peace unless the Board would enforce the law against trawling'.[94] That threat was reiterated in a boldly phrased complaint which Maclaurin addressed to the Crown Agent on 26 August.[95]

Meantime, the Board, after 'anxious deliberation', had decided that it was not prepared, of its own accord, to immediately re-enforce the law. This decision was communicated to Moncrieff on 29 August, and its express intention was to invite his intervention.[96] Moncrieff's response was immediate and perhaps more forceful than had been anticipated. He replied to Primrose next day, enclosing Maclaurin's 'grave and important' letter. His message was unmistakable. He, Moncrieff, would do his best to ensure that the public peace was maintained, 'whatever resolution the Board may arrive at'.[97]

In fact, no action was taken that year. The injustice of a sudden re-introduction of the law was generally recognised and objected to, principally on the ground that no allowance had been made for the trawl-fishermen to replace their nets with legal drift-nets. Two influential appeals reached the Board in September, the first from the Duke of Argyll — 'I think you must not enforce too rapidly down about Tarbert, or anywhere, when much capital has been invested in illegal nets . . .'[98] — and the second from Thomas Cleghorn, sheriff of Argyll.[99] Two petitions, together bearing 528 signatures, argued that many of the trawl-fishermen were in a state of poverty and therefore unable to acquire drift-nets.[100]

Intervention of the Curers

In the early months of 1860, a bill was in preparation to alter the fishery legislation. The time was right for the fish-curers to move behind the drift-net fishermen, and they effectively assumed control of the anti-trawling movement, turning the emphasis from passionate advocacy of violence to the equally passionate but no more subtle resort of public debate. Their intervention conferred a desirable respectability and, for that matter, credibility on the campaign.

The barrel bounty had been abolished in 1829, ending almost 80 years of government subsidisation of the herring fishery, but the principle survived in an economic bond between curers and drift-net fishermen. This bond was based on the contractual engagement of boats and crews by the curers, who offered a bounty to the fishermen for their catches, in addition to an agreed price per cran.[101] The firm of James Methuen, Leith, for instance, employed in the 1860s as many as 1,000 boats in a season, and advanced between £100 and £150 to each crew 'before it puts a net in the sea'.[102] Herring curing, susceptible to fluctuations in supply and demand, was thus an uncertain business. The continuing development of a competitive market in fresh herring, supplied cheaply and in great bulk by the trawl-fishermen of Tarbert and Ardrishaig, was recognised as an additional disruption of the trade.

This was acknowledged by B. F. Primrose as an important factor in the drifter-curer combination. 'The Merchant in Herrings,' he explained in April, 1860, 'generally has a large body of drift-net fishermen in his employ and pay. Both employer and employed look with aversion upon the whole trade in fresh fish, a trade in which, on the contrary, the consumers' interest directly lies.'[103] But, as he had argued in a letter to J. A. Moncrieff the previous year, the Board was 'powerless to move in the matter (because) the curers are with the fishermen, so that the Board would be running its head against the whole body for which it subsists'.[104] And that, cogently expressed, was the heart of the dilemma.

At a public meeting held in Glasgow on 27 March, both the suppression of trawling and the reconstitution of the Fishery Board had been demanded. James Methuen of Leith, head of the largest fish-curing business in Scotland,[105] emerged as the curers' chief spokesman. 'Nothing short of an explicit, stringent and peremptory Act will answer the desired end,' he insisted. 'The office of the Board should be solely and simply to put that Act in force, and they should be deemed incompetent for their office if they display neglect in such enforcement.'[106]

A committee of nine fish-curers, to organise the campaign, was appointed at the meeting, and it quickly went to work. The political possibilities of the event were not neglected, and extracts from the speeches were published and circulated among Members of Parliament. An anti-trawling petition was simultaneously prepared, and printed copies sent to every fishing town in Scotland. The petition included criticism of a proposal in the Herring Fisheries (Scotland) Bill – which had received its first hearing in the House of Commons on 8 March[107] – to grant the Fishery Board authority to suspend, whenever and wherever considered necessary, the prohibitions affecting nets and the methods of using them. This, the curers argued, would be tantamount to the abolition of the anti-trawling law.[108] (The offensive clause had originated with B. F. Primrose in the earliest stage of the bill's evolution.)[109]

The curers' opposition to the clause was well represented in the House of Commons, and the bill was rejected. Alexander Finlay, M.P., afterwards advised Primrose that the clause would be 'most strenuously opposed' in the Commons, and warned that if it were carried, which he did not anticipate, then the annual budget award to the Fishery Board would likewise be opposed, and the Board's very existence threatened.[110]

Pressure on Parliament was maintained. Two weeks after the public meeting a large deputation of curers, accompanied by fishermen representatives, including Martin Munro, met Moncrieff at his Edinburgh home. He received his intense visitors courteously, but the exchange of opinions was direct and forceful.

The delegation was determined that the bill should be withdrawn and replaced with a 'practical bill, drawn up with the assistance of practical men'. This demand Moncrieff dismissed. 'If you come forward to oppose my bill,' he warned, 'you weaken your case . . . This is not a new subject for me. I had occasion to study it well in 1851, and, therefore, I absolutely decline to withdraw it . . . I tell you that I won't begin again.' Significant concessions were, however, granted.[111]

Primrose's apprehension of the likely outcome of the forceful and increasingly organised lobby against trawling had already been expressed. 'The Act,' he wrote, 'could only be made more stringent by provisions so invasive of the rights and liberties of a British subject as Parliament ought never to grant and that no necessity could justify.'[112]

The curers followed up their initial approach to Moncrieff by appointing a delegation of four committee members to travel to London on 7 May.[113] On 8 May, the delegation conferred with Moncrieff and the Scottish M.P.s and repeated their recommendations. The response was immediate and favourable. A temporary Act would be passed providing for the repression of trawling for one year.[114] In the second draft of the bill, presented in the Commons two days later, the disputed clause was amended to allow the Board only a limited power to regulate the herring fisheries.[115]

The news was received by the trawl-fishermen – however and whenever it may have reached them – with a mute resignation, but John Campbell of Kilberry eloquently stated their case to Moncrieff. 'The Scotch Members,' he wrote, 'appear to treat the stopping of herring trawling for a year with the same coolness with which they would give orders to stop a steam engine for the same time. A steam engine may be stopt for a year, and, when fire and water are applied to it, go on as merrily as ever, but I don't think a trawler with his wife and children can live without food for the same space of time.' Campbell's revulsion to the announcement was the more intense for his mistaken belief that he would be required, as a justice of the peace, to try cases of trawling. In the sole instance of unreserved rejection of the law by a person in authority, he advised Moncrieff: '*Without delay* issue a new commission of the peace for this district, as I am convinced that there is not a magistrate between this and Tarbert who would not throw his commission into the fire.'[116]

On 13 August, 1860, after further amendments, Act 23rd and 24th Victoria, chapter 92, was passed.[117] Primrose was angered by the ultimate form of the bill, and when the debate had cooled sufficiently and the executioner's axe had lifted from the proximity of his own neck, he vented it passionately before the Board. 'It is not the Act of the Board,' he blustered. 'The Board was never officially consulted upon the bill in any of its stages. Nothing can have been more studiously ignored than the Board has been by all the Members of Parliament throughout the passing of this Act . . . The out-of-door opinions of self-constituted managers of the fisheries have been accepted instead, and made the basis of the legislation which has ensued.'[118]

By the new law, warrants to search for and seize illegal nets could be granted; a non-co-operative suspect could be arrested and detained for up to 24 hours for questioning; the penalty of conviction was defined, for the first time, as forfeiture of net and a fine of from £5 to £20; failure to pay could result in either imprisonment – maximum sentence 30 days – or poinding, that is the impounding of 'goods and effects' for the recovery of the fine and other expenses; forfeited boats and nets and other fishing gear could be sold by public auction or else destroyed.[119]

'They Seem Determined to Venture to the Last'

Despite an impressive appearance on paper, the new Act proved little more effectual than its predecessor. The main reason was that it lacked provision for the seizure of boats and catches.[120] The necessity remained of catching the trawl-fishermen actually working their nets, and the fishermen simply reverted to the practice of abandoning or concealing nets when alarmed.

That defect in the law was clearly demonstrated when, in a raid on Tarbert harbour in the early hours of the morning of Sunday, 23 September, 13 trawl-skiffs [121] with nets on board were seized by the crew of H.M.S. *Jackal* and removed to Greenock. The action was, as Primrose would later complain, 'absolutely illegal', and the Board was spared embarrassment only by the discovery that all of the skiffs were improperly named [122] and therefore liable to detention for a month by a clause in the Act of that year. [123]

The discovery of the seizures caused a stir of anger among the fishermen of Tarbert. They were 'going about in a half-raised state excited by drink, vowing vengeance on the crew of the *Jackal* and on all drift-net fishers'. Some of them, it was said, had 'gone off in their boats and maliciously cut the buoys off the drift trains'. [124]

The determination of the Tarbert fishermen to persist at trawling was, however, not long upset. Almost a month after the raid, with the restoration of the 13 skiffs pending, the fishery officer at Ardrishaig, George Thomson, reported that all the affected fishermen had again got trawls. 'I used my utmost endeavours,' he wrote, 'to convince the trawling classes that the law was now against them and would be enforced with the utmost rigour, of which they seem to be quite aware – yet, withal, they seem determined to venture to the last.' [125]

The boats – but not the nets – were restored to their owners on 23 October, having been exactly a month locked away in a Greenock boatyard. [126]

The Tarbert community was, at the time of the raid, already caught in a state of tension. Earlier in the month a crowd of villagers had tussled with three policemen and two fishery officers who had seized a net on the harbourside. [127] Four women – Catherine and Mary Law, Mary Hyndman, and Flora McBean – were each jailed for 30 days, and two men – James McLean and James Law – for 40 days. [128]

That the morale of the officers and men of the *Jackal* was consistently low throughout their term of duty on Lochfyne is evident in one report after another from that ship. In June, at the beginning of the fishing season, and while the bill was still passing through Parliament, Lieut. Francis Simpson, the then commander, had written to Primrose stating his need of more men and expressing hope that Primrose would back him up in a request to the Admiralty. He had five men in 'sick list', and assured Primrose: 'My crews are nearly all done up, what with the constant rain and exposure for so many hours. It is too much for them.' [129]

The morale of the force ashore was just as low as that afloat. The fishery officers, though augmented by a unit of seven fishery policemen, provided for in the legislation of that year, were frustrated time and time again in their efforts to seize trawls. William Gillis reported from Tarbert on 3 November that the trawlers continued to 'go out and come in in broad daylight'. [130] Later that month he applied to the Fishery Board for a revolver or cutlass for his 'personal protection against the fishermen'. [131] The Board rejected his request. [132]

On a November night a boat's crew from the *Jackal* had been surrounded by eight or 10 crews of fishermen when they seized a net. They were 'hooted and yelled at by the trawlers, and at one time were about to use their firearms in self-defence'. [133] By December it was admitted that no fishery officer would 'dare to take even six herrings out of a fishing boat for evidence . . . always being surrounded on the quay by 50 or 80 fishermen'.[134]

The year 1861 was not a happy one for the trawl-fishermen. A third anti-trawling bill was passed by Parliament, with the defects of the previous two amended. It was calculated to hit the trawlers hard, and in that object it succeeded. The Act received a premature baptism in the blood of a young fisherman shot dead on a hazy June night off the wooded shore of Otter. An officer and marine of H.M.S. *Jackal* stood trial on charges of culpable homicide, but were acquitted by jury. The face of opposition was set firmly, as never before.

Trawl-fishermen and the Fishery Board again came in for bitter criticism, at the annual soirée of the Glasgow fish-curers on 4 Feburary, 1861.

'Two of Her Majesty's war steamers are said to be continually at Lochfyne and the lochs, but we seldom hear of these vessels being far from Rothesay,' commented chairman James Dunn. 'At that place the evening enjoyments seem to be more attractive on land than hunting out trawlers at sea.' He assured the audience at the Glasgow City Hall that it was by the Board's 'perfect indifference' that trawling was allowed to continue to 'a greater extent than ever'.[135]

The tremors from Glasgow inevitably reached the House of Commons. James Dunn's claim that trawling continued was cited by one M.P. on 14 April. Moncrieff admitted that there had been some 'minor defects' in the previous session's legislation, and that a further bill might be necessary.[136]

The second anti-trawling bill within a year was presented by Alexander Finlay and David Mure, M.P. for Bute and a former Lord Advocate.[137] It became law on 1 August, and its provisions impinged not only upon fishermen, but also on fish-buyers and their customers, who could be fined or jailed for possessing herring caught by trawl-net, and their boats or boxes, baskets and creels liable to forfeiture. Legal right to seize fishing boats and catches was at last granted, with forfeiture the extreme penalty.[138]

The Killing of Peter McDougall

By the time the bill had passed into the statute books, Peter McDougall, a young Ardrishaig fisherman, was almost two months dead, and the officer and marine charged with his death already freed to resume their duties on board H.M.S. *Jackal.*

The time was about midnight on 6 or 7 June, 1861. A pair of Ardrishaig trawl-skiffs, the *Star* and the *Weatherside,* lay to a single anchor off the wooded shore north of Otter Spit. The fishermen had completed hauling their net for the third time that night. John Hamilton and Dugald McEwan were preparing the *Star's* net for the next haul, but the rest of the fishermen sat in the *Weatherside* puffing on their pipes and talking among themselves. The night was 'close, calm, and not very clear'.[139]

Suddenly, from the shore there came a cry and a gunshot. A ball was heard whizzing past the boats. Most of the fishermen immediately dived down. Several more shots were fired, but although close to the water's edge the two figures on the shore were unidentifiable in the haze. Duncan McBrayne, owner and skipper of the *Weatherside,* called to them, 'In the name of God what are you at all?' The answer was, 'Come ashore and we will damned soon show you what we are.'[140]

The sequence of events was rapid, and within four minutes of the first shot's being heard, the *Star*, rowed by Hamilton and Dugald McEwan, touched on the beach. Very few strokes brought the *Weatherside* to the shore, and, as the crew was about to get out, Neil McEwan noticed that Peter McDougall was still down. 'I thought he was hiding himself,' said McEwan. 'His head was leaning to the side next the shore, and his face was down, and his knees under him. I caught him to make him rise . . . as the firing had ceased. I at once found that he was shot, and on lifting his head I saw blood coming from a wound about his forehead. I was frightened and said, pretty loud, that he was killed.'[141]

Duncan McBrayne's attention was drawn to the dying boy. He saw blood on his face and put his hand down to touch him. By then, the two men had gone off in the direction of the wood. McBrayne jumped from the boat to pursue them, roaring as he ran, 'You have murdered one of our men!' He overtook them and they stopped. The officer, who seemed nervous, gave his name and said that he would get his boat, which McBrayne understood to be an offer of assistance.[142]

McDougall was still groaning a little when McBrayne returned to the shore and lit a match to look at him, but when the boats reached Ardrishaig the village doctor pronounced him dead.[143]

The account of the shooting, according to gunner Robert Hawton and the marine who accompanied him, William Parker, not surprisingly differed. Hawton, 31 years old and merely two years promoted from 'before the mast',[144] testified that, before going down to the shore, he had waited and watched until the fishermen had hauled their net and shot it again. He had hailed them, but 'they spoke to each other in Gaelic and just went on with their work'. He had twice ordered the marine to fire blank cartridges over the boat, but the fishermen did not respond. He said that he then ordered Parker to load again and fire wide of the boat, but claimed not to remember whether he had specified that the marine use blank cartridge.[145] Parker emphatically stated that the third order had been to send a ball wide of the boat.[146] Hawton had not believed that a fisherman had been killed, considering the fuss a 'decoy to get me out of the way while they secured their nets'. When he had returned to the scene with his boat, the skiffs and men were gone.[147]

The post mortem examination of Peter McDougall was conducted on 8 June, by which time he was stretched out on a plank and dressed for burial. He was described as 'above the ordinary height, stout, and well-made', and appeared to be about 19 years of age. The minutely detailed report revealed that 'it was one continuous wound that carried death, entering the forehead near the right eye, and carrying away a part of the forehead'. Daylight could be seen through the wound.[148]

On 15 July, 1861, Hawton and Parker appeared in the High Court, Edinburgh, on alternative charges of murder or culpable homicide. Both pled not guilty.

The central strain of evidence elicited from Naval witnesses concerned the legitimacy or otherwise of the use of firearms. The definitive judgement was expressed by Rear-Admiral William Ramsay,[149] who had served against smugglers during his career. 'The course we followed,' he said, 'was to fire a musket unloaded, and if the vessel did not heave to, to fire a loaded musket ahead, then over her, and if that did not do we fired into her. That is the only rule we have. There is no written law.' He concluded: 'Anyone acquainted with firearms must know that accidents happen very often indeed'. Lieutenant Edward Lodder of the *Jackal* also stressed the risk of accidents. 'Although I was firing a rifle wide of the boat', he said, 'it would not surprise me if it struck the boat or some one in it. If the ball strikes the water, it is apt to bound off in another direction.'[150] That evidence in support of the use of firearms was undoubtedly an influence of singular importance upon the verdict of the jury. It also persuaded the Solicitor-General, William Maitland, to withdraw the heavier charge of murder.[151]

George Young, for the accused, maintained that while they were engaged in supporting the law, the fishermen were a 'band of lawless men'. He insisted that 'where a man has no motive to serve of a personal kind, his acts must be presumed to be in the discharge of his duty'.[152] The jury's unanimous verdict of not guilty was received with loud applause.[153]

Poverty and Imprisonment

Some of the Tarbert fishermen, perhaps informed of the new strengthened legislation of 1861 and unhappily reluctant to endure another season of scheming and dodging, provided themselves with drift-nets. Many, however, and perhaps the majority, were unyielding, and the consequences to these men and their families were severe. For many fishermen, the prospect of incurring debt to equip themselves for drift-netting may have been more intimidating than that of continuing as trawl-fishermen, with the incumbent risks of loss of boats and gear, and of personal imprisonment. However the division of those prepared and those unprepared to begin the fishing season as legal fishermen was determined, the profit to either class would be negligible.

The subjugation of the trawl-fishermen was facilitated by the reduction of the market for illegally caught herring. The provision in the Act for seizure of suspected trawled herring exposed buyers as well as fishermen to prosecution. Six seizures of herring were achieved at Glasgow fish-market in November. One buyer was jailed for 22 days, and two others were each fined £20. The other three cases failed.[154] A 20-ton smack, the *Jane Gordon*, was forfeited by her owner, a Girvan curer.[155] Of the Tarbert fishermen, 26 would spend periods of 14 or 20 days in the jail at Campbeltown, one, Robert McFarlane, serving two separate sentences.[156] Four Ardrishaig fishermen, fined £10 each at Inveraray in July, were presumably more comfortably housed than any of their Tarbert counterparts. All four were spared imprisonment when poinding of their property – perhaps the greater misery – realised the fines.[157] Six men from Dalintober (once a distinct village, but now merged with Campbeltown) were jailed for 10 days in April.[158] In December, three Campbeltown fishermen were jailed for 30 days. A fourth, James O'Hara, had by then absconded to the West Indies.[159]

In 1862, the effects of the repression touched every corner of the Tarbert community. Families starved; payment of rents was continually neglected; debts accumulated; and the young men were going away. Evidence of these effects was heard in Tarbert by a Royal Commission appointed in August to enquire into the operation of the anti-trawling laws.

C

The Commission, spawned by the violent repudiation of these laws, [160] comprised Dr. Lyon Playfair, a scientist and member of the Fishery Board, Lieut.-Col. Charles Maxwell, and that pre-eminent intellectual, Thomas Henry Huxley. The commissioners lost no time in getting started, and that very month were on Lochfyneside interviewing fishermen, merchants, and others concerned in the issue. [161]

The report of the Commission, along with a selection of the evidence collected, was published in 1863. Its conclusions and recommendations will later be examined in some detail, but in the meantime personal accounts of the condition of the Tarbert population in 1862 may be relevantly quoted.

James Elder, for 40 years a baker in Tarbert, spoke of the great distress. 'I have known some families without a meal for 48 hours,' he said. Some had 'hardly fed themselves since the fishing began this year', and others he was feeding on credit 'in hopes of future payment'. He estimated that about a sixth of the population – 'young men, principally unmarried men' – had left. [162]

Archibald McCalman, merchant and banker, agreed that the population of the village had been reduced. Those fishermen able to raise sufficient capital had adopted drift-netting, but many of the best fishermen had been forced to emigrate, and others had become seamen in coastal and foreign-going merchant ships. [163]

James McLarty, a village merchant, was sure that the Tarbert fishermen were not earning at drift-netting a half of what, on average, they had earned at trawling. Their lack of money obviously concerned him, as part of his business was ship-chandlery. Most of the men who had bought drift-nets during the previous season had not yet repaid more than half their debts. [164]

Thus, in 1862, the will of the Tarbert fishermen was finally broken. 'Many of them,' reported the Fishery Board, 'have either joined the crews of drift-net boats, or have set up drift-nets for themselves.' The freedom of the drift-net fishermen from 'encroachments' had provided an impetus to boat-building, and many smacks were launched that year. The expectations of their owners were satisfied, for the Lochfyne fishery that year was 'the most abundant ever known'. It yielded 79,893 barrels, [165] and in the final quarter the fleet of 776 fishing-boats and 80 curing vessels was the greatest known to have congregated in the narrow upper loch. Early in July attempts had been made at Tarbert to resume trawling, but the seizure of two skiffs had acted as a check, and not another attempt was ventured throughout the remainder of the season. [166]

Revival and Triumph

In 1863, the Royal Commission report explicitly condemned the repressive legislation, and two years later a second commission delivered another body-blow. By 1866 the zone of defiance had extended north of Otter into the very communities which, seven years before, had been prepared to reject trawling with uncompromising violence. Naval and fishery officers were beginning to admit that the spread of trawling could not be halted except by overwhelming force.

In 1865 a rebellion was gathering force on Lochfyneside. High prices for herring at the beginning of the summer season, and the low yields at drift-netting, induced almost all of the Ardrishaig crews directly or indirectly to adopt trawling. Those not actually fishing watched the movements of the *Jackal* and her boats.[167] Nets were periodically seized.[168] In that year, too, the second Royal Commission delivered its report, and reaffirmed the conclusions of the first, that the anti-trawling laws had been both harmful and unnecessary.

Responsibility for the promotion of the Acts of 1860 and 1861 was laid squarely with the drift-net fishermen and the curers. The trawlers, reported the Commission of 1862, considered themselves 'sacrificed to a class who pursue the fishing in a different way, and dislike encroachments on their ancient methods'.[169] The commissioners conceded that the recent legislation may have been necessary to prevent disturbances of the peace, but reasoned that the Acts, in terms of their ostensible purpose of conserving the herring stocks, were 'altogether unnecessary . . . They are essentially Acts for protecting class interests'.[170]

The repeated claims of the drift-net fishermen and curers that the ultimate consequence of trawling would be the extermination of the herring, was challenged by the report of 1863. 'Our enquiries satisfy us that the fishery of Lochfyne has suffered no diminution by the operations of the trawlers,' the commissioners wrote. On the contrary, it could be regarded as a steadily progressive fishery.[171] Their confident conclusion was reached by ensuring that the periods of comparison were sufficiently long to correct the annual fluctuations. The gross yields from Lochfyne, based on Fishery Board annual returns, were recorded as:

Annual catch from   1839-43: 20,119 barrels
1844-48: 15,427 barrels
1849-53: 19,149 barrels
1854-58: 25,744 barrels
1859-62: 42,165 barrels[172]

Trawling had been 'an important means of cheapening fish to the consumer . . . and has thrown into the market an abundant supply of wholesome fresh fish at prices which enable the poor to enjoy them without having to come into competition with the curer'. The quantity of herring sold fresh already exceeded that sold cured. From 1854 to 1862, 129,000 barrels of Lochfyne herring had been cured, compared with 168,530 consumed fresh.[173]

The second Royal Commission, reporting merely two years later, had understandably little of significance to add to the conclusions of its predecessor. Most important, the herring fishery had not improved in 1863 and 1864, despite the suppression of trawling, which was admitted by the curers. The supply of herring had, in fact, failed to equal the demand, and large quantities of Norwegian cured herring had been imported.[174]

The trawl-fishermen's defiance of law would begin in earnest during the early stages of the summer fishery of 1866. To what extent their determination was fostered by the Royal Commission reports is impossible to estimate. Certainly, a repeal of the laws in 1864 had been hoped for by the Tarbert and Ardrishaig trawl-fishermen, who in February of that year almost unanimously petitioned Parliament on the basis of the Royal Commission report of the previous year.[175]

Alarming reports began to reach B. F. Primrose in August, 1866. The trawlers were re-emerging in force, and the prospect was not a comforting one. The resurgence began tentatively in 'some out of the way creeks about Skipness',[176] but by 25 August it was verging on the uncontrollable. George Reiach, assistant inspector of fisheries since December 1863,[177] was so anxious that he posted two separate letters to Primrose that day.

There was suddenly much to report, and unpleasant deductions to be formed: 'It is estimated that at least 40 trawl-nets are amongst the Lochfyne trawlers . . . and it is said that generally three crews are about each boat . . . The fishery constables and the *Jackal* crews were repeatedly threatened with violence and assaults by the trawlers and those who bought trawled herrings, and in several cases carried out their purpose to the serious danger of human life . . . Fire-arms have been resorted to but most fortunately no accident occurred . . . I am informed that many of the trawlers and all the (buyers) carry guns to sea with them at night . . . If trawling is to be suppressed, nothing but an overwhelming force can be effective . . .'[178]

By mid-September the trawlers were at work in great force, supplying smack buyers who transferred the catches to steamers at sea or in East Loch Tarbert, without entering harbour.[179] The crew of marine constables had, by then, refused to go to sea, having been surrounded and threatened one night by trawl-crews.[180] Recommendations for a large increase in the fishery police force and for its provision with firearms had been received by the Board, and rejected.[181]

Alexander Finlay, M.P., sensing that the wind of public opinion had begun to blow from another airt, altered his own opinion accordingly. His revised attitude to the laws which he had helped into the statute books was communicated in November to the Lord Advocate, but not without a tortuous attempt to justify their creation. The continuation of 'co-ercive measures' would, he wrote, be unwarranted, 'more especially as the opinion of the fishermen does not appear to be so strong against trawling as it was'.[182]

His conclusion certainly was correct. Late in the season of 1866 the fishermen of Lochgilphead and the upper loch 'became violators of the law quite as much as those termed trawlers'. These men argued that, even considering fines and forfeitures, the trawlers 'could make good earnings while theirs were very poor, trifling'.[183] Evidence to support that statement was abundant. Between 16 and 21 November no fewer than nine seizures of boats and trawls were made along the coast between Ardrishaig and Inveraray.[184]

One of the Glasgow fish-curers who had been prominent in the agitation against trawling had had a trawl made early in the fishing season and sent it to an upper Lochfyneside man who fished and cured for him.[185] A change had indeed occurred.

Early in January, 1867, Primrose heard from the Lord Advocate. He intended to prepare a bill to amend the existing fisheries legislation and to cancel the prohibition of trawling.[186] In that month, too, seven skiffs and 25 nets were recorded as seizures in the Kyles of Bute.[187] These figures are perhaps indicative of the extent to which trawling was going on, rather than of an increased efficiency on the part of the policing forces.

On 15 July, 1867, the amending bill was passed as Act 30th and 31st Victoria, chapter 52, and with it the restriction on fishing methods was repealed.[188]

By 7 August Primrose could assure the Lord Advocate that 'all is perfectly peaceable'. The previous week's fishery had been very successful and excellent hauls of fine herring had been taken at all the fishing stations. Orders had been issued to pay off four fishery constables, and the force would be further reduced and perhaps dispensed with altogether 'if no brawls spring up among the fishermen as the darker nights set in'.[189]

By 21 August, two smacks, 17 skiffs, eight trawl-nets, one sail, and 16 fish-boxes had been returned to their owners.[190] The contents of the Fishery Board's stores at Greenock, Tarbert, and Rothesay were auctioned off at the end of April, 1868, and realised £75 7/3d. The stores were then given up.[191] The 'Barracks' at Tarbert – a five-roomed harbourside building which had housed the fishery constables stationed in the village – was abandoned in May, and its free supply of coals and candles stopped. By then, only Sergeant John Allan remained in the village,[192] the last representative of a set of laws which had proved both ill-advised and unworkable.

REFERENCES AND NOTES

1. A. M. M. Turner, The Distribution and Development of Settlement in the Region Lying North-west of Upper Lochfyne (unpublished thesis). Quoted by A. Fraser, *Lochfyneside,* Edinburgh, 1971, 10.

2. D. C. MacTavish, *Inveraray Papers,* Oban, 1939, 87.

3. A. McKerral, *Kintyre in the Seventeenth Century,* Edinburgh, 1948, 146.

4. A. R. Bigwood, The Campbeltown Buss Fishery, 1750-1800 (unpublished M.Litt. thesis, University of Aberdeen 1972), 8. Extracted from Campbeltown Town Council Minutes, 2 February, 1744.

5. Particular Papers, mss. Inveraray Castle. Communicated to author by E. R. Cregeen, letter of 3 December, 1976. (An additional grievance was recorded in a petition sent in February, 1744, to the Third Duke of Argyll, from Campbeltown: 'In two seasons 1742 and 1743 . . . we were obliged to keep our larger boats on the east side of the land of Lewis, and to carry our small boats to the west side by land at a considerable expense, where we fished only in the open ocean.' *Ib.*)

6. *O.S.A.,* Vol. X, Campbeltown, 552.

7. Bigwood, *op. cit.* Table 5. Between 1761 and 1770, 26 buss bounties were paid at Inveraray, Tarbert, and Kerry, compared with 264 at Campbeltown during the same period.

8. *Ib.,* 13 and 32.

9. *R.R.C.* (1878), viii.

10. Bigwood, *op. cit.,* 13.

11. *Ib.,* Table 6: Number of shareholders in each buss, 1764-95. Table 7: Occupations of shareholders in busses, 1760-95.

12. *Ib.,* Analysis of Campbeltown buss papers: Table 21, 1764-5; Table 22, 1775-6; Table 23, 1784-5; Table 24, 1795-6.

13. *Ib.,* 56-7.

14. *Ib.,* 13-15.

15. *Ib.,* 110.

16. *Third Report of the Fisheries,* 1785, 48.

17. Bigwood, *op. cit.,* 110.

18. *Ib.,* Summary, 2.

19. *N.S.A.,* Campbeltown, 463.

20. *Ib.*

21. *Ib.*, South Knapdale, 264.

22. *Ib.*, Inveraray, 32.

23. B. F. Primrose, *Paper* (1852), 1.

24. B. F. Primrose, F.B.E. (1856), 106-8.

25. *E.R.C.* (1864), 1113.

26. *R.R.C.* (1878), vi.

27. Reports to H.M. Treasury in or since 1848, by J. G. Shaw Lefevre, on the subject of the Fishery Board in Scotland, 23. Pub. 1856.

28. *Ib.*, 18.

29. F.B.E. (1856), 56.

30. *E.R.C.* (1864), 1171. Oral tradition also has it that ring-netting originated by the attachment together of pieces of drift-nets. J. Weir, 28 November, 1975: 'The drifts they wir usin' wisna killin' the herrin' . . . So, they joined two or three barrels (of nets) together an' made a ring-net of it, an' hauled it to the shore.' D. McFarlane, 4 May, 1974, offered a similar version.

31. 1 December, 1837, quoted by J. M. Mitchell, *The Herring − Its Natural History and Importance,* 1864, 61.

32. Rept, 22 December, 1836, Inveraray, A.F. 7/104, 160.

33. *R.R.C.* (1878), xxv.

34. Rept., 21 December, 1839, Ardrishaig, A.F. 7/104, 217.

35. Rept., 10 May, 1842, Ardrishaig, A.F. 7/105, 16.

36. B. F. Primrose to J. Stewart commanding Fish. Bd. cutter, the *Princess Royal,* 22 May, 1849. The *Swift* of Tarbert case, A.F. 3 1/35, 417.

37. J. Stewart − above-mentioned case − 19 May, 1849, Tarbert, A.F. 37/1.

38. 17 May 1849, Inveraray, *ib.*

39. Nov. 1848, Ballimore House, by Strachur, *ib.* (That the trawlers were not 'regular' fishermen, as Campbell contended, was a fallacy which would be repeated by other opponents of the method, to discredit the Tarbert men involved in it, and to advance the case for its suppression.)

40. List of petitions in favour of trawling, A.F. 37/12.

41. 19 December, 1848, Stonefield Castle, A.F. 37/1.

42. J. Miller, general inspector of fisheries, rept. to Fish. Bd., 2 May, 1850, Leith, A.F. 37/2.

43. B. F. Primrose to Miller, briefing him prior to departure, 14 July, 1849, A.F. 3 1/36, 122.

44. B. F. Primrose, F.B.E. (1856), 574: 'The general inspector was against the trawl − one of the hottest of the lot.'

45. J. Miller, *op. cit.*

46. B. F. Primrose to Ld. Advocate, 19 April, 1851, A.F. 3 1/37, 375.

47. *Public Bills,* 1851, Vol. I, 193.

48. When the bill had passed successfully through the House of Commons and was about to be presented in the House of Lords by the Duke of Argyll, Primrose contacted John Richardson, one of the Government solicitors involved in the drafting of the bill, to express his hope that the Duke would 'stick to the Net clause and carry it'. He emphasised that it was 'by no means the same thing to the Board to have the Bill without it, now that the Bill has advanced to its last stage, and so much expectation has been raised in the trade that it will be obtained'. 7 July, 1851, A.F. 3 1/37, 453.

49. Owner of 12 trawl-skiffs and 12 or more trawl-nets. L. Lamb, assist. inspector of fisheries, 8 January, 1852, Tarbert, A.F. 7/85, 315.

50. B. F. Primrose, F.B.E. (1856), 459-60, and Minutes, 2 July, 1851, A.F. 1/18, 42-4.

51. *Public General Statutes,* 1851 (14th & 15th Vict.), 174.

52. *Ib.*

53. 27 December, 1851, A.F. 7/85, 302.

54. J. MacFie, 11 May, 1852, A.F. 7/105, 282.

55. Tarbert, 127 fish-curers and fishermen, 20 July, A.F. 37/4.
Ardrishaig, 88 fishermen and others, 23 August, List of Petitions in favour of trawling, A.F. 37/12.
Saltcoats and district, 122 fish-curers and others, 11 Aug., *ib.*
Skipness and Claonaig, 16 fishermen, 3 Sept., *ib.*
Arran, Saltcoats, and Irvine, 53 fishermen and curers, 28 September, *ib.*
Arran, 93 fishermen and others, 22 November, *ib.*

56. 20 July, 1852, A.F. 37/4.

57. Inveraray and district, 444 signatures, 31 December.
Ardrishaig and district, 313 signatures, 16 December.
East Otter, Kilfinan, and other places in Kerry, 189 signatures, 31 December.
Lochgair and district, 75 signatures, 31 December.
All A.F. 37/7.

58. F.B.E. (1856), 548-50.

59. G. M. Jackson, Fishery Journal of H.M.S. *Porcupine,* 19 June-2 July, 1853, 19 June, A.F. 37/10.

60. *Ib.* 20 and 21 June.

61. A.F. 37/9.

62. Precognitions, Turner/Rennie case, 5-6, A.D. 14 53/288. (Several word deletions and substitutions, and various reorganisations of the text – not indicated in the passages 'quoted' – were made to improve both the fluency and sense of passages.)

63. J. Johnstone or Brodie, Precognitions, 13.

64. C. McKeich, *ib.,* 8/3.

65. *Glasgow Herald* trial report, 30 September, 1853.

66. Precognitions, 127-160, *op. cit.*

67. *Glasgow Herald, op. cit.*

68. G. M. Jackson, to Fish. Bd., 22 October, 1853, Lochgilphead, A.F. 37/11.

69. 28 October, 1853.

70. Neil Fletcher, Precognitions, 32, *op. cit.*

71. *Ib.,* 8/5.

72. G. M. Jackson, letter dated 2 July, accompanying Fishery Journal, *op. cit.*

73. *F.B.R.* (1854), 5.

74. B. F. Primrose, F.B.E. (1856), 602.

75. *R.R.C.* (1863), 9 (quoted).

76. Confidence was such that in 1858 Government cruisers on the Scottish coasts received confidential instructions from the comptroller-general of the Coastguard not to interfere with trawl-fishing, a change of fishery laws being then 'in contemplation'. *Herring Fisheries of Scotland,* Fish. Bd. published statement, 20 April, 1860, 2, A.F. 37/141.

77. Letter to Ld. Advocate, J. A. Moncrieff, 22 March, 1859, A.F. 3 1/42, 112.

78. F.B.E. (1856), 522.

79. *Ib.,* 461.

80. Statistics of herring-fishing boats in the Firth of Clyde 1850 to 1859. L. Lamb, 8 May, 1860, Greenock, A.F. 37/19.

81. 'The trawling was for a long time confined to the broad parts of Lochfyne, but about four years ago the trawlers went above Otter Spit into the narrows. Formerly the coast in that quarter was thought too bold, but they got accustomed to it, and found it good trawling ground.' W. Dawson, W. Hamilton, and J. and D. Bruce, Ardrishaig, *E.R.C.* (1862), 5.

82. J. Fraser, chief constable of Argyll, *ib.,* 6.

83. J. McLean, *E.R.C.* (1864), 1182 and 1184.

84. L. Lamb to Fish. Bd., 30 October, 1858, Inveraray, A.F. 26/9, 55.

85. *Ib.,* 57, 8 November, 1858.

86. L. Lamb, 'Notes', 8 May, 1860, Greenock, A.F. 37/19.

87. L. Lamb, reports to Fish. Bd., 5 and 14 Feb. 1859, A.F. 7/86, 146 and 149.

88. B. F. Primrose to J. Matheson, Lews Castle, Lewis, 16 February, 1859, A.F. 3 1/42, 80.

89. L. Lamb, 25 July, 1859, Ardrishaig, A.F. 26/9, 89.

90. *Ib.,* 93-4. W. McFarlane, 6 August, 1859, Furnace. (Enclosure with rept. below.)

91. L. Lamb, *ib.,* 89-90, rept. of 11 August, 1859, Cumlodden.

92. *Ib.,* 95. Signed by J. Maclaurin, 8 August, 1859. (Enclosure with above rept.)

93. L. Lamb, *ib.,* 90, 11 August, 1859.

94. *Ib.,* 91.

95. J. Maclaurin to the Crown Agent, 26 August, 1859, A.F. 37/13.

96. A.F. 3 1/43, 95.

97. 30 August, 1859, A.F. 37/13.

98. 5 September, 1859, Inveraray, A.F. 37/14.

99. 14 September, 1859, Ardrishaig, *ib.*

100. *Ib.* Petitions identically phrased. The first carried 355 signatures, mainly of Tarbert fishermen and traders, but also those of Campbeltown, Skipness, Ayrshire, Bute, and Arran fishermen. The second was signed by 173 fishermen and others of Ardrishaig. Both dated 19 September, 1859.

101. *R.R.C.* (1878), viii-ix.

102. J. Methuen Jnr., *E.R.C.* (1862), 28.

103. To E. Ellice, M.P., 9 April, 1860, A.F. 3 1/43, 351.

104. 22 March, 1859, A.F. 3 1/42, 112-3.

105. Obituary of J. Methuen, *Glasgow Herald,* 4 September, 1862.

106. *The Trawl – Herring Fishing in Scotland,* 4-5, A.F. 37/197.

107. *Parliamentary Papers,* 1860, III, 593-9.

108. Petitions accompanying *The Trawl . . . op. cit.*

109. Draft of bill prepared by B. F. Primrose and passed on to David Mure, Ld. Advocate, substituting the existing prohibitory statutes with a clause granting power to the Board to form bye-laws. 22 June 1859, A.F. 37/197.

110. Letter, 22 March, 1860, London, A.F. 37/197.

111. *The Morning Journal,* 11 April, 1860, 5, A.D. 56 95/3.

112. To E. Ellice, 9 April, 1860, A.F. 3 1/43, 370.

113. B. F. Primrose to Ld. Advocate, 7 May 1860, A.D. 56 95/3.

114. *Glasgow Herald,* 8 and 9 May, 1860.

115. *Parliamentary Papers,* 1860, III, 601-8.

116. 11 May, 1860, Kilberry, A.D. 56 95/3.

117. *Public General Statutes,* 1860, 448.

118. Statement at Fish. Bd. meeting of 14 November, 1860, A.F. 37/197, 23-4 (hand written).

119. *Public General Statutes, op. cit.,* 448-58.

120. *R.R.C.* (1863), 6.

121. *Cintra* – Malcolm McDougall; *Elk* – John McEach; *Atlantic Scout* – John McFarlane; *Rover* – James Hay; *Elk* – James Campbell; *Flyer* – Archibald Smith; *Mars* – John Johnson; *Sarah* – Donald Bain; *Betsey* – Dugald McFarlane; *Fly* – James Law; *Catherine* – Dugald Thomson; *Vesta* – Archibald Black; *Kelpie* – John Carmichael. Receipts, A.F. 37/21.

122. B. F. Primrose to Ld. Advocate, 1 November, 1860, A.F. 3 1/44, 30.

123. *Public General Statutes, op. cit.,* 451. Owner's name and name of port to be painted distinctly in white or yellow Roman letters on the boat's stern. Letters to be 2 inches high and on a black ground.

124. R. Ballantyne, fishery officer, 28 September, 1860, A.F. 37/20.

125. 20 October, 1860, A.F. 37/21.

126. A.F. 3 1/44, 14.

127. W. Gillis and G. Thomson, fishery officers, 5 September, 1860, A.F. 37/22.

128. P. Wilson, fishery officer, 15 December, 1860, A.F. 37/25.

129. 19 June, 1860, A.F. 37/137.

130. A.F. 37/22.

131. J. Miller to Fish. Bd., 19 November, 1860, advising against supplying a weapon, A.F. 37/23.

132. Minutes, 28 November, 1860, A.F. 1/23, 11.

133. J. Miller to Fish. Bd., 28 November, 1860, Tarbert, A.F. 37/23.

134. J. Miller, 22 December, 1860, Tarbert, A.F. 37/25.

135. *Glasgow Herald,* 5 Feb. 1861.

136. *Hansard,* Vol. 158, Third Series.

137. *Parliamentary Papers,* 1861, IV, 639.

138. *Public General Statutes,* 1861 (24th & 25th Vict., cap. 72), 370-3.

139. D. McBrayne, *The Scotsman,* rpt. of trial, 16 July, 1861, and Precognitions, Hawton/Parker case, 2, A.D. 14 61/275.

140. D. McBrayne, Precognitions, 3, *op. cit.,* and A. McBrayne, *The Scotsman, op. cit.* (quotations).

141. N. McEwan, Precognitions, 25-6, *op. cit.*

142. D. McBrayne, *ib.*, 5-7.

143. *Ib.*, 9-10.

144. Lieut. E. Lodder, *ib.*, 64/3.

145. R. Hawton, *ib.*, 104-6.

146. W. G. Parker, *ib.*, 111.

147. R. Hawton, *op. cit.*, 107-8.

148. John Hunter, Dugald Campbell, and Hugh Jackson, 'surgeons', Precognitions, 114-18 (entire report), *op. cit.*, and J. Hunter, *The Scotsman, op. cit.* (quotation).

149. He had entered the Royal Navy in 1809, was promoted to commander in 1831, to captain in 1838, to Rear-Admiral in 1857, and would attain the rank of Vice-Admiral in 1864, six years before his retirement. W. Laird Clowes, *The Royal Navy,* London, 1903, Vol. VII, 569.

150. *The Scotsman, op. cit.*

151. *Ib.*

152. *Ib.*

153. *Ib.*

154. Minutes, A.F. 1/23, 172, 201, and 264.

155. G. Thomson, 30 Dec. 1861, Ardrishaig, A.F. 26/9.

156. Duncan McLellan, and Robert and Donald McFarlane;
Malcolm McFarlane, and Neil and John McMillan;
John McFarlane, Malcolm Smith, Edward Campbell, and Robert McFarlane Jnr.;
Archibald Smith, John Johnson, and Duncan Campbell (the *Solan,* G.K. 505);
John Smith, Alexander McDougall, and Dugald McFarlane (the *Mary,* G.K. 444);
Robert Thomson, John Law, and Donald Kerr (the *Catherine,* G.K. 503);
Duncan McAlpine, Neil Smith, Neil McMillan, and Robert McFarlane (the *Star,* G.K. 440);
Donald McCallum, Lachlan McLullich, and John Guille.
A.F. 37/157.
James McCaog, member of Campbeltown crew – refer to 159 – possibly in an advisory capacity.

157. Neil and Dugald McEwan, Hugh McKellar, and Robert Campbell, A.F. 37/157.

158. John and Duncan Martin, James Robertson, John and Peter Carmichael, and Campbell Stewart, *ib.*

159. Duncan O'Hara, Alexander McLean, and William Davidson, *ib.*

160. B. F. Primrose, *E.R.C.* (1877), 2.

161. *E.R.C.* (1862), Lochfyneside evidence, dated.

162. *Ib.*, 11.

163. *Ib.*, 12.

164. *Ib.*

165. *F.B.R.* (1862), 4.

166. Rept. for 1862, A.F. 26/10.

167. Lieut. H. McN. Dyer, H.M.S. *Jackal,* 4 August, 1865, Ardrishaig, A.F. 37/40. He commented: 'It is not easy to elude their vigilance. If they are in earnest, the boats are always hailed in Gaelic, and not being able to answer in that tongue at once betrays us.'

168. On 1 July, trawls discovered and seized on the Ardrishaig road, at Silvercraigs, and on Eilean Mór, off Silvercraigs. An Ardrishaig skiff, the *Margaret,* owner Robert Dawson, had already been detained. J. Murray, fishery officer at Ardrishaig, repts. 22 June, 1 July, and 3 July, A.F. 26/10.

169. *R.R.C.* (1863), 7.

170. *Ib.*, 31.

171. *Ib.*, 30.

172. *Ib.*, 8.

173. *Ib.*, 31.

174. *R.R.C.* (1865), xliv-xlv.

175. J. Murray, 24 February, 1864, A.F. 26/10, 100.

176. G. Reiach, assist. gen. inspector, 9 August, 1866, A.F. 7/88, 17.

177. Minutes, 23 December, 1863, A.F. 1/24, 83.

178. A.F. 7/88, 21-2 (first letter).

179. G. Reiach, 11 September, 1866, *ib.,* 25-6.

180. G. Reiach, 15 September, 1866, *ib.,* 26.

181. Minutes, special meeting, 26 September, 1866, A.F. 1/25, 35.

182. 28 November, 1866, A.D. 56/97.

183. J. Murray, rept. for 1866, 21 January, 1867, Ardrishaig, A.F. 26/10, 243.

184. 16 November. The *Mary* of Lochgair, G.K. 383, with trawl on board (Donald Fletcher, Robert Mitchell, and John Currie);

   The *Janet* of Furnace, G.K. 885 (Neil McCallum and John Cameron);

   Trawl belonging to William McKellar, West Otter. 21 November. Small boat, with trawl on board, belonging to Peter Crawford, Inverae;

   Trawl concealed beneath boat on shore at Achagoil;

   Trawl, belonging to Robert Munro, in net-store at Achagoil;

   Trawl sunk in two fathoms of water off Sandhole.

   Both rpts. J. Murray, 19 and 22 November, A.F. 26/10, 229-33.

185. G.Reiach, 17 November, 1866, Greenock, A.F. 7/88, 45.

186. B. F. Primrose to J. B. Nicolson, private secretary to Ld. Advocate, 14 January, 1867, A.D. 56/96.

187. A.F.·3 1/45, 789-93.

188. *Public General Statutes,* 1867 (30th & 31st Vict.), 426-30.

189. A.D. 56/96.

190. G. Reiach, rept., 21 August, 1867, Ardrishaig, A.F. 7/88, 111.

191. Minutes, 20 May, 1868, A.F. 1/25, 317.

192. *Ib.,* 311.

2

Traditional Community Ways

Housing

THE village of Tarbert, whence ring-netting was pioneered, is built on the eastern
shore of the narrow neck of land which links Knapdale, in mid-Argyll, to the
Kintyre peninsula, and from that geographical feature derives the name, An
Tairbeart, the isthmus. The village in the mid-nineteenth century was characteristically West Highland in appearance, as described by the author 'Gowrie'
(William Anderson Smith), who lived there in 1867. His account is uniformly
sympathetic, though not unaffected by occasional disgust, conditioned, it may have
been, by that very sympathy: 'Facing the beach there is a line of white-washed
modern-looking houses, but beyond – inside this cleansed epidermis – the whole
town is composed of miserable hovels, apparently thatched about the
commencement of the century, but for many a year in the habit of growing much
heavier green crops than I saw anywhere in the cultivated grounds of the
neighbourhood. The rain has painted the fronts of these huts a dirty green, with
colour drawn from the reeds and grass above, giving them a most filthy appearance.
One would think that neither cleanliness nor any other good thing could proceed
from such holes, yet the people don't look so dirty as you would expect, and they
seem healthy . . .'[1]

In such Lochfyneside crofting-fishing communities as Low Kames, thatched
dwelling-houses were occupied into the earliest years of the twentieth century. The
thatch, or *tugha*, was cut on a hillside a mile distant. The men went by moonlight,
because 'they wirna supposed tae be going there for it', and carried a bundle of the
rushes (*sguab thugha*) home on their backs. These were stored until a quantity
sufficient for thatching had been amassed. The thatch, when laid on in a series of
strips – working, bottom to top, from a ladder – was pegged to the underlying
layer of turf (*sgrath*) with rows of hazel-rods. Each rod, or *sgolb*, had first been
cleaned with a knife, and sharpened at each end. Similar rods – differentiated,
however, by the name *pinneachan* (pins) – were used for pinning in the eaves the
netting which covered and secured the thatch from high winds. Re-thatching would
be done every second or third year.[2]

The croft-house of the Campbell family at Silvercraigs, four miles south, was
slated in the nineteenth century, at some time long adrift of living memory, but the
thatching of the byre continued well into the present century. The house comprised
a kitchen and a bedroom, with a loft above, which was reached from the kitchen by
a wooden ladder connected to a trapdoor. The bedroom contained two large beds,
and the kitchen a box-bed, under which was kept a home-made *hurlie* (a moveable
bed). Most of the children bedded down in hurlies in the loft, which was partitioned
to allow room for the storage of nets and other fishing gear.[3]

4. A former dwelling-house at East Kames, Lochfyneside, abandoned during the latter half of the nineteenth century. Converted into a byre c. 1890, the thatched roof was replaced with corrugated iron. Photograph by Lachlan Paterson, Carradale, November, 1979

The cooking pots and kettle were hung above the fire on an iron chain (*slabhraidh*), which was suspended from a bar, or *bolt*, built into the gable of the house. In many houses that bar was the stock of an old anchor. The height of the pot-hook above the grate could be adjusted by raising or lowering it a link on the chain, and, when cleaning the chimney, the chain could be detached completely from the bar. In the houses at Low Kames the occupants themselves built the fire-place. Sticky blue clay was taken from the ebb and packed between stones or − if obtainable − bricks at the sides of the fire-place. To form a grate, three or four iron ribs were then embedded in the clay, which would solidify like cement. Peat was said to burn longer in a brick-and-clay fire-place, which heated less than one made of stone. In the early nineteenth century the fire − or *gealbhan* − was contained in an iron creel (*creathal*) placed upon a flagstone on the earthen floor, with the smoke going up through an 'opening' (*fosgladh*) in the thatch. A few of these drystone houses − 'They would see, through the wall, the cabbages they wir gonny eat' − still exist, in a ruinous state, above the shore at Low Kames.[4]

By the end of the nineteenth century the houses were illuminated by paraffin lamps. Earlier in the century, however, the standard lamp was the cruisie, or *cruisgean*, an open metal dish with a spout or *strup*, into which a wick of several skinned rushes, tied or woven together, was laid. The fuel commonly used was home-produced cod-liver oil. Each family had also metal moulds in which candles − usually of mutton fat − were made.[5]

Food and Physique

The staple diet of all the fishing communities of Kintyre and Lochfyneside was fish, and especially herring, eaten fresh in summer and autumn, but pickled in barrels before the onset of winter. The fish would invariably be accompanied by potatoes, which many families grew for themselves. Most crofter-fishermen on the shores of Upper Lochfyne owned a cow, for which their plots would provide sufficient grass (p. 54). Milk was very much part of their 'ordinary food'.[6] Oats, the fourth important element in their diet, were eaten as *brochan* (porridge) or as *aran-coirce* (oat-cake), and could be purchased at 15/9d per boll (140 lbs.) in 1859.[7]

Such families as the Campbells and the MacVicars, however, grew their own oats. The seed was hand-sown, from a sheet knotted around the neck, and the field would be harrowed immediately afterwards using a borrowed horse. The crop was reaped with a *corran* (sickle) and tied in sheaves by women and children following the reaper. A sheaf was termed a *sguab*, four or six of which together formed an *adag*. After about a week's drying, the crop was stacked into a *mulan*.[8]

The flail (*sùist*) was, within living memory, replaced by a threshing-stick (*maide-bualaidh*) among the crofter-fishermen of Low Kames. The operation there was an outdoor one, the oats being beaten on a board – perhaps a plank out of a skiff – laid on an old sail or a tarpaulin. The grain (*sìol*) was later shaken in a riddle (*ruidil*), formed of perforated sheepskin stretched and laced to a wooden frame. The remaining chaff (*moll*) would blow away, and 'weed seed' pass through the skin. Seed for the next season's sowing would be selected by hand. Quern-stones for grinding oats, though remembered, ceased to be used at some stage of the nineteenth century, and oats were instead milled.[9]

The main fertiliser was seaweed (*feamarach*).[10] The Campbells could gather ample supplies in the bay at Silvercraigs, and, as the field lay by the shore, the wrack had simply to be forked into barrows and wheeled away for spreading.[11] The MacVicars, however, when seaweed was scarce around Kames, had to go off in a small boat and gather elsewhere. The weed was cut from the rocks at low tide using old sickles, then carried to the foreshore in creels and left in heaps to drain before loading into the boat. With a 'good ebb' 30 to 35 creel-loads might be gathered by a single family.[12]

The potato grounds at both Silvercraigs and Kames were, within living memory, dug by spade (*caibe*), though further back the distinctive *cas-chrom* or *cas-chaibe*, as it was also known at Kames, was employed.[13] The crop was also dug by hand, using a fork (*gràpa*), and the potatoes gathered into baskets, hand-made of hazel and willow rods. The crop was preserved in a pitted heap (*dùn buntàta*) covered over with bracken or rushes and an outer crust of firmly packed earth.[14]

Hay was also grown on the crofts, and small quantities of turnips and other vegetables. While the men were absent from the crofts, such as at the Minch herring fishery in May and June, the women would attend to the necessary tasks.[15] Some Tarbert fishing families maintained crofts during the nineteenth, and to an inconsiderable extent into the twentieth century, but, as 'Gowrie' emphasised: 'Everything is fish; the very little patches of cultivation, recovered at great labour from the rocks and hills, belong to the fishermen, and you often see a man in the fields throwing his sail, with the yard and mast attached, over a stack of corn to preserve it from the weather. When the stacks are thatched and completed too, the universal covering is an old herring net, bound neatly round them . . .'[16]

The crofters at Kames and elsewhere would buy a sheep from a nearby farm, kill it, and preserve the flesh in brine for winter. Many crofting families kept a pig, perhaps for six months, and then killed it for winter food. Its flesh, too, would be pickled, but was later rolled and hung in the kitchen.[17] Butter was churned on most crofts, and scones and oat-cakes were produced almost daily. At Silvercraigs the large oat-cakes were toasted before the fire after removal from the griddle, and then hung in a flat home-woven basket from a kitchen rafter.[18]

The diet of the crofter-fishermen and their families in the early years of the present century included not a few foods which are seldom, if ever, eaten now. Of these may be mentioned *brochan-càil,* curly kail and oatmeal cooked in water,[19] champed potatoes, eaten with milk and perhaps cheese, especially when money was scarce,[20] and nettle-soup, a spring medicine taken to 'clean the blood'.[21] Other elements in the people's diet, which have since been rejected, such as dried saithe and cod-head soup, will be referred to later.

Village stores retailed a variety of foodstuffs which, though basic by today's standards, would, in the main, have been unobtainable − except occasionally − by the majority of the fishing population in the mid-nineteenth century. Munro's Store, Furnace, sold beef salted by the barrel at £2 5/8d, and a sheep weighing 60 lbs. could be bought for £1 8/-. The sale of hams, costing 17/- each, increased after 1865. Casks of butter were priced at £1 5/3d, and a cheese at 6/9d. Treacle, which was very cheap, was commonly used, but syrup only occasionally.

Towards the end of the century, salt beef was no longer ordered in large quantities; instead, minced collops, mutton, and tins of tongue − at 1/2d each − were bought. The appearance of bakeries in the latter half of the century discouraged bulk-buying of meal and flour, and many people instead bought bread, rolls and scones, and, from time to time, sweet products such as buns, fruit-cakes, tarts, and sultana-cakes. New Years were celebrated with an increasing variety of luxuries as the century approached its end. In 1886, apples, oranges, lime-juice cordial, lemonade, almonds, and sweets were purchased by the population of the Furnace district.[22]

The most important food was, however, fish, and not least of all because it was obtainable at 'only' the cost of physical toil. Without a supply of herring, especially in pickle, to sustain them throughout winter, the fishermen and their families would face starvation. During the six weeks preceding mid-October of 1835, merely six barrels of herring were caught in Lochfyne, and many fishermen 'had it not in their power to preserve a few herrings for the use of their Families'. The result was near-starvation, without prospect of improvement, as there was no evidence that the 'tempestuous' weather would abate, or that herring would appear.[23] Such deprivation returned to the lives of the Tarbert fishermen and their families 16 years later, in 1852, with the introduction of the anti-trawling legislation. Their hardship was intensified by the failure of the potato crop, and the majority of them was reported by July to be in 'absolute starvation' (p. 11).

Herring taken ashore for salting were called, in Tarbert, *sgadan taighe* ('house herring'). The fishermen would fill with herring a leg of their oilskin trousers, tied at the bottom, and carry the load home on their shoulders. Herring when salted were termed *sgadan saillte* or *sgadan goirt*.[24] In the crofting communities cod and saithe were immersed in tubs of brine and then laid in rows on the roofs of the houses to dry. Cod, especially, would later be hung in the loft, beside the chimney, where they remained dry.[25]

Small saithe, caught at evening with rod and line from rowing-boats, were also salted at Silvercraigs and Kames. In order to put into use more than one rod per person, the rods were secured at the stern by means of a removeable beam, known at Kames as a *batag* and at Silvercraigs as a *sgathach*. Though serving the same purpose, they differed in design, a noteworthy point – the communities were not five miles apart – in this age of mass-produced, standardised utensils. The *batag* was a single beam laid across the stern of the punt (*geòla bheag*), overlapping each gunwale by several inches for security. Brushwood was lashed to the underside of the beam using light rope, and the sharpened ends of the fishing rods – *slatan-iasgaich* – were shoved in.[26] The *sgathach* was formed of two pieces of wood into each of which was cut a series of notches. When the separate boards were placed together, the corresponding notches formed holes into which the rod-ends could be inserted.[27] In both cases, one or two – usually two – persons sat aft on the beam tending the rods, while an oarsman kept the boat in slow motion through the dusky waters.[28] These evening trips along the shores were happily remembered, having had a recreational as well as an economic value.

At Kames, cod-heads would be boiled by the potful. The flesh was picked off and eaten, while the juice, with salt added, provided a soup.[29] At Silvercraigs, and in other crofting-fishing communities, cod-livers were melted in a large stone jar on the hob. The oil, when strained clear, could be applied to seaboots to keep the leather supple (p. 109), and was also taken as a medicine.[30]

Shellfish were also gathered and eaten, especially winkles (*faochagan*) and razor-fish (*spooties*). The latter, which could be collected in worthwhile quantities only by an exercise of patience and skill, were immersed in boiling water to facilitate shelling; after shelling, they would be fried in butter. Winkles were gathered for sale in Ardrishaig and provided a little additional income to the household.[31]

In springtime, when food was scarce, some Tarbert families might 'walk the White Shore' to gather shellfish. The *maorach-bàn* – 'white shellfish', smooth-shelled, deep-burrowing bi-valves similar to the cockle – would be raked up at ebb tide and carried home in baskets. They could be boiled or stewed – 'That's what they wir livin' on in the springtim'.'[32]

Many country folk watched the offshore waters for such appearances of herring as gulls and gannets distinctively feeding, and fish breaking the water. The repeated contention, in the mid-nineteenth century, that the trawlers were 'in general, not genuine fishermen, but interlopers, consisting of tradesmen, small farmers, farm servants and other landsmen',[33] was erroneous, but there were undoubtedly opportunist fishermen in the coastal villages north and south of Tarbert. As late in the century as 1881, trawl-nets were being maintained in the west Kintyre farming community of Rhunahaorine for occasional use. In December of that year, a farm-manager there, Charles McKinven, having seen 'an unusual appearance of herring', alerted a crew, and 50 *maize* (about 25,000 fish) were taken. [34]

There were more methods of catching herring than trawling and drift-netting. In 1864, Angus Brown of Tarbert described 'jigging' for herring with a hand-line weighted at the end, and with four 15-inch (0.38 m.) lengths of wire attached horizontally, all bearing an unbaited hook on each end. The jiggers could be used effectively – by rhythmical jerking – as deep as 15 fathoms (27.43 m.), and could provide a single crew with as many as 600 herring. [35] In 1866 the curiously named 'blooncer-net' made its appearance on Lochfyne. It was a bag of netting, 10 to 12 feet (3.05-3.66 m.) long, attached to an iron ring of 4 or 5 feet (1.22 or 1.52 m.) diameter. When towed astern of a boat it was 'said to be more useful in taking herrings than the drift-net where the shoal is dense, but not equal to the trawl net'. [36] And in Gareloch and the Clyde estuary, a year later, 'large quantities of excellent herrings' were caught using small lap-nets and ordinary quarter-cran baskets. [37]

Saithe, a once-prolific coastal fish, was eaten in all fishing communities, though, in some, probably as a stand-by food. Certainly, a distinct aversion from the fish has prevailed in most parts of Kintyre during the present century, which might just be attributable to connnotations with poverty. The use, in the nineteenth century, of small-meshed trawl-nets to capture the young of the species – referred to in Fishery Board records in connection with Tarbert, Carradale, and Campbeltown – is briefly noted on p. 189.

E. Estwyn Evans, in *Irish Folk Ways,* noted that the saithe 'has different names in Gaelic for every stage of its growth to demonstrate its former importance and popularity', [38] and that is true for Kintyre. Gaelic names survive vigorously: 'cuddin' or 'cuddie' (*cudainn*) is the smallest stage, and is followed by 'gleshan' (*glasan*), 'peuchtie' (*piocach*), and, finally, the largest stage, 'stainlock', a word of probable Norse origin.

Most crofting-fishing families had their own well, on the bottom of which might be spread *chuckies,* shining-white shore pebbles. In most wells could be found an old, solitary trout, which served to keep clean the water, particularly of grubs and flies in summer. These trout would become quite tame, and appear from cover to accept worms or oatmeal dropped in for them. Water could be carried from a well in either a bucket or a *stòp* (stoup). [39]

The fine physical condition of the Lochfyne fishermen, and particularly those of Tarbert, was not infrequently noted by natives of Argyll as well as by impressionable visitors. John Campbell of Kilberry informed readers of the *Glasgow Herald* on 20 June, 1860, at a time when the public image of the Tarbert fishermen was at its most wretched: 'When I see a strapping, powerful man, tidily dressed in seaman's attire, I safely conclude he is a Tarbert fisherman.' Another Argyll landowner, William Campbell of Dunmore, was even more determined to convince readers of that hostile newspaper of the admirability of the Tarbert fishermen: 'I assure you, on the faith of having had opportunities of observing men in many parts of the world, that I have never seen a more muscular, full-sized, full-fed, in short, a finer body of young men.' [40]

The extraordinary extent to which they employed oars on the trawl-skiffs undoubtedly developed their physique, as did the severe physical demands of their work. The denunciation of the upper loch drifters, by the Ardrishaig and Tarbert trawlers, as 'a lazy set of men', will be examined in the next chapter, but that trawl-fishing was the more uncomfortable and exhausting of the two occupations is a certainty.

Almost 40 years later, an anonymous writer in the *Glasgow Evening Times* attributed the fishermen's physical development to dietary factors. 'Great strong men are the Lochfyne fishers,' he wrote, 'and it is doubtful if their marrow can be

5. Uncovered, the MacVicar well at Low Kames, Lochfyneside, which continues to supply the two houses in constant occupation there. Three wells originally served the communities at East and Low Kames, but the other two gradually filled in through disuse. Photograph by Lachlan Paterson, Carradale, November, 1979

matched on any other part of the Scottish coast . . . Tall, deep-chested and shaggy, no better proof could be had of the nutritive properties of fish as a daily diet.'[41] Dugald Mitchell was more qualified than most to comment. A native of Tarbert and a doctor there, he described the Tarbert fishermen in similar terms, but added: 'He cannot, however, be said to be long-lived. The exposure he endures, the spurts of hard work, and the quick cooling down in the cold night air seem to tell upon him speedily, and at a comparatively early age he . . . is forced to give up work. For a Tarbert fisherman to reach the age of seventy is quite an exceptional circumstance, and few grey heads are to be found in their midst.'[42]

D

The discomforts in the lives of the trawl-fishermen were numerous. One, the early practice of wading into the sea to haul the nets ashore, will be referred to later in this chapter, in an extended context, but there were other (unavoidable) afflictions. In summer-time, fishermen were frequently subjected to irritation of skin and eyes caused by the stinging tentacles of the jellyfish *Chrysaora isosceles* (known to fishermen, in Scots, as *scowders*, or 'scorchers', and in Gaelic as *sgaldrachan*, or 'scalders').[43] These jellyfish, which bear 24 long tentacles, were frequently netted in great numbers and left slime and parts of their tentacles on nets and ropes. The work of shaking the nets, and even the flurrying of herring taken on board, spread the irritation to the men's faces. The stinging properties of the tentacles were effective even after the jellyfish had disintegrated into powder in a net stored away for a period of many months, as the author knows from personal experience.

Fishermen were prone to rectal illnesses, such as piles, which can be caused by chronic constipation. Some fishermen habitually passed the entire working week of five or six days without excreting anally, and would − or could − do so only when they returned home at the week-ends.[44] Meal and water mixed and health salts were taken by the skiff fishermen of the present century to relieve constipation (p. 106), and some believed that a walk ashore, during the week, helped.[45]

The disease fistula was specifically referred to by a Commissioner of Fisheries, Lord John Murray, in a letter to the Fishery Board in 1849. The rather bizarre remedy which he suggested was the cutting of holes in the thwarts on which the fishermen sat to row. The advice of four Scottish medical authorities was sought by the Board, and the opinion of three was that such a resort would 'rather aggravate the evil'. The disease, they agreed, 'proceeded from disordered bowels', and was 'increased among fishermen by the difficulty they experienced in relieving nature while at sea in their open boats'.[46]

Fuel

Peat was the main fuel of the poorer people − most fishermen included − who lived along the shores of the Kilbrannan Sound and Lochfyne, with wood an equally inexpensive alternative. Coal, for the minority in the nineteenth century which could afford it − principally in the larger communities such as Inveraray, Ardrishaig, and Campbeltown − cost between 12/- and 14/- per ton during the 1840s, and was shipped from Glasgow and Ardrossan.[47] Smacks, which were frequently redundant fishing craft (p. 58), brought coal to all the sizeable fishing communities until the 1920s. Where a suitable pier or jetty was lacking, they would be anchored in a bay, and the cargo discharged into rowing-boats, thence into carts on the shore.[48]

On certain parts of the coast, such as at Silvercraigs, peat deposits had long been worked out by the end of the nineteenth century. Both the fishing families there, the Campbells and the MacEwans, brought coal from Greenock in their skiffs in April, before the start of the fishing season. The 10-ton cargo would be discharged into punts, rowed ashore, and then carted or barrowed the short distance to the

house, outside of which it would be piled into a *gualag* (coal-heap). That supply
would be supplemented by drift-wood gathered at Otterferry, on the opposite shore
of the loch, and loaded into a small boat, and by withered whins, burnt the previous
spring. Fired heather (*falaisg*) was cropped for kindlings.[49]

Peat was cut annually by the crofting-fishing population of the Kames district.
The account which follows is much simplified. Initially, the turf would be removed
from the bank (*bac-mòine*) using a spade. Then the cutting began, each peat (*fòid*)

6. The coal-smack *Euphemia* of Carradale, after discharging a cargo at Port Rìgh, c.
1912. The *Euphemia,* which was owned and worked by the brothers Gilbert and James
MacIntosh, carried coal between Troon and Carradale until 1916, when Gilbert died and
the vessel was sold. Built in Ardrossan during the late nineteenth century, her capacity
was 32 tons. On her foredeck, the crew is seated to a cup of tea. The elderly man in the
foreground is the Port Rìgh boat-builder, Matthew MacDougall. Per M. MacDougall,
Port Rìgh

being sliced out vertically in a single action of the winged peat-iron (*torbhsgian*).
The peat was tipped from the iron on to the bank, to be handled or barrowed out
and spread to dry. After an adequate drying period − perhaps eight or nine days,
depending on weather − the peats would be fitted together into *teinnteanan*, being
four or five peats upright, with a single one laid longways at the head. When dry,
the peats were built into stacks (*cruachan*) on the hill. (In most areas where carts
could be taken, peats would be removed to the habitation and stacked outside or in a

shed.) The stack was broadly based, and tapered to a single row of peats at the top. The peats were inclined slightly, as they were laid, so that the finished stacks would 'shed the rain'. Loads of peats would be carried from the hill in home-made hazel creels, perhaps a load in the morning and another in the evening. A pile of peats − covered by a piece of sail, perhaps − would be accumulated on the croft to ensure a continuing supply of fuel should weather conditions prevent a tramp to the hill.[50]

Earnings

Fishing has always been, and remains, an economically uncertain occupation. Although trawl-netting did not guarantee more regular wages than drift-netting, its greater productivity occasionally put massive amounts of money into the hands of successful fishermen. John McLean of Tarbert, in the mid-nineteenth century, once saw £550 divided between 16 men.[51] That wage of about £34 each for a week's work may be compared with the near-contemporaneous 12/- weekly wage earned by shoemakers, tailors, and wheelwrights. Only masons earned more than £1 per week among the non-professional class of workers, and even the highest-paid school-teacher in Inveraray could, in 1843, expect a salary of only £39 16/8d, though he was provided with a free house and a potato garden, and had the right to graze a cow on the town's common pasturage.[52]

A large catch of herring usually benefited many more fishermen than those directly responsible for it. From one trawl-net were taken 1,500 *maize* (three-quarters of a million) herring, loaded into 13 smacks and skiffs, before H.M.S. *Jackal* 'scared away' the fishermen.[53] The payment to the crews of these carrier boats was either £5 − or £1 or more to each man, a satisfactory week's wage − or one half of the profits of the load, a less certain reward.[54]

The sale prices of herring fluctuated according to the quantities available on the Lochfyneside markets, but one set of 1860 prices, for winter herring, reveals that trawled fish of a 'fair size' sold at 1/7d per 'long hundred' of 120 fish, and drift-net herring at from 1/6d to 2/-.[55] Drift-net herring fetched, on average, 1/- more than trawled herring, and occasionally more; but, infrequently, trawled herring would sell for 1/- and 2/- more per long hundred.[56]

'The Tarbert Man must have his Dram'

The operation of working from the shore was evidently an uncomfortable one. Laurence Lamb's report in 1851 of accounts received from three Tarbert fishermen indicates as much: 'They stated that the fishermen engaged in (trawling), having to wade into the sea, are almost constantly wet to the neck, and in this state they come on shore and go direct to the dram shop where they will remain until it is time to go to sea again. This conduct they will continue for days and in this manner nearly the whole of their earnings are squandered.'[57] The author's great-great grandfather John Martin, a Dalintober fisherman with practical experience of both trawling and drift-netting, gave a similar account in 1864: 'The trawlers are in a great way subject to getting very wet and draggled with dragging the herring and nets, whereas the drift-net fisherman can put on a suit of oil-cloth and keep himself almost dry.' He added cautiously: 'No doubt they very frequently take a glass of spirits or whatever they choose when they come in.'[58]

Serious drunkenness among the trawl-fishermen was evidently prevalent about that time, though the tendency would hardly have been of recent origin. John Stewart, commander of the *Princess Royal* fishery-cutter, had reported drunken squabbling in September, 1846: 'I am sorry to say that the fishermen at Ardrishaig and Tarbert have frequently quarrels among themselves. These arise from the large sums of money which they realise and spend mostly on spirits. I am often called upon to settle these disputes, and our interference has always been the means of restoring order.'[59]

Years earlier, before trawling had begun on Lochfyne, Lord Teignmouth, who visited Tarbert in 1827 and 1829, reported 'no less than twenty public houses' in the village. The superintendant of a Tarbert distillery informed his Lordship that the fishermen carried whisky to sea with them. 'Sir,' he said emphatically, 'the Tarbert man must have his dram, let the world sink or swim.'[60]

Heavy drinking was by no means restricted to trawling communities. Lochgilphead, with a population of between 500 and 600,[61] had, in 1844, 23 inns and 'alehouses', selling 5,600 gallons of whisky annually. That number was, however, an 'improvement' because earlier in the century, of the 90 houses in the village, more than 30 had been licensed to sell spirits and ales.[62] 'The almost universal connection between herring-fishing and whisky-drinking makes (fishing) rather a curse than a boon to the people,' wrote a mid-Argyll land-owner, Sir John Orde, in 1844, his eyes on Lochgilphead.[63]

Unlike the drift-net fishermen, whose wages were relatively meagre but more regular, the trawl-fishermen with their 'sudden gains' could afford to 'get on a spree and remain so for a week', as Sergeant John Kennedy of the fishery police, a detached but understandably critical observer, commented in 1862. He described the trawlers as 'a wild set of men'.[64] A Tarbert merchant and banker, Archibald McCalman, expressed a similar opinion: 'The trawlers are a younger and more disorderly set of men, while the drift-net fishers, being older, are steadier.' He admitted that the trawlers were 'given to drink', and offered the ingenious explanation that 'the intermarrying here produces a good deal of this disorder'.[65] The chief constable of police for Argyll, James Fraser, declared that the Inveraray fishermen were, as drinkers, 'as bad as any fishermen in the loch', but added that 'they do not commit as ferocious assaults as the Tarbert men'.[66] Charles Mactaggart, procurator-fiscal at Campbeltown, was in no doubt that trawling led to the 'demoralisation of the people', and described Tarbert, in the middle of fishing, as 'perfect pandemonium'.[67]

Until recent years, an ash tree stood on the northern shore of Camus na Ban-tighearna (The Lady's Bight), and was named Mrs Black's Tree because, according to local tradition, when the fishermen were working along the shore between Tarbert and Skipness, the vigorous Mrs Black would walk across the hills to the bay in her long billowing skirts, with skins of illicit whisky concealed beneath them. The fishermen would put the bows of their skiffs to the edge of a flat tidal rock on the shore beneath the tree, and Mrs Black would go down to them and 'peddle the whisky over the bow' [68]

David McFarlane of Tarbert remarked with some amusement: 'That wis a common thing in the old days, four men goin' away wi' a good *scud* in them; but by the time they rowed doon tae Laggan, they wir sober. The oars wid soon take it oot ye.' He remarked, too, that some of the herring-buyers, who conducted their business from smacks amid the fleet at nights, carried a jar of whisky with them. 'There wis,' he said, 'a certain element that wid go where they wid get the dram.' [69] 'Gowrie' described the joviality of a trip back to Tarbert after a successful night's fishing: 'So . . . we set off home amid stories and good humour − for both smacks handed round the glass liberally.' The fishermen were 'reckless, merry, and good-humoured, all kindly attention to the stranger they knew nothing of . . .' [70]

Clothing

The attire of the fishermen, as described by 'Gowrie', was 'sou'westers and fishing-boots, with guernsey frocks (woollen jerseys) and thick blue trousers; others in blue shirts and heavy waistcoats'. [71] Material with which to make clothing could be purchased from a general store such as Munro's of Furnace. There, in 1860, plyding or plaiding, a coarse woollen cloth − 'not the same with flannel, but differing from it in being tweeled' − was available, white, checked, or dyed, with blue the most popular colour. Clothes of that material 'served in all weathers . . . and were not easily torn'. Red or blue flannel, or 'open woollen stuff of varying degrees of fineness' were used both to make petticoats and to line the men's waistcoats. The women's frocks and skirts were made of guernsey and stocking yarn respectively, both of which materials were knitted from wool. [72]

Prior to the mid-nineteenth century, oilcloth garments and seaboots were by no means standard wear. Thick jackets and short boots − on Lochfyneside, *peitean cùrainn* and 'blootchers' (bluchers), respectively [73] − served, inadequately it may be supposed, as protection from wind and rain. When the grandfather of J. T. McCrindle of the Maidens, Ayrshire, returned home from a six-week trip to the Lochfyne winter herring fishery, one year in the nineteenth century, his jacket was almost falling apart on his back − near-continual rain had rotted the sewing. [74]

Colin McKeich of Tarbert's upper attire on the night that he was wounded (p. 11) consisted of a linen shirt worn under a flannel shirt. [75] In his boyhood in Tarbert during the late 1880s and the early 1890s, Hugh McFarlane regularly saw his father return from a week's summer fishing with the sleeves of his blue shirt stained white by the scales and *sgoll* (scum) of herring. He remembered these flannel shirts as having been about a quarter-of-an-inch thick − 'lik' a blanket' − with a collar split at one side, and four or five buttons attached. A waistcoat was worn over the shirt, and the 'rig' was completed by an expensive silk Navy-style scarf, specially knotted. The most popular hats in Tarbert at that time were round, wide-brimmed 'ministers' hats', but other fishermen wore the unmistakable 'Balmoral' bonnets, and some, who spent seasons on yachts, or who periodically sailed 'deep sea', returned to the village with smart caps, complete with shiny scoops. [76]

7. A group of Dalintober fishermen, c. 1910. Flat caps are already in vogue, though three of the elderly men retain 'cheese-cutters', and one a high soft hat. Waistcoats are worn by the oldest of the men, in retirement, while the active fishermen wear the conventional working dress of the time, dark-blue hand-knitted jerseys and thick dark-blue trousers. The group includes representatives of the major Dalintober fishing families of the period 1885-1935: MacMillan, McKinlay, McGeachy, and Martin. Two informants quoted in this work — Duncan McSporran and Henry Martin — were, as young fishermen, present in the group.

L.-R: Andrew McKinlay; James MacEachran; Campbell Martin; Michael Brodie; John McKinlay; James Finn; Neil Bell; James Smith; Angus MacMillan; Alexander Mac-Millan; Neil MacMillan; Malcolm McGougan; Donald MacKay; John Taylor; Duncan McGeachy; Duncan McSporran; Henry Martin; John McGeachy; Archibald Taylor.
Seated in front: Donald McCallum; Alex Finn; James McKinlay; David Paterson.
Seated on wall: William McKinlay and William Martin.

Per W. Anderson, Campbeltown

The End of the Gaelic Language

The community at Tarbert, in the middle of the nineteenth century, was conditioned by the Gaelic ethos. Deculturalisation was, however, already under way in schools. Most of the children in the Tarbert district in 1840 'understood the English, as greater pains are taken with them in school'.[77] Some of the older villagers were probably able to communicate in English, but for many this was a difficult effort to make, as Gaelic was their first and true language. When Colin McKeich, two days after his wounding, was visited at his home by an officer of H.M.S. *Porcupine,* his communication with the officer had to be assisted by a travelling clock-maker, who was 'better able to speak English'.[78] For the next generation, genuine bi-lingualism would be possible, but in that unself-conscious attainment the wasting of its inherent cultural character, perhaps 13 centuries and 40 generations mature, had irreversibly begun. One generation more and the language would be finished.

The association of the English language with educational and social advancement encouraged, to a considerable extent, a hostility to Gaelic in that section of the community which had successfully 'gone over' to the professionally more useful of the two languages. The Rev. John McArthur, minister of Kilcalmonell and Kilberry, the parish immediately south and west of Tarbert, may relevantly be quoted. 'The Gaelic is the vernacular language of the parishioners,' he wrote in 1843, 'but the English is displacing it, and the sooner it overmasters it the better.'[79]

Another factor detrimental to the survival of the language was the increasing presence of English-speakers in the village, largely a consequence of the operation of steamer services from Glasgow. Tarbert was, in fact, one of the first harbours in Europe to be visited regularly by a steamer, and that vessel was the forerunner of them all, the *Comet*. Her route, for several weeks between Glasgow and Greenock, was extended in September, 1812, via Tarbert and the Crinan Canal, to include Oban, Port Appin, and Fort William.[80] By 1840, no fewer than three steamers visited Tarbert daily throughout the summer, and in winter at least one.[81]

Almost one-sixth of the 131 signatures on a Tarbert pro-trawling petition of 1852 are recognisably non-Gaelic. Of that number, perhaps six are Irish, and probably belonged to fish-curers, but the others include curiosities, Lightbody, Elder, Arbuthnot, and Warren among them.[82]

The author 'Gowrie' lived in Tarbert in 1867, and his observations of the village and experiences at sea with the fishermen constitute a substantial part of the resultant publication, *Off the Chain, Notes and Essays from the West Highlands*. Though the awareness was denied him, his presence there as an English-speaker within a community historically predisposed to the Gaelic language represented, in its early stage, the imposition of an alien culture, with a proved potential for predominance. Paradoxically, perhaps, it is 'Gowrie's' account one must consult for descriptions of the physical character of Tarbert during that middle period of the nineteenth century. The vision of the stranger in an alien country is almost invariably clearer, even if uncomprehending of the *essences* of appearances, for familiarity breeds, if not contempt, indifference.

Hugh McFarlane of Tarbert, who was born in 1884 of Gaelic-speaking parents, did not himself acquire fluency in the language, and his knowledge was largely concentrated in fishing terminology. His explanation, 90 years later, of the Gaelic decline, is tragically familiar, but nevertheless deserves quotation, if only for its implicit condemnation of linguistic passivity: 'It wis dyin' oot gradually aal the time, barrin' when the ould wans wid get thegether . . . The incomin' crowd wir all speakin' English and, not tae offend them, they spoke in English too. It jeest died away from that.'[83] 'Gowrie' was grateful for such consideration, and remembered the fishermen 'delicately apologising for talking Gaelic among themselves when he could not understand it'.[84] For the next generation there would be no need for such apologies, there being no need to use Gaelic.

In both Inveraray and Campbeltown the anglicisation process was far advanced by the mid-nineteenth century. The settlement of south Kintyre by predominantly Scots-speaking families from Renfrewshire and Ayrshire was conducted by the Campbell House of Argyll on behalf of the Scottish government, which was

8. Young boys, with strung herring carried on switches between them. Campbeltown Old Quay, c. 1900. Per W. Anderson, Campbeltown

concerned to subject unruly Gaels in Kintyre and elsewhere. Major colonisation began in 1650,[85] and continued, intermittently, until about 1860.[86] Relations between the native Gaelic population and the 'sober, hardworking' settlers were characterised by hostility from the beginning. In 1843, almost two centuries after that first extensive plantation, a minister at Southend, 10 miles from Campbeltown, could record that the Lowlanders 'have rarely amalgamated themselves by inter-marriages with the Highlanders', and that they continued to worship apart from their 'Highland brethren'.[87] The conclusion of one local writer, Latimer MacInnes, was that 'there was no intention or desire on the part of these Lowlanders to be absorbed in the native Gaelic-speaking population, probably because they regarded themselves as superior stock'.[88]

The decay of the Gaelic language was probably retarded, however, by an influx in the early eighteenth century of Gaelic-speakers from rural Kintyre and the Highlands.[89] By the end of that century, two-thirds of the townsfolk were 'Highlanders'.[90] But, inevitably, the language succumbed, and Kintyre has become yet another withered limb of the once-powerful *Gaidhealtachd*.

With the Lowland lairds and barons there had come to Kintyre numerous tenants and cottars to whom the farms were sub-let by the tacksmen.[91] Many of the descendants of these land-workers removed to Campbeltown and adopted fishing, as an anti-trawling petition signed by more than 200 Campbeltown and Dalintober fishermen in 1852[92] evidences.

9. After the exertion. A skiff has been launched in spring-time, c. 1898, at Port na Cùile. Among the Carradale fishermen is a single Skyeman. Per D. McIntosh, Carradale.

The fishing community at Carradale, the most considerable in Kintyre after Campbeltown and Tarbert, had little contact with the Lowland settlers, whose lands lay principally within the fertile plain of Laggan, to the south. The Carradale fishermen were more familiar with Skye crofters, who arrived annually at the beginning of June and fished with them until September, when they returned home for the sheep-smearing. The communication problem was of quite another order. James McIntosh, who was born in Carradale in 1868, remembered that the Skye Gaelic was 'not too easy understood',[93] a complaint which undoubtedly cut both ways.

The tradition of employing hired hands from the Western Isles had originated with the smack fishery. The petition against trawling subscribed to in Campbeltown and Carradale in 1852 was, when submitted to the Fishery Board, accompanied by a note explaining that there were 18 boats from the district 'gets their Crews from different Parts in the North hilands each Boat requiring from 5 to 7 Men, and as these Men depend principally upon the Success of the drift Nets, it is Natural to infer that they would Support the enclosed Petition'.[94]

With the decline of the Campbeltown smack and lugger fleets in the 1880s, which will be examined in the next chapter, the Skye fishermen ceased to go there, but, in Carradale, crew shortage for the trawl-boats ensured the continuation of the annual influx until about the end of the century. During winter and spring, many of the Carradale skiffs were laid up, but when the Skyemen arrived by steamer, clothed in blue flannel shirts and smelling strongly of peat, the entire fleet was able to go to sea, some boats manned wholly by native fishermen, but many with two natives and two Skyemen.[95]

Examination of that petition of 1852 reveals that most of the prominent names in Carradale then, prevail still. There were, for instance, six Campbells, five MacDougalls, five McMillans, and five Patersons.[96]

The village of Ardrishaig (locally, *An Rudha*), the final of the four important post-1867 trawling communities, clustered into existence after the opening of the Crinan Canal in 1801.[97] By 1840, a population of about 400 was settled at that eastern entrance of the canal.[98] A community of Ayrshire fishermen gathered there, and their presence was surely sizeable to judge by the recurrence of such names as Bruce, Law, Hamilton, and Dawson in the documents of the Fishery Board. No information on the motivation for their settlement is, however, available. They may have been encouraged to go there to stimulate the development of the community – *The New Statistical Account* referred in 1841 to the Ayrshire fishermen's having 'brought in a good style of skiff with a single lugsail'[99] – or they may have heeded the establishment of the village and acted for themselves. Ayrshire fishermen were no strangers to Lochfyne; they had been fishing there since the sixteenth century, and probably earlier.[100]

William Hamilton told the Royal Commission in 1862 that he had fished for 20 years at Ardrishaig, and, before then, for 10 years at Ayr,[101] which dates his settlement in the village to about 1842. Gaelic was then 'generally spoken, but gradually disappearing' in the district.[102]

By the end of the eighteenth century, English was established as 'the prevailing language' in Inveraray,[103] and that state of development was not reversed. In the villages between there and Ardrishaig, however, Gaelic continued predominant. At Furnace in 1883, despite the presence in school of the children of quarry-workers from north Wales, north Ireland, and Aberdeen, the school-master could note that 'a few Gaelic-speaking pupils have incredible difficulty in either speaking or reading the English language'.[104]

On a bright morning in June, 1975, the author sat with an old man in his cottage at Minard, listening to memories. His talk was repeatedly broken by the growl of motor traffic, mainly tourists coming and going on the lochside, intent upon appearances. In that very room, 80 and more years before, his father, James Turner, had regularly sat before the window, at a small table, to divide among his two crews the money earned for the month's work. Sam Turner and his brothers received a penny from each of the seven men, and thought it 'a fortune'.

'If ye wir standin' at that corner at the week-end,' said Mr Turner, indicating the road outside, 'there wirna a word o' English spoken. Every word wis spoken in Gaelic.' He remembered dark nights when the herring fleet was at work in the narrow loch, and the flaming torches on the boats created the impression of a 'wee city'; and still nights, when the Gaelic of the fishermen carried from shore to shore.

'Are there any Gaelic-speakers left in the village?' the author asked him. 'No, they are all dead. There's no' a Gaelic-speaker in the whole village. In fact, the village here is full o' strangers. All the houses are practically sold tae holiday people. Empty all winter . . .'[105]

Superstition

Analysis of the bases of superstition, and of the psychological processes whereby patently illogical beliefs come to exercise a rigid influence on human behaviour, is outside the scope of this book, indeed, outside the capability of the author to undertake. A few conspicuous factors may, however, be noted. Few fishermen were willing to discuss superstition in terms of its significance in their own lives — a reluctance which itself tends towards superstition — but preferred a general, almost impersonal approach to the subject; or else, reversed the direction of the enquiry by citing instances of benefits which resulted from deliberate breaching of superstition. Two examples of that trait are later presented, *viz.* white stones, and rabbits.

'Fishing as an occupation is very dangerous and economically chancy, and as a result fishermen have always been extremely superstitious,' wrote C. L. Czerkawska in *Fisherfolk of Carrick*. 'Basic insecurity and superstition seem to go hand in hand . . .' The period between a fisherman's leaving home, and actually stepping on board his boat, was, as she states, 'a time of preparation for the state of being at sea, and many taboos belong to this transitional period'.[106]

'Taboo' persons achieved their notoriety by the accident of physical deformity, by outlandish appearance, or, in the case of women, by extreme old age. The latter were especially feared, perhaps because of an enduring unconscious association of female decrepitude with supernatural power. One such woman, acknowledged as kindly, but nevertheless regarded uneasily by fishermen, lived close to Dalintober quay. One Monday morning an uncle of the author's, John Martin, turned around the corner of her house with his oilskins under his arm. 'Well, John,' she said from her doorway, 'are ye goin' tae the fishin'?' — 'Well,' he replied, 'A was, but seein' A've met you, A can go away back hame.' And he did.[107]

That certain women possessed power to influence a fishing trip — such as by control of sea winds — is a popular notion in Gaelic folklore. That notion is perhaps preserved in the Tarbert *buitseach* (witch), which was applied to a person who was considered capable — wittingly or otherwise — of exercising bad luck on fishing. An equivalent word was *bòcan*, which is literally a hobgoblin or bogle. The author, as a boy in Campbeltown, was familiar with the word. It was a customary tease-word if a night's fishing had been unproductive. 'A doot we've a bockan aboord,' a fisherman would remark, looking meaningfully at the young passenger.

A Campbeltown fisherman, John McIntyre, admitted to superstitious apprehension whenever he would encounter an eccentric local tramp who slept in closes or in the corners of alleys. If he happened to be seen and greeted by the tramp, he immediately returned home and sprinkled salt over his left shoulder, before setting off once more.[108]

Persons who could exercise good luck on a fishing venture were, significantly, uncommon, and the hopeful impulse which they generated in the fishermen was often spontaneous. As the crew of the *Senga* of Tarbert lounged on her deck on a breezy summer's evening, discussing the state of the weather, and unable to decide whether or not to venture out, an Irish stone-knapper's wife appeared on the quay and asked if they would be going out. They confessed to being 'in the swithers', and she at once ordered them out. That night they filled the boat with herring, and were paid £1 a basket. 'They wir thinkin' that she knew they wir goin' tae get them,' remarked Hugh McFarlane who, as a member of the crew, was present that evening.[109]

Red-haired women of any age were unwelcome on board fishing boats, and would be prevented from casting off mooring ropes in the evening, [110] but Angus Martin of Dalintober attested that his accidentally encountering a red-haired neighbour on a Monday invariably coincided with a successful week's fishing. [111] Ministers and priests were also shunned on working days. [112]

The Devil, it would seem, had a special interest in the Tarbert men. A group of camper-fishermen survived a visitation on a night last century. 'There were a crowd up at the Maol (Dubh), up Ardrishaig way,' said John Weir, 'an' they wir in in a wee bight up there, makin' thir tea through the night, when a voice hailed them from the road. They didn't pause tae consider, but they hoofed it away − "It's Himsel'!" . . . "Himsel'" was the Devil. They were very superstitious. Maybe a tramp goin' past gied them a hail.' [113]

The Gaelic superstition of the unluckiness of hearing, before breakfast, the first cuckoo-call of the year, survived among Tarbert and Carradale fishermen until the present century. One old Carradale fisherman, Archie Paterson, heard the ominous notes of the cuckoo coming from a tree at Cour, while he rested on the stern of an anchored boat, awaiting breakfast. 'Ye bugger,' he spluttered, 'if I had a gun I'd go up an' chip the bloody heid off ye.' [114] The stock remark passed to the luckless person in such a circumstance was, 'The cuckoo shit on ye.' [115]

In accordance with a rare superstition relating to ownership of fishing craft, a fisherman might decline to part with a boat which he no longer required. 'I knew a man that was like that,' said Donald MacVicar of Kames. 'And he left the boat lyin', a good boat, an' it lay on the shore till it rotted. He didn't want tae sell it. He never told me why.' [116] Ernest Marwick, in *The Folklore of Orkney and Shetland*, comments on the superstition and explains it in terms of a belief that 'they would remain alive until the ancient disused boat in which they had fished so often began to fall to pieces'. [117]

Hatch-boards, when removed from a boat's hold, were stacked one on top of another, painted sides uppermost. To turn a board was considered unlucky. [118] Whistling at sea was certain to agitate the older fishermen intensely, and they would reprove the offender for 'whistling up the wind'. [119]

A superstitious observance forbade the turning of a boat 'against the sun' in harbour. The custom was rigidly enforced. As one fisherman remarked, 'If ye turned against the sun, ye could turn right back.' [120] The superstition had − and has − wider applications in society, and Isobel Cameron in her essay *Superstitions and their Origins* reckons the observance of *deiseal* − that is, the way of the sun − to be a survival of a tenet of the Druid religion. [121]

It was believed by many of the crofter-fishermen of Torrisdale that the presence of a 'boutrie bush' (elder) close to their cottages afforded protection from 'evil spirits'. Said James Campbell: 'It may not be beside the door, but it's no' far away.' [122] Rowan was also associated with power, and in a time out of mind Tarbert fishermen would use only tillers made from that wood. [123] A piece of wool, carried to sea in a pocket, was considered lucky by some fishermen. [124]

Taboo words, most of which probably originated on the East Coast of Scotland, [125] were numerous, and substitutes generally served when referring to the objectionable things. Thus, salmon became 'billies'; rats were 'long-tails'; rabbits were 'bunnies', and pigs 'curly-tails', or, in Campbeltown, 'doorkies'. A story of the refusal of a fisherman's young son to speak the word 'pig' was told by George Newlands of Campbeltown. At school Willie was asked to repeat the lesson: 'P-i-g . . . pig', but he resolutely and persistently recited: 'P-i-g . . . doorkie', until the teacher, exasperated, asked: 'Now what's the reason?' He replied, immediately and with relief: 'My daddy'll no' let me say that!' [126]

White-handled knives and white stones were considered unlucky in both Kintyre and Ayrshire. At the appearance of a white stone in a boat's ballast, the old Tarbert fishermen would expostulate: *'Tilg i, tilg i!'* (Throw it, throw it!). [127] Some men intentionally defied the superstition, such as the Ayrshire fisherman who deposited a *chuckie* among the ballast, and at the cleaning of the boat after a successful season produced it and triumphantly displayed it to his skipper. [128]

Another superstition which was frequently defied was that relating to rabbits. Many crews carried guns to sea with them, principally for the shooting of sea-birds (p. 102), but on the Hebridean islands, especially, they hunted rabbits to provide fresh meat for themselves and to relieve the monotony of week-ends spent at anchor in remote bays. [129] Some of the Tarbert fishermen even took ferrets on board their boats, and would go ashore along Lochfyneside and the Kintyre coast when the moon was bright and 'nothing doing'. The superstition was occasionally totally discredited, as on a night when a pair of Tarbert boats were filled to the gunwales with herring soon after a poaching venture. The rabbits, which had been strung on a boat-hook in the hold, were hanging like 'droont mice' when the hold was finally emptied. [130]

REFERENCES AND NOTES

1. 'Gowrie', *Off the Chain: Notes and Essays from the West Highlands,* Manchester, 1868, 64.
2. D. MacVicar, 2 February, 1977 (S.S.S.), and 17 September, 1978 (H.D.S.G.).
3. Mrs J. (Campbell) MacBrayne, 22 February, 1977 (S.S.S.).
4. D. MacVicar, 2 February, 1977 (S.S.S.) and 17 September and 29 October, 1978 (H.D.S.G.).
5. D. MacVicar, 16 February, 1977 (S.S.S.).
6. *N.S.A.,* North Knapdale, 639.
7. A. Fraser, *Lochfyneside,* Edinburgh, 1971, 57.
8. D. MacVicar, 16 February, 1977 (S.S.S.) and 17 September, 1978 (H.D.S.G.).
9. D. MacVicar, 16 February, 1977 (S.S.S.) and 29 October, 1978 (H.D.S.G.).
10. D. MacVicar, 17 September, 1978 (H.D.S.G.).
11. Mrs J. MacBrayne, 22 February, 1977 (S.S.S.).
12. D. MacVicar, 16 February, 1977 (S.S.S.).
13. D. MacVicar, 15 November, 1978 (H.D.S.G.).
14. D. MacVicar, 2 February, 1977, and Mrs J. MacBrayne, 19 January, 1977 (both S.S.S.).
15. Mrs J. MacBrayne, 19 January, 1977, and Ms M. Campbell, 27 January, 1977 (both S.S.S.).
16. *Op. cit.,* 65.
17. Mrs J. MacBrayne, 19 January, 1977; Ms M. Campbell, 27 January, 1977; D. MacVicar, 16 February, 1977 (all S.S.S.).
18. Mrs J. MacBrayne, 22 February, 1977 (S.S.S.).
19. As above.
20. As above, 19 January, 1977 (S.S.S.).

21. Mrs M. MacVicar, 2 February, 1977 (S.S.S.).

22. A. Fraser, *op. cit.,* 57-8.

23. A. Sutherland, fishery officer at Inveraray, rept. to Fish. Bd., 19 October, 1835, A.F. 7/104, 137.

24. *S. taighe* – H.D.S.G.

S. saillte – H. McFarlane, 10 December, 1976; M. MacDougall, 1 March, 1978 (both H.D.S.G.); Ms M. Campbell, 27 January, 1977 (S.S.S.).

S. goirt – H. McFarlane, 10 December, 1976, and G. C. Hay, recorded 11 May, 1979.

25. D. MacVicar, 2 February, 1977, and Mrs J. MacBrayne, 22 February, 1977 (both S.S.S.).

26. D. MacVicar, 29 October, 1978 (H.D.S.G.).

27. Ms M. Campbell, 23 October, 1978 (H.D.S.G.).

28. D. MacVicar, 29 October, and Ms M. Campbell, 23 October, 1978 (H.D.S.G.).

29. D. MacVicar, 2 February, 1977 (S.S.S.).

30. Mrs J. MacBrayne, 22 February, 1977 (S.S.S.).

31. D. MacVicar, 2 February, 1977, and Mrs J. MacBrayne, 19 January, 1977 (both S.S.S.).

32. H. McFarlane, 24 February, 1978 (H.D.S.G.).

33. B. F. Primrose, *Paper* (1852), 4.

34. *Campbeltown Courier,* 17 December, 1881.

35. *E.R.C.* (1864), 1186.

36. J. Murray, fishery officer at Ardrishaig, ann. rpt. for 1866, A.F. 26/10, 243.

37. G. Reiach, assist. gen. inspector of fisheries, to Fish. Bd., 23 April, 1868, A.F. 7/88, 175.

38. London, 1957, 226.

39. Mrs J. MacBrayne, 19 January, 1977; Ms M. Campbell, 27 January, 1977; and D. MacVicar, 16 February, 1977 (all S.S.S.).

40. 29 May, 1860.

41. *A Night with the Herring Fishers,* 1901 (cutting incompletely dated).

42. *Tarbert Past and Present,* Dumbarton, 1886, 115.

43. H.D.S.G. (Tarbert). D. MacVicar, 17 September, 1978, gave *sgaldainn* (H.D.S.G.).

44. A. Martin, 14 May, 1976.

45. As above.

46. B. F. Primrose, to Prof. Miller, transmitting Murray's suggestion, 25 May, 1849, A.F. 3 1/35, 428.

Primrose's consultation with medical authorities reported to meeting of Fish. Bd., 6 June, 1849, A.F. 1/16, 192-3.

47. *N.S.A.,* Saddell and Skipness (which included Carradale), 451; South Knapdale, 278 (in each of these parishes the main fuel was peat); Campbeltown, 468; Inveraray, 43.

48. H. Martin, 3 May, 1974, and Duncan MacKeith, retired farmer, Saddell, 10 March, 1977 (S.S.S.).

49. Mrs J. MacBrayne, 19 January, 1977, and Ms M. Campbell, 27 January, 1977 (both S.S.S.) and 23 October, 1978 (H.D.S.G.).

50. D. MacVicar, 2 February, 1977 (S.S.S.) and 16 September, 1978 (H.D.S.G.).

51. *E.R.C.* (1862), 13.

52. *N.S.A.,* Inveraray, 31 and 39.

53. J. McLean, fisherman, Tarbert, *E.R.C.* (1864), 1184.

54. J. McMillan, fisherman, Tarbert, *ib.,* 1177, and G. Thomson, fishery officer, rpt. to Fish. Bd., 20 October, 1860, A.F. 26/9, 132.

55. W. Gillis, fishery officer at Tarbert, to Fish Bd., 3 Nov. 1860, A.F. 37/22.

56. L. Lamb, assist. inspector of fisheries, to Fish. Bd., 13 April, 1860, Greenock, A.F. 37/18.

57. 29 December, 1851, Tarbert, A.F. 7/85, 304.

58. *E.R.C.* (1864), 756.

59. To Fish. Bd., 30 Sept. 1846, Skipness, A.F. 37/1.

60. Quoted by D. Mitchell, *Tarbert in Picture and Story,* Falkirk, 1908, 86.

61. *N.S.A.,* South Knapdale, 263-4. In 1831.

62. *Ib.,* Glassary, 699.

63. *Ib.*, 689.
64. *E.R.C.* (1862), 9
65. *Ib.* 12.
66. *Ib.*, 6.
67. *Ib.*, 17.
68. D. McFarlane, 24 January, 1975, and H. McFarlane, 7 June, 1974.
69. 4 May, 1974 and 24 January, 1975.
70. *Off the Chain, op. cit.*, 60.
71. *Ib.*, 65.
72. A. Fraser, *op. cit.*, 59.
73. D. MacVicar, 29 October, 1978 (H.D.S.G.).
74. J. T. McCrindle, 29 April, 1976.
75. Precognitions, Turner/Rennie case, A.D. 14 53/288, 7.
76. H. McFarlane, 30 October, 1974.
77. *N.S.A.*, South Knapdale, 264. 'Pains' may be interpreted literally. D. MacVicar: 'Ye never heard anybody hardly, speakin' English, except in the school. If you were heard talkin' Gaelic there, ye got a clout on the jaw without bein' spoken tae.' 2 February, 1977 (S.S.S.).
78. C. McKeich, Precognitions, 8/7, *op cit.*
79. *N.S.A.*, Kilcalmonell and Kilberry, 410.
80. D. Mitchell, *Tarbert in Picture and Story*, 72, *op.cit.*
81. *N.S.A.*, South Knapdale, 275.
82. A.F. 37/4.
83. 30 November, 1974.
84. *Off the Chain*, 61.
85. A McKerral, *Kintyre in the Seventeenth Century*, Edinburgh, 1948, 85.
86. L. McInnes, *Dialect of South Kintyre*, Campbeltown, 2 (undated, but collection was undertaken in 1920s and 1930s).
87. *N.S.A.*, Southend, 428.
88. *Op. cit.*, 3.
89. A. R. Bigwood, The Campbeltown Buss Fishery, 1750-1800, 9 (unpublished M.Litt. thesis, University of Aberdeen, 1972).
90. *O.S.A.*, 1794, XXXV, 546.
91. A. McKerral, *op. cit.*, 80-1.
92. A.F. 37/7.
93. 3 June, 1958, recorded by E. R. Cregeen.
94. P. McCallum, fisherman, New Quay Head, Campbeltown, 3 January, 1852, A.F. 37/9.
95. J. McIntosh, 3 June, 1958, recorded by E.R. Cregeen.
96. A.F. 37/7.
97. *Argyll Estate Instructions*, ed. E.R. Cregeen, Edinburgh, 1964, 45.
98. *N.S.A.*, South Knapdale, 270.
99. *Ib.*, Glassary, 691. Boats were previously 'wherry or schooner-rigged'.
100. D. C. MacTavish, *Inveraray Papers*, 87. Refer to p. 4 of this book.
101. *E.R.C.* (1862), 5.
102. *N.S.A.*, Glassary, 689.
103. *O.S.A.*, Inveraray, 1791, V, 303.
104. A. Fraser, *op. cit.*, 92.
105. S. Turner, 10 June, 1975.
106. Glasgow, 1975, 41.
107. A. Martin, 16 June, 1974.
108. 15 July, 1976.
109. 11 June, 1976.
110. Ms M. Campbell, 27 January, 1977 (S.S.S.), etc.
111. 16 June, 1974.
112. H. McFarlane, 11 June, 1976, and J. McIntyre, 15 July, 1976.

113. 27 June, 1975.

114. H. McFarlane, 11 June, 1976.

115. H. McFarlane, noted from. Also Calum Bannatyne, retired shepherd, 17 March, 1977, Campbeltown (S.S.S.).

116. 2 February, 1977. Also referred to by Ms M. Campbell, 27 January, 1977 (both S.S.S.).

117. London, 1975, 93.

118. A. Martin, 16 June, 1974.

119. R. McGown, 29 April, 1974. A continuing, if rarely asserted, superstition.

120. A. Martin, 16 June, 1974. It was the custom among Tarbert fishermen, before leaving harbour, to put their boats in a complete circle with the sun. G. C. Hay, noted 9 May, 1979.

121. A. MacGregor, *Highland Superstitions*, Stirling, 1946 (reprint), 7-8. From foreword.

122. 1 January, 1958, Carradale. Recorded by E. R. Cregeen.

123. G. C. Hay, noted 9 May, 1979.

124. Ms M. Campbell, noted 30 December, 1977.

125. I. F. Grant, *Highland Folk Ways,* London, 1961, 275.

126. 7 February, 1975.

127. H. McFarlane, 5 April, 1976.

128. J. T. McCrindle, 29 March, 1976.

129. J. McIntyre, 15 July, 1976.

130. H. McFarlane, 30 October, 1974.

E

3

The Ascendancy of Ring-Netting

Lochfyne

TO represent the hostility to trawling of the drift-fishermen as the mere carpings of a 'sleepy set of men',[1] entrenched in backwardness and resentful to the point of envy of the progressive and generally more successful fishermen of Tarbert and Ardrishaig, would be less than just. There seems little doubt that they recognised in the advance of trawling the ultimate destruction of their way of existence.

A 'proprietor' in the Inveraray district, Patrick Forbes, informed the Royal Commission in 1864 that the local fishermen were, with few exceptions, economically dependent on the fishing industry, though some had crofts. 'Every small cottage you see is mostly a fisherman's house,' he said. 'They have . . . potato ground attached, and they work that. They have their house and enough grass perhaps for a cow, and as much ground as will enable them to plant what the cow's manure will produce.'[2]

The Tarbert and Ardrishaig fishermen were contemptuous of the capabilities of the Inveraray men, even as drift-net fishermen, the occupation to which they clung so tenaciously. 'The Tarbert men would get more fish with six barrels (of nets) than the Inveraray men would do with a hundred,' boasted John McMillan of Tarbert in 1864. His explanation of the Inveraray men's reluctance to adopt trawling was equally hostile: 'It was because they were not qualified for the work. They cannot do it. They are too lazy.'[3] The collective opinion of four Ardrishaig fishermen two years before had been: 'The men there are lazy and will not follow the fish.'[4] These allegations were probably not devoid of truth, but there is the counter evidence of the involvement of the upper loch fishermen in the Lewis herring fishery. On a day in May, 1862, for instance, 35 boats, crewed by fishermen from Inveraray south to Otter, sailed for the north Minch.[5] (The number of these boats which belonged to Inveraray cannot, however, be established.)

The upper loch fishermen were not, in fact, completely without experience of trawling in the early years of the method's existence. Martin Munro, the fishermen's leader in 1859 (pp. 13-16), had 'once practised it himself'.[6] (He was, incidentally, still about Inveraray in 1877, when the Royal Commission of that year visited the town, but, to judge by his evidence, the young man's anger was spent.[7]) John McNichol and Alexander McArthur, both upper loch fishermen, tried trawling in the 1850s, the latter for two seasons. His trawl was a converted drift-net.[8]

The geographical isolation of Inveraray cannot be ignored when seeking explanations of the progressive decline of the fishing fleet there from 1870 onwards. The main fishing grounds of the Clyde were within a half-day's sail of both Tarbert and Campbeltown, but Inveraray's location close to the head of the longest sea loch in the British Isles opposed its fishermen's complete and successful involvement in the Clyde herring fisheries. The successive elimination of Lochgilphead and

Ardrishaig and the numerous other smaller fishing communities on the lochside to the north may partly be explained by the absence of harbours suitable for the larger class of motor-boats which began to be introduced in the 1920s.

'It is observable,' wrote an anonymous newspaper commentator in 1878, 'that the drift-net boats at present fishing have an unusual number of old men, whose sons have gone away to other employments and left Lochfyneside. Everywhere good boats lie rotting in the sun, and with the population going these will not be replaced.' The drift-net fishermen of the upper loch were alleging that disturbance of the herring shoals in the lower reaches by Tarbert and Ardrishaig trawl-fishermen prevented the fish from 'finding their way as they used to do into the upper waters'. By then, however, these fishermen were 'somewhat heedless of the loss and disappointment occasioned by trawling, since they regard the fishery as doomed'.[9]

A year before, John McKellar of Inveraray had remarked: 'There are not now 40 boats inside of Otter Spit, not one third of the number that there used to be.' He himself had acquired a trawl-net, 'to get a share of what was going'.[10]

By 1883 the Inveraray fishing fleet had been reduced from 50 boats employing 171 men to 21 boats and 36 fishermen, which suggests that crew shortage would have prevented some of these boats from putting to sea. Fishery officer William Jeffrey commented that, 'Owing to the failure of the herring fishing in Upper Lochfyne, Inveraray has almost ceased to be regarded as a fishing place; the few boats which now belong to it are oftener down about Ardrishaig and Tarbert than about Inveraray.'[11]

The fleet, by 1900, had been reduced to 10 second-class boats − 18-30 feet (5.49-9.14 m.) of keel − employing 18 men, and that pattern of decline was evident also on the eastern shore of the loch. From Cairndhu to Newton, 15 second-class boats were registered for fishing, employing 30 men, but the fishermen were 'simply occasionals, fishing only when the herrings come to their doors, and do not follow the fish all over the loch'. From Otter to Ardlamont, the figures were 6 and 12 respectively.[12] These statistics reveal that between 1890 and 1900, 10 boats disappeared from these two fleets, and that 48 fewer fishermen were employed.[13]

The comparative productivity of the drift- and trawl-net fleets indicates impressively the main cause of the decline of drift-netting. In 1899, for instance, in the Clyde generally, 328 trawl-boats grossly outfished a greater drift-net fleet of 404 boats, landing 54,059 crans of herring, valued at £70,325, compared with 15,965 crans of drift herring, valued at £18,009.[14]

There is no evidence that the drift-net fishermen on the eastern shore of the loch turned seriously to trawling, other than a brief reference in the annual report of the Fishery Board for 1902 to trawls having 'come more into use' among the fishermen based between Cairndhu and Newton.[15] Eight years later, in 1910, the decline was almost complete. Inveraray supported a fleet of just 4 second-class boats, employing 9 men; from Cairndhu to Newton, the figures were 7 and 18, and from Otter to Ardlamont, a 14-mile coast, only 1 boat, with a crew of 3 men, fished.[16] Four years later, the combined fleets of Inveraray and the eastern shore of Lochfyne totalled 5 boats, employing 13 men.[17] The brink of ruination had been reached, and there would be no recovery.

The decline on the western shore of the loch affected most severely those communities which had continued to reject trawling, but was generally less dramatic. The fleet of 4 second-class boats which had fished from Kenmore in 1890 was gone by 1914; the fleet of 8 Furnace boats was also gone; the reduction at Crarae was from 7 to 3; at Minard from 12 to 5; at Castleton from 18 to 6; at Lochgair from 13 to 2; and at Lochgilphead from 41 to 9.[18] Of all these places, only Minard would retain a fishing community until the end of the 1930s – a single pair of trawl-crews, at that.[19]

The drift-net fishermen had complained as early as 1877 that scarcity of herring in Lochfyne was forcing them to carry more nets. John Bruce of Ardrishaig told the Royal Commission that in his youth 10-ton boats carried seven barrels of nets, whereas the four-ton boats now carried 12 barrels.[20] A fisherman of Kames, Archibald McEwan, complained that 'formerly no boat need leave Lochfyne, but since the trawl they require to go (elsewhere)'. The population had been 'scattered' and the fishery was such that if the head of a family were to be 'lost' the only resort for his dependants would be the work-house. In the past, even old and infirm men were able to 'maintain themselves comfortably by fishing quite close to their homes'.[21] John McKellar of Lochgilphead had no doubt that the spread of trawling had ruined the Lochfyne herring fishery. 'When trawling was legalised,' he said, 'all the fishermen got trawls, and they fished Lochfyne out; they then went to the Kyles of Bute and fished them out; they then went to Loch Goil and Loch Long and fished them out. They got money for the fish at the time and did not care for the future.'[22]

The fishing industry on the Kyles of Bute, that arm of the sea which reaches around Ardlamont from Lochfyne and embraces the northern portion of Bute, had declined correspondingly. The fishermen there maintained their opposition to trawling, and in 1897 petitioned Parliament for its abolition, but the appearance on their coasts of 'immense shoals of herring' later that year caused an alteration of opinion,[23] and by 1904 the Fishery Board could report that trawling was 'yearly becoming more popular'.[24] Despite two prosperous seasons in succession, 1903 and 1904, there had been a decrease in the numbers of men and boats engaged in fishing. The principal cause of the unexpected regression was that the younger men had shunned regular fishing and were, instead, seeking employment on yachts and in the merchant navy. One conclusion only could be drawn from the statistics, that fishing by district crews was 'rapidly becoming a thing of the past'.[25]

The fishing fleets of the Isle of Bute and Kyles declined from a total of 115 second- and third-class boats, employing 132 men, in 1900, to 52 boats, employing 51 men, in 1914.[26]

Campbeltown and Kintyre

The Campbeltown fishing industry, like that of Lochfyneside, was centred on drift-netting, and much capital was invested in boats and gear. The Campbeltown fishermen were, however, able to manage a gradual and successful transition from drift-netting to trawling, and this was achieved by their prompt accommodation of trawling – after its legalisation – in the scheme of their operations. Complete commitment to trawling would, however, take a generation to develop.

The Campbeltown fishermen, like their drift-netting counterparts on Lochfyneside, visited the north Minch annually in early summer, a tradition which originated in the early eighteenth century and which was strengthened by involvement in the buss fishery, after the introduction of the bounty scheme in 1750 (pp. 4-5).

Investment in the Minch fishery was strong in the 1860s. In 1863, five large half-decked fishing vessels were built in Rothesay for Campbeltown owners, and five were under construction.[27] On a single day in May of the previous year, 29 Campbeltown boats passed through the Crinan Canal on passage to Lewis. There were also nine boats from Carradale and four from Skipness.[28]

The Mackerel Luggers

In the early 1870s the interest of some Campbeltown fishermen was attracted to the annual mackerel fishery at Kinsale, on the south coast of Ireland, and that interest was translated by 1876 into a fleet of fine luggers. These boats, the earliest of which arrived in 1872, are not to be confused with the traditional smacks. The lugger was built on a much larger scale, with a keel length of 48 feet (14.63 m.), compared with 24 feet (7.32 m.), though in 1862 a few smacks of 35 and 36 feet (10.67 and 10.97 m.) were constructed for Lochfyneside owners.[29] The lugger was, moreover, completely decked, and lug-rigged, 'as English boats are',[30] whereas the smack was only partly decked and cutter-rigged.[31]

By 1875, the Kinsale mackerel fleet numbered 10 vessels, of which eight had been specially built in St Ives, Cornwall, at a cost of more than £400 per boat, including sails.[32] The keel length of these vessels was, as already stated, about 48 feet (14.63 m.), giving an overall length of 52 and 53 feet (15.85 and 16.15 m.). Their beam was about 15 feet (4.57 m.), and their draught 7 feet (2.13 m.).[33] They were equipped with both herring drift-nets and mackerel drift-nets, each fleet costing £200, which brought the total capital investment to more than £800,[34] which may be compared with the £35 or £36 necessary to purchase and equip a trawl-skiff.[35]

In 1876 four new luggers were added to the fleet, all of them constructed at St Ives. A local boat-builder, however, began that year the making of a lugger based on the design of the imported craft.[36] The fishery, which customarily began in March and continued until May or June, had been successful that year, and gross earnings per boat ranged from £300 to £400, which allowed for a 'fair profit' to the owners.[37] A fleet of 375 boats, all registered first-class – 30 feet (9.14 m.) and more, keel length – was engaged in the Kinsale fishery, 50 of them Irish and the others principally from the Isle of Man, Lowestoft, and Campbeltown. At the beginning of the season prices ranged from £3 to £5 per 127 fish (counted as 100) and rarely fell below £1. The buyers maintained a fleet of 'six powerful steamers and a dozen smart Jersey cutters' to carry fish to the English markets. Weekly catches by Campbeltown boats at the beginning of the season ranged from 800 to 2,000 mackerel, later increasing to from 1,000 to 11,000.[38]

At the end of May, 1876, 31 smacks sailed for the herring fishery at Howth, having delayed to await the arrival of Skye crews by steamer from Clydeside. The extent to which the principal Campbeltown boat-owning families had invested in smacks may be judged by the numbers credited to them. Of that fleet, 9 belonged to the McIsaac family, 8 to the Cook family, 4 to McMillan and McKeith, and 3 to John Martin.[39] These families also monopolised ownership of the lugger fleet, which would finally increase to 20 vessels.[40]

By 1895, the mackerel fishery was in serious decline, owing to a succession of bad seasons which ruined several boat-owners,[41] and diminishing numbers of luggers

10. A smack, with trawl-skiffs moored alongside, at Campbeltown New Quay, c. 1890.
Per W. Anderson, Campbeltown

made the annual trip until, in 1900, only one vessel ventured to Kinsale, and returned in June having scarcely cleared the expenses of the trip.[42] By that time most of the luggers had been sold or converted into coal-carriers. Three fishermen, Dugald Robertson, Alexander McMillan, and the author's great-grandfather, John Martin, faithfully maintained a lugger each until the First World War, but these did not put to sea within living memory, except when taken to Dalintober beach to have the mussels and seaweeds scraped from their bottoms. They were finally sold.[43] Some luggers became homes for destitute families. They were set up along the Kilkerran shore of Campbeltown Loch, supported by an upright on the seaward side of the hull.[44] The forerunner of the fleet, the *Europa*, which had been built in

1872 by William Painter of St Ives, was, ironically, the last survivor. She was broken up in 1929, having spent the last 20 years of her existence trading, principally in coal, between Campbeltown and the Ayrshire ports.[45]

The operational range of the Campbeltown luggers – each crewed by eight men – was extensive, and though it is by their involvement in the Kinsale mackerel fishery that they are now remembered, they also worked the Minches, around the Orkneys and Shetlands, and into the North Sea.[46]

The smacks, too, were disposed of, and ended their times similarly. The fleet of first-class boats – a category which included both luggers and smacks – declined from 29 in 1890 to 4 in 1914. The number of second-class boats also declined during that period, from 171 to 83. The 323 fishermen employed in 1914 represented a reduction of 306 on the corresponding number in 1890. That downward trend is observable also in the statistics for Carradale and Torrisdale – 56 boats employed 230 men in 1890 and, in 1914, 30 boats employed 122 men.[47]

The drastic decline of the Lochfyneside fishing fleets had its parallel in Kintyre. The fleet of 24 second-class boats, mainly drift-netters, employing 135 men from the Kilbrannan Sound villages of Skipness, Claonaig, Cour, and Grogport, declined to a single boat, employing a couple of men, during the period 1890-1914.[48]

The Trawl-skiffs

The trawl-skiffs of the original class were small and lightly built, and were propelled by four oarsmen. Their swiftness and manoeuvrability were decisive factors in the fishermen's evasion of detection during the most difficult years of the trawl-net's prohibition. Their lightness was necessary because the fishermen preferred to haul their nets directly to the shore. 'When the net is hauled on shore,' said Ardrishaig fisherman Duncan McBrayne in 1861, 'the men land from their boats which they heave on the beach.'[49] The smacks used in drift-netting were therefore unsuited to shore work.

These early trawl-boats ranged in length from 20 to 25 feet (6.10-7.62 m.) and were about 6 feet (1.83 m.) in beam.[50] Such a vessel would cost, in 1852, from £12 to £20.[51] They were open from stem to stern and the fishermen got about principally by rowing, though the boats were generally equipped with a small lugsail, and a few, like the *Pelican* of Tarbert in 1869, with a jib.[52] That a lugsail was not carried habitually is evident in Colin McKeich's statement that on the night that he was wounded, in June 1853, 'As one of the other boats in company with us had no sail, we gave her a tow.'[53] That circumstance was, however, probably rare. The 13 trawl-skiffs seized in Tarbert harbour by the crew of H.M.S. *Jackal* in 1860 (p. 18) were all but one provided with mast and sail. Each was equipped, in addition, with four oars – one of which, in three cases, was broken – a rudder, and a kedge-anchor.[54]

By 1862 the trawl-skiffs were not, as a rule, ballasted, which often resulted in deterioration of the quality of a catch. 'When the boats have no ballast, and the fish get into the water at the bottom of the boat, they lose the scales,' said Tarbert fisherman John McMillan.[55] The lack of ballast need not necessarily have damaged the catches, because the water could have been allowed to drain off the herring as they were basketed aboard, and the fish, by frequent use of the pump, properly preserved. That, however, was impracticable 'when the crews of the men-of-war are always ready to pounce on the trawlers and prevent them handling their fish quietly'.[56] As early as 1819 the Fishery Board had decided that, without special legislation, the improving of Scottish fishing boats – notably the furnishing of pumps and floors – would be impossible.[57] The ballasting of the skiffs seems to have been more general by 1867, when 'Gowrie' described the Tarbert boats at evening 'visiting the adjoining rocks to take in ballast and await their friends'.[58]

In the event of a heavy catch of herring, the greater part of the ballast would be dumped overboard to increase a boat's carrying capacity.[59] There is a tradition that jettisoned ballast-stones fouled the inshore fishing grounds and caused damage to nets subsequently set in the area.[60] With the introduction of larger classes of trawl-skiffs, permanent stone ballast, boarded over, became practicable.

Living from Home

Some crews, when fishing at a distance from Tarbert, had with them a smack, which provided shelter after a night's work. These vessels were of the kind used at drift-netting, for which occupation they undoubtedly were built, but the decline of interest in drift-netting at Tarbert would have limited their use in that industry. Large open skiffs were also used as shelters, but sleeping arrangements, as described by 'Gowrie' in 1868, were crude and hazardous: 'When there has been no fishing for some time . . . and the main body of fish has gone up the Kyles, or even up Lochfyne, then the boats go and make a stay at those places, not returning for weeks if the takes are good. During these trips the crews sleep in the large open skiffs, in which they visit Stornoway in the beginning of the season. They are sailed up when the wind is fair, a sail spread over them, and a fire kept alight in a small open portable grate. On their periodical return to obtain a change of clothes and put themselves to rights, their suits are always singed and spoiled, from drawing in towards the warmth in their sleep.'[61]

There is the information from John Weir of Tarbert, who received it from his grandfather, of Tarbert skiffs being taken to Loch Seaforth, Lewis, on the decks of the herring-carrying steamers which followed the fleets.[62] And Hugh McFarlane of Tarbert recalled sailing north, more recently (about 1900) on the herring-steamer the *Petrel* to Loch Eishort, Skye, as one of a drift-netting crew, and their open skiff was lashed to the deck of the steamer for the passage. Accommodation on the fishing ground was provided in a dismasted sailing-ship, the *Coquette*, which also served as a curing base.[63]

Tents also served as shelters. The *Campbeltown Courier* of 22 April, 1876 reported fishermen's tents pitched near Crinan. A fleet of Tarbert boats 'of the small open or skiff kind' was lying overnight there, having passed through the canal on its way to the Lewis drift-net fishery. The report concluded: 'Crews when away from home take with them a tent, which being pitched ashore forms a comfortable sleeping place in rough weather.'[64]

The Camper Fishermen

When the trawl-fishermen of Kintyre began to camp ashore on a large scale cannot be established with certainty. Fishery Board documents from 1848 onwards contain no references to either huts or tents as fishermen's habitations, although there occur numerous references to the employment of smacks to lodge the fishermen.

There can be little doubt, however, that the practice – considered generally – was very old. Legislation passed in 1705, during the reign of Queen Anne, allowed fishermen 'for their greater conveniency to have the free use of all ports, harbours, shoars, fore-lands and others for bringing in, pickleing, drying, unloading and loading the same upon payment of the ordinary dues where harbours are built . . .'[65] That legislation would seem to have been made available to fishermen in charter form. John McCrindle, grandfather of Maidens fisherman J. T. McCrindle, kept, in a seven-day clock at his home, a copy of the 'Queen Anne Charter', until it finally yellowed and disintegrated with age.[66]

In 1775 a more comprehensive Act was passed. It granted free use of the shores below the highest water-mark and of 100 yards (91.44 m.) of foreshore on waste or uncultivated ground, for landing nets and casks, and erecting sheds for curing, etc.[67]

Disputes between fishermen and unsympathetic landowners inevitably occurred. In the only such controversy in Kintyre of which there is record, William Fraser of Skipness prosecuted John McMillan and several other fishermen, in 1848, for trespass on a piece of ground.[68] The conclusion of the case is, regrettably, unknown.

The advent of large-scale camping in home waters may be attributed to two dissociated factors. First, the disappearance of smacks from the Tarbert fleet through disrepair or sale – by 1890 not a smack remained there.[69] The Tarbert trawl-fishermen, whose interest in distant fisheries was minimal, would not have re-invested in such expensive craft solely to provide periodic accommodation on the fishing grounds. Second, the general adoption of trawling by fishing communities previously opposed to the method, in particular Campbeltown. Many of these fishermen invested in skiffs, but could not have afforded smacks.

In 1862 the procurator-fiscal at Campbeltown, Charles Mactaggart, informed the Royal Commissioners that, 'The drift-net boats here are not the property of the fishermen in many cases, but the property of men with capital, who pay their fishermen wages.' Mactaggart reasoned, with remarkable foresight, that, 'If the men were reasonably provident, the trawling system would encourage property among them.'[70] Just such a development occurred, as the local registers of fishing boats confirm, and as the account of one Campbeltown fisherman, David McLean, will illustrate: 'The fishin' industry in Campbeltown was owned by just a clique. Gradually, the men thought it wis time that some o' them wir branchin' out. Any well-doin' men that had a successful season an' knew how tae take care o' their money got boats o' their own an' they were independent o' these fellows.' His own father had begun as one shareholder among several in the ownership of a pair of skiffs, but finally acquired full ownership, having taken possession of the shares of departing crew-members.[71]

Trawl-fishing in the Campbeltown district, during the period of the method's prohibition, had been confined to winter, and was not engaged in by more than three or four crews at any time.[72] After the repeal of the legislation, however, many Campbeltown fishermen acquired trawl-nets. Indeed, a renewal of anti-trawling agitation among the fishermen of Lochfyne in 1874 drew an angry reaction from one Campbeltown fisherman. In a letter to the *Campbeltown Courier* he wrote: 'We have been taught trawling from the Lochfyne district, and have fitted up nets and boats at great expense, and can now outstrip our instructors, and hence the cry . . . We are determined not to give up trawling till all are satisfied or receive compensation for our nets and boats.'[73]

The crews attached to smack-owning families trawled in open boats, but, as was the case in Tarbert, lodged on board the larger vessels. Archibald Cook, a member of one of the major boat-owning families, explained in 1877: 'It is the habit at Campbeltown to go out with two skiffs. The men in the skiffs catch the fish . . . and

11. Kintyre and Lochfyneside ring-net skiffs, with accompanying smacks, beached at Loch Ranza, Arran, 1880s. When fishing in that area, crews customarily moored their vessels in the sheltered inlet behind Loch Ranza Castle and caught a steamer home, returning on Monday morning to resume work. Per Messrs. Morrison/Macdonald (Paisley) Ltd.

place them on board a large boat, which is lying at anchor, and in which the men live. These large boats are employed in the Irish fisheries and afterwards become the houses in which the men live in the trawling season . . . for four months. The fish are stored in the large boats till the buyers take them.'[74] Another Campbeltown fisherman, Dugald Robertson, stated that the Irish fishery prevailed from April until August, and that on the fishermen's return they adopted trawling.[75] The fishermen also used their smacks, in the absence of herring-steamers, to carry fish to market.[76]

If one were to trust the scantiness of the documented evidence, then the conclusion might be that the custom of camping ashore had not existed on a considerable scale. Oral tradition, however, bears the last threads of the story, which is of a simple and companionable existence on the shore; but understandably the rough edges of that story have been rubbed away. When in later years the old men and the young gathered in the warmth of the little 'dens' and lit their pipes as one and spoke to forget the howl of the wind across the Sound and the cracking of the rigging in the night above them, the old men would have told of good times.

These camper-fishermen have gone, every one of them, and lie in the little graveyards above the changing seas. The Tarbert men are in Cill Aindreis; the Carradale men in Brackley and Waterfoot; the Campbeltown men are in Kilkerran, and the Dalintober men in ruined Kilchousland. From their sons, and the other men who later sailed with them, the little gathering of information has had to be made.

The total written evidence of camping on the Kintyre shore is contained in two reports which appeared in the *Campbeltown Courier* in May, 1880: 'Upwards of 70 pair of trawling skiffs are meantime pursuing the fishing. Of these about one-half are belonging to Carradale, Tarbert and Arran, the crews of which lodge in tents on the shore.' A week later the second report revealed that the campers were on the move northwards, following the herring shoals: 'The strangers and a few local crews have gone up the Sound, taking with them their lodging houses, and pitching in the locality of Carradale.' By June the grassy bays between Carradale and Skipness – with such names 'sweet to the mouth' [77] as Grianan, Sunadale, Cour, Crossaig, Eascairt, and Claonaig – would lie under canvas. Huts were erected in later years, [78] presumably by fishermen more prepared to station themselves.

In time, particular families became associated with certain bays and bights along that shore, and these families remained by their own pitch with a tenacious faith, as a story told by David McLean illustrates. The crews of his father's pair of skiffs had sought herring along their 'beat' at the Minister's Head for three weeks and had not marked a fish. They resolved one night to shift, and were sailing north when, in Mr McLean's words, 'My Father didna know whether it was intuition or not, but they turned back, an' they got fifty poun' a man for thir shot, an' that wis a year's pay tae most o' the folk at that time.' [79]

Memory of these traditional associations of place-name with family-name survives in Campbeltown only, and in small proportion at that, it may be assumed. As the McLeans favoured the Minister's Head, so the Brodies favoured Cour; the McKays went to Dearg-Uillt; the O'Haras to south of the Minister's Head; [80] the McGowns to Eascairt [81] . . . and the rest is lost. The recital, by a Carradale fisherman, of the names of four of the principal camping families from Campbeltown has an almost jingle-like quality: 'There wir "Teedelty", Brodelty, Cook, an' "The Junk"' (Neil McLean, Brodie, Cook, and John McIntyre). [82]

On the Shore

In the morning, after any catch had been disposed of to a steamer, the fishermen would return to the shore and either moor their boats close to the rocks, [83] or heave out the stone ballast and drag them on to the beach over rollers purposefully left ashore. [84] In the case of moored boats, they would undoubtedly be constantly alert to any shift in wind direction. If the wind changed and was blowing in upon their boats, they would remove them to another *cùil* (nook) to prevent damage. [85]

Cooking was managed over a drift-wood fire or the *griosach* (brazier)[86] which they carried on board during the night for brewing tea or boiling a meal of herring and potatoes in the 'fish-pan'. (The risk of fire on board the skiffs was minimal, the braziers being positioned on the bare ballast stone.[87]) If the coals in the brazier were not catching alight, one of the men might '*birl* (whirl) it roon his heid tae make it take'.[88] 'Gowrie' described, with some amazement, a similar practice: 'I have seen a fisherman hold one of these grates, flaming, over the boat's side for five minutes, to let the breeze blow through it, and raise it to a fiercer heat. He held it by a handle a few inches over the lighted coals, and how his hands retained their hold I cannot imagine.'[89] An old man, John O'Hara, with whom John McWhirter had fished,

12. Fishermen's hut, with open boat upturned on the foreshore. The photograph, probably taken in the Kyles of Bute, belongs to the period c. 1880-94, when the photographer Matthew Morrison was active. Per Messrs. Morrison/Macdonald (Paisley) Ltd.

described to him the system of cooking at his family's station. The O'Haras built their fire beneath a slab of the slaty rock found on that part of the shore, and fried the herring, in their own fat, on that slate.[90]

For the first six months of his working life, before the introduction of part-decked skiffs, James McIntosh of Carradale took a 'piece' with him to the fishing every night.[91] These pieces, or sandwiches, were wrapped in big, fresh handkerchiefs which would be tied beneath the thwarts to keep them dry during the night.[92] John Weir of Tarbert was told that 'usually when they'd had a busy night, when they come in in the mornin' the pieces wir still below the beam'.[93] Oatcakes were eaten at sea by many of the Lochfyneside fishermen. If a catch of herring yielded fish containing *melts* (the roe of the male), some of these might be removed and eaten –

perhaps raw – with oatcake (*bonnach èineachan*).[94] Hugh McFarlane recalled an account of spearing flounders in Skipness Bay, which he heard from an old fisherman, Johnny Hyndman: 'They wid put on the pot an' go doon an' get a *gad* (strung quantity of fish), pickin' the big wans. Johnny said: "A'd go doon in the moarnin' an' just pick whoot A wis waantin'. The pot wid be boilin'; just gut them, an' intae the pot."'[95]

One story, which typifies the period, was told by two fishermen, Henry Martin of Dalintober and David McLean of Campbeltown. Their versions were substantially the same, but the latter's was the more directly sustained of the two and merits full quotation: 'I remember hearin' a tale. A storm came away an' the men went in earlier than they intended, maybe aboot three in the mornin'. An' some o' them took a pail an' went away up an' dug a feed o' potatoes oot the farmer's field. So, durin' the day the farmer wis passin' wi' his dog, an' the herrin' wir fryin' an' the potatoes boilin', an' while they wir chattin' the young fellow cried dinner was ready. An' they invited the farmer intae the hut. Oh, he was delighted, an' he went in, an' "Oh, my God," he says, "the herrin's lovely, an' that's the best potatoes I ever ate." – It wis his own potatoes – "What kin' o' potatoes is that?" – "Oh," says wan o' the men, "that's the Early Walkers."'[96]

The last camper-fishermen were probably two crews of elderly Dalintober fishermen who 'went up the shore' in small line-skiffs. They were then 'kinna old tae get a berth at the fishin''.[97] That would have been about 1900, and the custom had ended, on a significant scale, for at least 10 years.

Smacks continued to serve as living accommodation for crews until the provision of forecastles in skiffs became general about 1885, so that camping was not an experience common to all fishermen. Nor was the practice of smack and skiffs operating in company confined to the summer fishery in the Kilbrannan Sound. In February and March smacks were stationed with the fishing fleet on the Ballantrae Banks during the great annual fishery on spawning herring there. In March, 1881, a gale caused 10 skiffs lying off Girvan to 'break away from their moorings . . . or be cut away from the smacks to which they were attached'. Two boys, left alone on board the smack *Mary Cook*, suffered the alarm of being trapped alone on board the vessel to weather the two-day storm. The crews of the skiffs had been unable to return to her and had to run to Stranraer.[98] Those crews without smacks, especially the Tarbert fishermen, lodged in houses at Girvan during the 'Banks' fishery. Each fisherman took with him, in a *kist* (chest), sufficient food to sustain him throughout the six- or seven-week-long industry. All the foodstuffs that would 'stand', such as butter, cheese, sugar, and sea-biscuits, were carried across to Ayrshire.[99] Winter weather conditions undoubtedly precluded camping ashore.

(A brief note on the last fishermen to camp on the Kintyre shore may be of interest, though contextually irrelevant. Crews of fishermen from the Rhinns of Islay customarily arrived at the Mull of Kintyre in summer, to fish for cod and saithe. They stationed themselves in crude shore huts built of stone and turf and thatched overhead. The interiors of the huts were lined with dry bracken, on which they slept. Catches – taken with hand-lines during slack water – would be cleaned, split, salted, and spread to dry on the rocks around their huts. The tradition ceased about the time of the First World War.[100])

Week-end Fishing

If the wind suited a passage home by sail, then the fishermen would set off from their station about the end of the week, though some might remain on the fishing ground for several weeks without relief. They would, therefore, in the latter case, spend a week in every month at home, a respite which spanned the 'week of the moon',[101] when nights were bright and fishing tended to be unsuccessful (p. 176). The fishermen, especially those from Campbeltown, did not, however, pass their week-ends in idleness. They were actively at fishing even on Sundays, much to the distress of those church-going people who witnessed their indiscretions. Sunday fishing had been prohibited by law in 1815,[102] so that the indignation excited by the practice had legal as well as moral principles to support it.

From a letter written by a citizen of Tarbert, who chose not to identify himself, and published in the *Rothesay Express* in June, 1883, the public annoyance aroused by the issue may be judged: 'The usual quiet of the Sundays here is again being intruded on by numbers of the Campbeltown fishermen, who, not being content with the week's fishing, are taking advantage of the Sabbath day for the prosecution of their work. On Sunday last numbers of them were observed trawling during the day near the mouth of the harbour. Visitors here were naturally wondering what they were doing, as it is understood we have no Sunday fishing in this district. The fish were bought by O'Brien's boats; the other buyers did not encourage the Sunday work. Can the Fishery Board not step in and prevent this, what we believe to be illegal fishing, for the good of all concerned?'[103]

The Fishery Board did, indeed, intervene in the following year, by stationing a Naval cutter, the *Neptune,* in the Kilbrannan Sound. The Board's annual report for 1884 claimed that the practice had been completely stopped,[104] but the Campbeltown men were at it again during the summer season of 1885. Reporting to the Board that year, George Brook F.L.S., examined the argument that to prevent week-end fishing would be 'unfair to those who come from a distance, and who do not go home at the end of each week', but decided that he 'fail(ed) to see any unfairness in it, (as) some of the fishermen prefer to return to their homes in the middle of the week, so that they may return to make the earliest catches on Monday morning'.[105] In 1886 the practice had evidently been curbed. At least, H.M.S. *Jackal* failed to make any detections, though 'the electric light was used at intervals'.[106]

The week-end activities of Lochfyne fishermen near Fort William had caused much offence to the native population in the previous decade. John MacDonald, a fisherman of that town, complained to the Royal Commission in 1877 of 'strangers from Ardrishaig' violating the Sabbath. He reasoned: 'When a Scotchman is away from home he is not so particular about the Sabbath as while he is at home.'[107] The Commission noted in its report 'a strong feeling among Scotch fishermen that the absolute forfeiture of nets, equivalent in some cases to the ruin of a fisherman, is not too high a penalty for the offence of Sabbath-breaking'.[108]

Certain of the Tarbert fishermen were not averse from disregarding the week-end prohibition law when it urgently suited them to do so, if one may judge from a story told by Hugh McFarlane. A Tarbert skipper, Archie *'Mór'* (Big) Campbell, and his crew had rowed along the coast on a Sunday evening to their station north of Skipness. They were sitting in their tent puffing at pipes and talking to pass time until the day turned, and considerations of law and religion could be disregarded. 'This Erchie Mór wis there wi' the (only) waatch. He went oot at the back o' eleven, an' went for a wee *dander* (stroll) doon. There wis the whale an' the *buckers* (porpoises) ootside . . . An' Erchie's listenin' there. He heard the whale. Now, the herrin's here. Aheid wi' the waatch! He came back up. "Come on," he says. "Boys, the time's near up." − "Oh no, Erchie." − "Aye, there's the waatch!" he says. "The whale an' the buckers is oot there at Port a' Chruidh, an' they're no' leavin' at all." They filled the boats − a big "touch"[109] − an' away home in the early moarnin'.'[110]

In 1889 a more restrictive law was passed which prohibited fishing between sunrise on Saturday and an hour before sunset on Monday, from 1 June until 1 October.[111] It also introduced the prohibition of daylight fishing, and many Kintyre and Lochfyneside trawl-fishermen were fined £1 or 10/- − hardly a deterrent − for the offence.[112]

The Floating Market

The development of trawling had been assured by the advent of rail transport, because the fish could be conveyed fresh to market throughout the year. The practice of discharging trawled herring into boxes to be loaded on to the regular 'luggage steamers' at Tarbert and Ardrishaig prevailed until the mid-1860s. Unless discovered actually at work with trawl-nets, or with trawl-nets on board their craft, the fishermen were immune from prosecution. The Act of 1861, however, provided for the seizure of suspected trawled herring, and for the prosecution of both fishermen and buyers (p. 19), therefore the landing of catches in harbour became a potentially dangerous practice.

By 1866, however, a new and safer arrangement had evolved, as a report by George Reiach, assistant inspector of fisheries, explained: 'Great facilities are afforded (the fishermen) by the smacks which move about and take the herrings from them during the night and fill them into boxes, which are shipped on board the steamers at sea, or in the bays without coming near harbours.'[113] Smacks had been employed on a small scale before then, but the transfer of catches at sea to steamers was a significant development which would eventually eliminate these sailing-craft.

In 1869, special herring-carrying steamers were operating in the Minches, thus 'opening up new fisheries at places which, until the introduction of steam, were too distant from markets to allow the herring fishery to be prosecuted remuneratively'. The early summer Minch fishery was particularly cited: 'Great quantities of herring are now caught and carried to Liverpool and other ports for consumption in the large towns of England. These fish are in general but slightly sprinkled with salt, the rapidity with which the voyages are made enabling them to be got to market almost in a fresh state.'[114]

Five years later, specially chartered steamers were introduced on Lochfyne for the transportation of fish to the Clyde railheads.[115] Their effect on the fishing industry was described in 1877 by John McGeogan, for 15 years a curer at Tarbert: 'The introduction of steamers here has given a great stimulus to the fresh fish trade. The trawl-caught fish are in Glasgow market at seven a.m. The drift-net fish not before four p.m. The trawl, therefore, confers a great benefit on the Glasgow consumers, and on the consumers in other places. The fish are ready at nine, to go even to Billingsgate.'[116] The reason for the great discrepancy in times of arrival at Glasgow market was that the trawl-fishermen could have herring caught and loaded aboard

13. At Campbeltown Old Quay, c. 1900, the herring-steamers *Nightingale*, *Rob Roy*, and *Talisman*. The long-serving skipper of the *Rob Roy* was the well-known Neil Hyndman, a native of Skipness. Per W. Anderson, Campbeltown

the attendant steamers several hours after darkness, whereas the drift-net fishermen did not begin to lift their nets until dawn, and might spend hours hauling them and clearing them of entangled herring.

The drift-net fishermen voiced objections to the damage caused to their fleets of nets by the steamers. Said John Bruce of Ardrishaig: 'There is a class of steamers running about night and day after the trawlers which carry away the buoys and nets. These boats buy herrings, but they do an enormous amount of destruction to the herring fleet. The boats are perhaps 20 times a night across the same track.'[117] Drift-nets, being set in trains, often in mid-channel, were understandably more prone to damage than the relatively small trawls which were usually worked close inshore.

In 1880, the *Campbeltown Courier* reported, seven 'screws' (in Gaelic *sgriubhachan-sgadan*[118]) – so called because of their means of propulsion – were 'kept moving about the Sound picking up the catch'. Sailing-vessels continued to operate, purchasing for both the fresh and the curing markets,[119] but their disappearance was imminent. The reconstitution of the Board of the British White Herring Fishery in 1882 – with a new title, the Fishery Board for Scotland – resulted in more detailed annual reports than had hitherto been issued, and from these reports may be ascertained the increase in the importance of the 'floating market', as the buying-steamers were collectively termed by the fishermen.

In 1883, six, and at times eight steamers accompanied a fleet of 431 boats at work in the Kilbrannan Sound, whereas 12 steamers were on Lochfyne attending the smaller fleet of 261 boats, most of which were trawlers, hence the discrepancy.[120] In 1884, 20 swift steamers were present on Lochfyne and the Kilbrannan Sound, which more than trebled the size of the fleet which had operated two years before.[121] The greatest number of steamers recorded was 21, in 1885,[122] but the size of the fleet fluctuated according to the state of the fisheries on Lochfyne and the Kilbrannan Sound, and the steamers periodically shifted to the Minches, when more profitable operations promised there.

At Sea on a Herring Steamer

The trading practices of the sea-going herring-buyers, or agents, were described in an anonymous account of *A Night with the Herring Fishers*, published in the *Glasgow Evening Times* in 1901.

'During the summer season Glaswegians frequently read in the local papers paragraphs to the effect that "there was a heavy fishing in Kilbrannan Sound last night, and the Campbeltown steamer was in consequence delayed for over an hour shipping them for the market". As likely as not their slumbers in the early hours of the morning were disturbed by the loud-tongued bell and the strident voice of the hawker yelling "Herring, best Lochfyne herring, new come in, new come in". But very few have any idea of the exciting scenes which take place at the fishing ground in Lochfyne when a cargo has been secured, say about 3 a.m., and the rush to catch the Glasgow market is made by the little light-draught steamers owned by the various fish merchants of Glasgow.

'These vessels have been cruising up and down the loch from sunset to sunrise, the only period during which the fishermen may cast their nets. They have been darting hither and thither, wherever their captains think catches may have been made, they being guided to the fishing boats by means of torches when night falls thick and black, and blots out the peaks of Arran and the undulating slopes of the Kintyre peninsula. . . .

'Keen hands (the fishermen) are at a bargain. Having secured a good haul of herring, they will not begin to trade until the buyer of each steamer in the vicinity has been pulled alongside in the row-boat. Then the huckstering begins, one buyer being pitted against the other, thereby running up prices in a way to warm the heart

of the men sitting knee-deep amid tens of thousands of herrings dragged inboard
. . . The bidding may start at five or six shillings per box, and, if fish be scarce, may
not stop till as much as thirty shillings has been reached. On the other hand,
however, a couple of skiffs may have a heavy haul in some sequestered spot with
only one steamer in the vicinity. As there is no competition for them the buyer gets
them at his own figure – possibly at two or three shillings per box. This is regarded
as a "thief's bargain", and great is the joy of the purchaser when coming to Glasgow
he finds that fish are scarce, and will in the open market fetch thirty shillings. It is a
cash trade pure and simple. As soon as all the herrings are packed into boxes, these
are counted and a wad of bank-notes representing £100 or whatever the sum may
be, changes hands.

'Steamers which fail to secure a sufficiency of boxes to warrant their running to
Glasgow usually make for Fairlie Pier, whence they are conveyed by rail, the
owners being apprised by telegram of the extent of the consignment. But should the
run to Glasgow be made the vessel is put close in to Gourock Pier, in passing which
the man on the bridge shouts out the number of boxes on board, information which
is wired to the city by a railway porter, who stands by for that purpose.'[123]

The Campaign against Corruption

Although the determined grip of the Glasgow fish-curers on the Clyde herring
industry had been broken by 1880, and a regular supply of fresh and inexpensive
fish was available to the Glasgow public, the fishermen were less than satisfied with
their share of the benefits of the economic reorganisation of the industry. Their
principal objection was to the buyers' exploitation of the lack of a legal measure for
the sale and delivery of herring on the Scottish west coast.

A petition, organised in 1882 by John Martin of Dalintober, and signed by 223
Campbeltown, Dalintober, and Carradale fishermen, requested the Fishery Board
to introduce a 'just measure', such as the quarter-cran basket, stamped to
authenticate it.[124] The Board's reply to Martin was unhelpful, and in 1883 agitation
among the fishermen resurged, and meetings were held in Tarbert and Carradale.[125]

In 1888, the Argyll and Bute Fishermen's Association was formed to secure the
fishermen's independence of the sea-going buyers, and steamers were chartered to
run catches to an agent appointed in Glasgow.[126] Perversely, however, the Fishery
Board in the following year introduced a regulation which recognised the quarter-
cran measure as the legal one, and the Fishermen's Association was consequently
left with between 15,000 and 20,000 boxes exceeding that standard. The Board
refused to grant a special brand for that type of box, which held about three-eighths
of a cran, and was popularly termed the 'Glasgow Box'. The Association continued
to use the boxes, countering official objections by pointing out that as the content of
each box was known to the buyer, there was no infringement of the law.[127]

The Association seems to have failed in the early 1890s, having run into financial
difficulties, and dealings with the agents of the Glasgow fish-merchants were
resumed on the fishing grounds.

14. One of the last of the herring-carrying steamers, the *Watchful,* with a deck cargo of boxes off Fairlie, 1930s. Per R. Miller, Campbeltown

In 1902 the fishermen re-asserted their grievance that the system of selling catches to the agents by the Glasgow Box was unfair. The dispute was based on two allegations. First, that they had no guarantee that the boxes were strictly uniform in size, and second, and more serious, that the appropriation of spilled fish by the agents and crews of the steamers, to be sold for their personal profit, was grossly abused. In the discharging of a catch, the fishermen, usually in darkness, and in all weathers, passed the fish on to the steamers' decks in small *spale-baskets.* [128] The steamers' crews were entrusted with supervising the filling of the boxes. [129] The fishermen complained that boxes were filled until they overspilled, and that a third or more of the fish handed on board the steamers was scattered over the decks and afterwards collected as 'scran', and boxed. [130] (That there is no evidence of the steamers' skippers being involved in the swindle is a point worth stressing.) [131]

In lower Lochfyne, in the spring of 1902, the fishermen were united in their refusal to sell catches except in quarter-cran measures. The agents refused the demand, and some crews dumped their fish overboard rather than yield. The entire Kintyre and Lochfyneside herring fleet remained in harbour for several days, while a deputation of fishermen debated the conflict with the fish-merchants of Glasgow, employers of the agents. The merchants tried to secure a compromise, but the fishermen were adamant in their insistence on the abandonment of the Glasgow Box. Unable, however, to endure the economic consequences of a long dispute, the fishermen returned to work almost immediately afterwards, and 'business was conducted on the old lines'.[132]

A Fishery Board enquiry later that year recommended that the quarter-cran measure should not be interfered with, but that the Glasgow Box be recognised as 'a measure customary in the trade', and, as such, legalised as an alternative, and branded to satisfy the fishermen of the uniformity of the boxes. The report also recommended that one fisherman should be posted on the deck of the steamer to count the number of boxes filled and to ensure that the lids were properly fastened.[133]

A 'very bitter feeling' was raised among the fishermen when the report became known. They insisted that the proposed legalisation of the Glasgow Box would deprive them of control over the quantities of herring going on board the steamers, and allow continuation of the 'scran' abuse. Exclusive adoption of the quarter-cran measure would, on the contrary, force the buyers to 'box their herring in a more careful manner, and both the seller and the buyers would be on equal terms'. The recommendation that a fisherman be deployed to check the boxing of the herring was dismissed as 'quite unworkable'. The entire crew was required for collecting and handing up the herring from the skiff, 'for time is valuable'.[134]

The issue in 1918 of a Ministry of Food order, under the Defence of the Realm Act, that sale of herring was to be by weight only, forced the buyers to adopt the official quarter-cran basket, which contained approximately seven stones.[135] Thus, what the fishermen failed to achieve by agitation, was achieved for them by the outbreak of a war, from which many of them would not return.

REFERENCES AND NOTES

1. J. McMillan, fisherman, Tarbert, *E.R.C.* (1864), 1179.
2. *Ib.,* 1156.
3. *Ib.,* 1179.
4. W. Dawson, W. Hamilton, J. and D. Bruce, *E.R.C.* (1862), 5.
5. G. Thomson, fishery officer at Ardrishaig, to Fish. Bd., 17 May, 1862, A.F. 26/9, 230.
6. *The Trawl — Herring Fishing in Scotland,* 6. A.F. 37/197.
7. *E.R.C.* (1877), 119.
8. *E.R.C.* (1864), 1116-7.
9. Cutting, incompletely dated.
10. *E.R.C.* (1877), 113.
11. Inveraray harbour report, 13 October, 1883. A.F. 38 154/2.
12. *F.B.R.* (1900), 118-9 (App. A).
13. *Ib.* (1890), 67 (App. D).
14. *Ib.* (1899), Part III Scientific Investigations, 16.
15. *Ib.* (1902), 113 (App. A).
16. *Ib.* (1910), 52 (App. A).
17. *Ib.* (1914), 52 and 54 (App. A).
18. *Ib.* (1890), 67 (App. D), and *ib.* (1914), 52 (App. A).

19. The boats *Sireadh* (Duncan Munro) and *Clan MacNab* (James MacNab). A. Fraser, *Lochfyneside,* Edinburgh, 1971, 25.

20. *E.R.C.* (1877), 125.

21. *Ib.,* 122.

22. *Ib.,* 125.

23. *F.B.R.* (1897), 233.

24. *Ib.* (1903), 246.

25. *Ib.* (1904), 237.

26. *Ib.* (1900), 118-20 (App. A), and *ib.* (1914), 52-4 (App. A).

27. D. Corner, fishery officer at Rothesay, 1 April, 1863, A.F. 37/37.

28. G. Thomson, fishery officer at Ardrishaig, 17 May, 1862, A.F. 26/9, 230.

29. G. Thomson, 1 May, 1862, *ib.,* 230. The largest boats previously used on Lochfyne were 23 to 24 feet (7.01-7.32 m.) of keel.

30. A. Levack, fishery officer at Campbeltown, rpt. for 1874, A.F. 22/7, 31.

31. B. F. Primrose, *Paper* (1852), 1.

32. A Levack, rpt. for 1875, *op. cit.,* 70.

33. Addition to the Fleet of Luggers, article in *Campbeltown Courier,* 7 October, 1876.

34. A. Levack, rpt. for 1875, *op. cit.*

35. J. Martin, fisherman, Dalintober, *E.R.C.* (1864), 755, and D. Bruce, fisherman, Ardrishaig, *ib.,* 1147.

36. A. Levack, rpt. for 1876, *op. cit.,* 104.

37. A. Levack, to Fish. Bd., 1 June, 1876. *op. cit.,* 91.

38. *Campbeltown Courier,* 25 March, 1 April, and 28 June 1876.

39. *Ib.,* 27 May, 1876.

40. A. Levack, to Fish. Bd., 11 May, 1877, *op. cit.,* 119.

41. *F.B.R.* (1898), 228-9.

42. *Ib.* (1900), 270.

43. A. Martin, 16 June, 1974.

44. D. McLean, 3 June, 1974.

45. *Campbeltown Courier,* 25 January, 1930. Original owners of the *Europa* were Alexander Montgomery and John MacLean, who was also skipper. She was sold in 1877 to the McMillan family of Campbeltown, and remained in its possession until the end.

46. *Ib.*

47. *F.B.R.* (1890), 67 (App. D), and *F.B.R.* (1914), 50 (App. A).

48. *Ib.*

49. Precognitions, Hawton/Parker case, A.D. 14 61/275, 14/3 (marginal annotations).

50. Details of the 13 Tarbert skiffs seized in September, 1860 (p. 18), including precise length measurements. A.F. 37/20. Also public auction notices relating to forfeited skiffs,e.g. of 28 May, 1862, the *Stork* and the *Dart* of Campbeltown and the *Lady* and the *Hero* of Tarbert (dimensions completely detailed). A.F. 37/159.

51. B. F. Primrose, *op. cit.,* 2.

52. Register of Fishing Boats, Inveraray District, Vol. 2, 2, Campbeltown Fishery Office. Vessel 21 feet (6.40 m.) of keel, first registered 1869, owner D. McAllister.

53. Precognitions, Turner/Rennie case, A.D. 14 53/288, 4.

54. Inventory submitted to Fish. Bd. by J. Miller, 29 September, 1860, Leith, A.F. 37/20.

55. *E.R.C.* (1862), 13.

56. L. MacDonald, Greenock, *ib.,* 19. For 40 years a fish-curer, and with experience also as a drift-net fisherman.

57. *F.B.R.* (1819).

58. *Off the Chain,* Manchester, 1868, 55.

59. D. McLean, 5 December, 1974, and (H.D.S.G.) H. McFarlane, 27 May, 1978.

60. D. McLean, 5 December, 1974.

61. *Off the Chain,* 20-1.

62. 27 June, 1975.

63. 5 April, 1976.

64. Report, Death by Drowning.

65. *Acta Parliamentorum Anne,* September XXI, M, DCC, V.

66. J. T. McCrindle, 29 April, 1976.

67. Cap. XXXI, *An Act for the Encouragement of the White Herring Fishery.*

68. Minutes, 18 October, 1848, A.F. 1/15, 401.

69. *F.B.R.* (1890), 67 (App. D).

70. *E.R.C.* (1862), 18.

71. 3 June, 1974.

72. C. Mactaggart, *E.R.C.* (1862), 17.

73. 18 April, 1874.

74. *E.R.C.* (1877), 138.

75. *Ib.,* 137.

76. A. Levack, 21 October, 1868, Campbeltown, A.F. 22/5, 230: 'Every pair of skiffs and trawls have a large boat in which they sleep and if need be run fish to market.' Also A. Martin, 16 April, 1974; H. McFarlane, 7 June, 1974; J. Weir, 28 November, 1975.

77. G. C. Hay, poem 'Kintyre', *Wind on Loch Fyne,* Edinburgh, 1948, 2.

78. R. McGown, 29 April, 1974; R. Morans, 1 May, 1974; D. McLean, 3 June, 1974; J. McIntosh, 3 June, 1958, recorded by E. R. Cregeen.

79. 3 June, 1974, and 17 Feb. 1975 (composite).

80. J. McWhirter, 10 Dec. 1974.

81. R. McGown, 29 April, 1974.

82. R. Conley, 11 June, 1975.

83. J. McWhirter, 10 December, 1974.

84. D. McFarlane, 4 May, 1974.

85. J. McWhirter, 10 December, 1974, and (H.D.S.G.) Hugh McFarlane, 27 May, 1978, the word *cùil.*

86. H.D.S.G. (Tarbert).

87. H. McFarlane, 27 May, 1978 (H.D.S.G.).

88. D. McFarlane, 4 May, 1974.

89. *Off the Chain,* 21.

90. J. McWhirter, 10 December, 1974.

91. 3 June, 1958, recorded by E. R. Cregeen.

92. R. Conley, 11 June, 1975, and J. Weir, 28 November, 1975.

93. 28 November, 1975.

94. D. MacVicar, 29 October, 1978 (H.D.S.G.).

95. 28 November, 1975.

96. 3 June, 1974. H. Martin's version, 3 May, 1974.

97. D. McSporran, 30 April, 1974.

98. *Campbeltown Courier,* 12 March, 1881.

99. H. McFarlane, 20 April, 1974.

100. Alastair Beattie, retired shepherd, Southend, Kintyre, 30 March, 1977 (S.S.S.).

101. D. McSporran, 30 April, 1974.

102. *R.R.C.* (1878), xxxiv.

103. Reprinted in *Campbeltown Courier,* 14 July, 1883.

104. *F.B.R.* (1884), xlvi.

105. Report on the Herring Fishing of Loch Fyne and the Adjacent Districts during 1885, *F.B.R.* (1885), 59. (Brook held post of lecturer on Comparative Embryology at the University of Edinburgh.)

106. *Campbeltown Courier,* 9 October, 1886.

107. *E.R.C.* (1877), 91.

108. *R.R.C.* (1878), xxxiv.

109. Etymology doubtful, but pronunciation represented exactly by English 'touch'. Has the meaning 'gain', usually financial.

110. H. McFarlane, 30 October, 1974. Told to him by one John 'Pud' McFarlane, who was present. Transcript edited and re-arranged in the interest of clarity.

111. Herring Fishery (Scotland) Act, 5th section, *F.B.R.* (1889), xxi.

112. For examples, *F.B.R.* (1909), 180. In Kilbrannan Sound, May and June 1909:
John McCaig, the *Glengarry,* T.T. 115 − £1.
Malcolm Johnston, the *Rolling Wave,* T.T. 33 − £1.
George Bruce, the *Cardross Castle,* T.T. 192 − 10/-.
Robert McGregor, the *Nightingale,* T.T. 162 − 10/-.

113. 11 September, 1866, Greenock, A.F. 7/88, 25-6.

114. *F.B.R.* (1869), 5.

115. J. Mitchell, *Tarbert Past and Present,* Dumbarton, 1886, 110.

116. *E.R.C.* (1877), 132.

117. *Ib.,* 125.

118. D. MacVicar, 17 September, 1978 (H.D.S.G.).

119. 29 January, 1881.

120. *F.B.R.* (1883), xliv.

121. *Ib.* (1884), xxviii.

122. *Ib.* (1885), xliv.

123. *Campbeltown Courier,* 2 September, 1901 (reprinted).

124. 13 February, 1882, to Fish. Bd., A.F. 37/56. Fishery officer's appended comments, 17 February: 'This matter is becoming yearly more and more a grievance among fishermen. I have heard them complain of some buyers who frequent Girvan, having baskets which in some cases would measure a fourth and a third more than the cran measure.' *Ib.*

125. *Campbeltown Courier,* 28 March, 1883.

126. *F.B.R.* (1888), 24.

127. L. Milloy and D'Arcy W. Thompson, Report to the Fishery Board for Scotland as to the Herring Fishery in the Firth of Clyde, 1903, 8-9, A.F. 56/794.

128. A shallow basket fashioned from *spales,* or laths.

129. C. Milloy and D. W. Thompson, *op. cit.,* 3.

130. Dalziel Maclean, to Secretary of State for Scotland, 25 August, 1903, A.F. 56/1457.

131. On that point fishermen are agreed.

132. *F.B.R.* (1902), 248.

133. C. Milloy and D. W. Thompson, *op. cit.,* 11-12.

134. D. Maclean, *op. cit.*

135. *F.B.R.* (1918), 72.

4

The Skiffs

THE decline of the practice of lodging ashore and in smacks began in the early 1880s with the introduction of the Lochfyne Skiff, which replaced the open boats traditionally employed at trawling. The length of the open skiffs had increased by the mid-1870s from the 20-25 feet (6.10-7.62 m.) range to a 25-30 feet (7.62-9.14 m.) range, but by the beginning of the next decade the standard length would exceed 30 feet (9.14 m.)[1] The Lochfyne Skiff was not, however, distinguishable by its increased dimensions, but by the incorporation of a small forecastle in its design. The available evidence[2] indicates that the first of the part-decked skiffs were the *Alpha* (C.N. 85) and the *Beta* (C.N. 84), launched at Girvan in the spring of 1882 for Edward McGeachy of Dalintober, whose family origin was on the island of Gigha, off the west coast of Kintyre.[3] Both boats had 25 feet (7.62 m.) of keel.[4] The value of the innovation was quickly recognised. Many fishermen had small forecastles added to existing craft, which were first 'stroked' (had one or more strakes or 'strokes' added), a modification which produced a dumpy type of vessel, later referred to by Ayrshire fishermen as the 'Penny Bank'.[5]

Between 1882 and 1885, numerous part-decked skiffs, such as the *Star* of Campbeltown, were built. She was launched in 1884, and was the first of her type in which John O'Hara sailed. His brief experience of trawl-fishing before then had been as a camper.[6] Hugh McFarlane went to sea as a boy in the family boat, the *Britannia*, which was of the original Lochfyne Skiff build. Constructed at the Port Bannatyne yard of James Fyfe in 1893, she was 32 feet 5 inches (9.88 m.) overall length, had 9 feet 2 inches (2.79 m.) of beam, 5 feet 4 inches (1.63 m.) of draught, with a net tonnage of 7.11.[7] The forecastle, at about 11 feet (3.35 m.) long, was evidently not as cramped as the very earliest forecastles. David McLean remembered his father's telling him that 'their room wis so crowded that if a man wanted tae turn in his bunk, he'd tae come out'.[8]

By 1896 the design had found favour beyond Kintyre and Lochfyneside. In Loch Broom, Sutherland, 'skiffs of the Lochfyne build are considered the most suitable craft for the loch herring fishing', reported the Fishery Board. Six such craft had been bought, '(and) with successful fishing in future, there is every probability of fishermen investing further in this type of boat, which, being partly decked, affords some means of shelter, and is a vast improvement on the small open boats hitherto used'.[9] Some – at least – of these skiffs were probably bought second-hand from Argyll owners, for by then had begun the phasing out of the fleets established more than a decade before. In 1902, 1903, and 1904 another building boom was experienced in the Lochfyneside, Kintyre, and Clyde boatyards.

In 1902, for instance, 17 second-class skiffs – that is, of 18-30 feet (5.49-9.14 m.) keel – were built in the Inveraray district, which included Tarbert and Ardrishaig, at an average cost of £97 each. That output represented the greatest, for that class of boat, of any fishing district in Scotland. Five skiffs were built in the Campbeltown district, each at an average cost of £9 less.[10] By 1904, the average cost of nine skiffs built in Inveraray district was £122, whereas at Campbeltown the average cost of an unspecified, but unquestionably smaller number of boats was £140,[11] a curious increase of almost 40 per cent on the average cost of two years before.

In the first decade of the present century the trawl-fleets attained maximum development. In 1905, 96 second-class boats sailed from Campbeltown, 36 from Carradale, 96 from Tarbert, and 60 from Ardrishaig.[12] The major boat-builder during that period was Archibald Munro of Ardrishaig, but there were productive yards in the other main fishing centres of Kintyre and Lochfyneside. In Tarbert alone there were five builders: Dugald Henderson, Archibald Leitch, Archibald Dickie, Duncan McTavish, and Thomas Fyfe. Another member of the expansive Fyfe family, Robert, was building at Ardrishaig, and at Port Bannatyne, near Rothesay, James Fyfe was active in the family trade. Robert Wylie's boatyard at Campbeltown, in the south, has already been mentioned. The yard furthest north was Donald Munro's at Blairmore, near Inveraray.[13]

Most of the boatyards maintained only a small labour-force which produced two or three skiffs annually. Dugald Henderson, whom Hugh McFarlane remembered as having employed a single assistant,[14] averaged one 35-foot (10.67 m.) skiff annually.[15] Duncan McTavish, however, employed four carpenters, including two of his sons.[16]

At least two skiffs were built at Ardrossan for Campbeltown owners by John Thomson, who had served a seven-year apprenticeship at the Fyfe boatyard in Fairlie.[17] The work-force at Ardrossan included John Thomson's wife, Isabella, of whom it was written, 'She could strike the water-line or use a hammer with the best of them.' She started work at six a.m. with her husband, and her carpentry was interrupted only by her return home at eight a.m. to attend to her family of 12 children. That done, she resumed work in the boatyard.[18] She died in September, 1944, having outlived her husband by 18 years.[19]

The skills of boat-building were also mastered by a few fishermen, notably Matthew MacDougall of Port Rìgh (King's Port) near Carradale. A natural craftsman, he began building boats at South Dippen during the late nineteenth century, his output during that period including a full-sized lugger for the Kinsale mackerel fishery.[20] When his fishermen sons were sufficiently experienced to manage the family skiffs, he devoted himself to boat-building, relying on their off-season assistance. A series of fine skiffs was built during the first decade of the present century, including the *Isabella,* the *Clan MacDougall,* the *Maggie MacDougall,* the *Mhairi,*[21] and the *Clan Matheson.* Pitch-pine for the planking was imported from Clydeside,[22] but the knees he cut himself from larch trees felled at Port na Gael, Saddell. These were pulled to the edge of the ridge by a mare – borrowed from the nearby farm of Whitestone – and then rolled on to the shore below. As many as possible would be loaded into the family ferry-boat, and the remainder lashed in bundles and towed to Port Rìgh.[23] Each vessel was built in an open field beside the family house, and, prior to launching, a section of the roadside dyke would be removed. Using ropes, rollers, and much manpower – supplied by local fishermen – the skiff would be manoeuvred at a 'half-sideways' angle down the bank and into the bay.[24] Some of the Dunure, Ayrshire, fishermen were likewise prepared to try their hands at the craft. In 1903, three skiffs, designed and built by fishermen, were launched at the village.[25]

The Lochfyne Skiff ranged in overall length from 32 to 37 feet (9.75-11.28 m.), though a few boats of 40 feet (12.19 m.) were ordered just before the First World War. The *Lady Charlotte* and the *Lady Edith*, built for Campbeltown owners in 1914, were fully 40 feet,[26] and were considered by some fishermen at that time to be unworkably large. The limitations of even 35- and 36-foot (10.67 and 10.97 m.) craft were described in 1908 by David Robertson of Campbeltown. 'Our boats are a small type, and unfit for any open fishing,' he wrote. 'We have to lie in port in rough weather. We only fish five nights per week, weather permitting, and consider it a fair season if we go to market 60 or 80 times per year.'[27]

15. The boat-builder Matthew MacDougall, with his son Alasdair, at work on a partly completed skiff behind his home at Port Rìgh, c. 1905. Per M. MacDougall, Port Rìgh

The Lochfyne Skiff was divided into three sections: forecastle, hold, and stern sheets. Two stout cross-beams strengthened the divisions. The forward beam was termed the 'break of the deck', and the bulkhead partitioning the forecastle from the hold was built down from it. The 'pump-beam', so called because the hand-pump for discharging bilge-water was mounted to it, traversed the after end of the hold, and aft of it, extending to the very stern-post, was the stern sheets, in which the net was laid.

The forecastle was decked over entirely, and within it the crew ate, rested, and slept. More commonly called the *den*, and with aptness, being a dim and smoky quarter when occupied, it was furnished sparsely, but sensibly, and wholly in accordance with the limited space.

Ballast

A stack of sand-bags (Gaelic, *pocaichean gainmheach*[28]) was carried on the skiffs, perhaps a dozen or two piled athwartships in the hold. The bags served as ballast, in addition to heavy stones deposited below the platform. The bags could also be positioned on the weather side, when tacking, to keep the boat 'on her feet', as the fishermen said.[29] When the tack was to be reversed, the two men stationed in the hold would transfer the bags to the opposite side. As Duncan McSporran of Dalintober remarked, 'The time used tae pass quicker if ye wir makin' a passage.'[30]

16. Late nineteenth-century skiffs leaving Tarbert on a summer's day of little wind. Per I. Y. Macintyre, Tarbert

If a catch was especially heavy – 'the full o' the boats', as such abundance was described – then the cord binding the mouth of each bag would be cut and the sand or gravel coped overboard.[31] Some fishermen were less willing to empty bags, dissuaded by the later necessity of having to refill them. Lightly built forecastle doors could be supported by a stack of bags inside the forecastle, if the hold was to be filled to capacity and there was a possibility that the weight of the catch might push the door in.[32]

The bags had various other occasional uses. They might be used for trimming a boat which was heavily loaded. Some boats lay deep at either the bow or stern, according to their partitioning, and a few bags could be placed fore or aft on the boat to stabilise her.[33] When a skiff grounded unexpectedly, she might be refloated immediately by propping a few bags on her bow to lift the stern and so free the keel.[34] Carradale crews, when taking a boat into Waterfoot for the week-end, would, if the tide was small, heap sand-bags on the bow 'tae put her by the heid', so that she cleared the sandy shallows at the mouth of the burn.[35]

The number of sand-bags necessary could be reduced by the fitting of a wooden 'ballast-box' across the hold. The box, which was a fixture, varied in height and breadth according to the build of the boat and her ballast requirement. These two factors were related. The deeper – or 'sharper' – boats, which could carry sufficient ballast below the platform, might not have benefited by the addition of a box, but it was generally adopted on the flatter-bottomed boats.[36] A foot-and-a-half (0.46 m.) square was the average size of the box, and it would be filled with stones and covered along the top with either a removeable board[37] or a nailed board.[38] The box enabled the skiffs to be kept cleaner. Herring were liable to slip in between stacked bags and rot undetected.[39] The bags, which were tarred when new, also rotted, and gave out their content, which was ordinarily coarse sand, but, when sand was unavailable, gravel. Sand, however, was more retentive of water bucketed over it to increase its weight, a common practice.[40] On some boats bags continued to be carried, in reduced numbers, after the introduction of ballast-boxes.[41]

The stability of a loaded skiff was improved by the longitudinal division of the hold into compartments. Stout planks, termed 'shifting-boards', were slotted into bearers both fore and aft of the ballast-box, thus creating four pounds into which the catch could be distributed. The boards stood about 2 feet (0.61 m.) high, but if an even higher partition was needed an extra board could be jammed in and more fish contained. The compartments were especially valuable on a breezy night, preventing the mass of herring from running unchecked from one side of the boat to the other with the motion of the sea.[42] The boards were not fixtures and could be prised up when basketing herring aboard to allow part of the catch to run from the port into the starboard compartment.[43] When discharging herring, the boards might be lifted out entirely, to afford the fishermen greater working freedom.[44]

The main ballast was, as already stated, deposited beneath the platform. Correct ballasting improved a skiff's performance under sail. The old class of fishermen took their seamanship very seriously and determined always to get the utmost performance from their craft. The Lochfyne Skiff carried about two tons of ballast.

The ballast stones were of a special kind – heavy, black, and sea-rounded, 'lik' cobble-stones'.[45] The Campbeltown fishermen gathered these in Black Rock Bay, at the mouth of Campbeltown Loch. They would tow a punt out with the skiff and load the stones into the punt.[46] The Tarbert and Carradale fishermen knew various shores where suitable stones could be gathered, though perhaps only at ebb tide. Certain Tarbert fishermen would go to Ardrossan Bay especially to ballast a new skiff; the stones there were of an exceptionally heavy kind.[47]

The stones were packed along the bottom of the boat, between the timbers in the hold section, to a height of a foot (0.31 m.) or more, and smaller stones set in to raise the ballast level with the boards and provide these with support. Pig-iron bars, having more compact weight, later replaced stones as the main ballast on carvel-built boats, though a layer of stones was ordinarily laid across it before fitting the platform boards. There was an opinion that clinker-built boats behaved better with exclusively stone ballast. 'It seemed tae be livelier wi' the stones, air goin' through them', said John Weir.[48] John McWhirter recalled that 'Young' John McKay's *Annunciata* carried about a ton of pig-iron bars, under a layer of stones. The iron had been obtained from a cargo-steamer at anchor in Campbeltown Loch. The fishermen had approached her to offer her crew fish, thinking that she was a coal-boat, and anticipating a lucrative barter, but she was laden with iron, some of which was stowed on her deck, and so McKay accepted a quantity of iron instead.[49]

As sand-bags harboured rotting fish, so too did stone and, to a lesser extent, iron ballast. The platform boards had to be maintained in good order, as cracks and gaps allowed fish to slip beneath. But scum and scales could not be prevented from running into the bilges, and the water would be *'scooshing'* from side to side with the motion of the boat, 'stagnant an' stinkin'', as Hugh McFarlane put it.[50] When the bilges were absolutely clogged, the fishermen would raise the boards, heap the ballast along the sides of the boat and wash it down with buckets of water.[51] The Ballantrae Banks fishery increased the problem, as herring spawn would slither down between the boards and into the bilges.[52] The skiffs were beached and thoroughly cleaned and overhauled in the month of May, and the ballast would then be heaved out on to the shore or quay and freshened. It might be cleaned once or twice again in the year, depending on the extent of the fishing and the consequent dirt.

Cement had been introduced on many skiffs by 1910. It was poured into the boat's bottom to the level of the 'floorings', though some fishermen added a quantity of net-bungs (corks) to the mixture to reduce its dead weight.[53] The replacement of stone and iron ballast by cement ensured cleaner bilges. A few bucketsful of water sluiced over the boards carried all the dirt aft to the pump-beam to be bilged out.[54]

Rowing

With the advent of the bigger class of skiffs, which rode higher in the water, a longer type of oar was required, and so came into use the 'sweep', which ranged from 20 to 25 feet (6.10-7.62 m.) in length. Two of these sweeps would be carried, and a smaller − perhaps 18-foot (5.49 m.) − oar in addition. The main oarsmen sat facing aft, one on the break of the deck, his oar extending over the port side, another on the forward (or mast) beam, his oar over the starboard side, while a third − usually the boy − stood on either side of the pump-beam, facing forward, with the small oar over the starboard side.[55] The steersman would have his rest at the tiller, and then give it up to one of the oarsmen and himself take over that oar. The boy, however, was not generally relieved, and when he became weary would simply haul his oar across and sit to rest. On a calm night, when oars would surely be required, the fire in the stove might be extinguished after supper had been cooked and eaten, so that smoke would not be flying in the face of the forward oarsman.[56]

Hugh McFarlane remembered that some of the older fishermen with whom he sailed in his youth had their preferred oar and would be satisfied with none other. The preferred oar was usually the most supple, and would bend on the stroke; the stiffer-wooded oars tended to be harder to pull on. Occasionally an oar would be cracked in use, but the offender was usually an inexperienced hand who had given it a quick uncertain jerk.[57] The belief in the personal oar possibly had its origin in the early trawl-skiffs. The entire crew then rowed in unison, so that each man might have had an oar for his exclusive use. An account of the equipment seized along with the 13 Tarbert trawl-skiffs in September, 1860 (p. 18) reveals the practice of marking oars with the initials or even the full name of the owner. Twelve oars out of 52 were so marked, for instance: 'A.S.F.', 'J.McE.', 'James Law', and 'John Black'.[58]

That the Tarbert men were an excellent class of oarsmen seems a certainty. David McFarlane's contention that they 'made a fool o' the navymen, rowin' them here an' rowin' them there',[59] is well substantiated in official accounts. The Tarbert men had, of course, been, in the main, inveterate trawl-fishermen since about 1840, and had developed the method of fishing with light, four-oared skiffs. John McWhirter recalled his admiration of the Tarbert men in his earliest years as a fisherman: 'They could pull fae Skipness tae Campbeltown an' ye widn't know the boat wis in the waater. They wir the greatest hands yet. Jeest tippin' her along at thir leisure, an' ye winna hear the oar goin' intae the waater.'[60] The section of the oar which rested between the thole-pins – or 'pootacks'[61] – was protected by a binding of leather, about a foot (0.31 m.) in length, a common provision, of course. The Tarbert men habitually greased that leather, which facilitated rowing and eliminated creaking and squeaking.[62]

When not in use the oars were laid along the starboard side of the vessel, with the sail and yard, and perhaps secured by a lashing. The oars were coated with varnish each spring to keep them 'respectable', as Hugh McFarlane put it. He remembered that new pine oars, crudely hewn in Norway, arrived in Tarbert by steamer, to be shaped and smoothed by a local carpenter.[63]

On frosty winter days or during the long calms of summer, the fishermen might 'put on' the oars and pull for 20 or 30 miles – and occasionally longer distances – to be home for a week-end. Hugh McFarlane remembered pulling from Ayr to Tarbert when a young man. With a distance of more than 35 miles ahead of them, the crew set out on a Saturday in July. The day was windless and when the slow, still evening fell to darkness they had reached the Garroch Head, the southernmost point of Bute. They were there crossing a busy shipping track, so the third oar was 'put in' and the man stood forward with a paraffin torch to keep watch and be ready to signal to any oncoming steamer. They reached Tarbert on Sunday morning, just as the church-goers were walking by the harbour to worship.[64] John Conley rowed from off the Ayrshire coast to Carradale on a Saturday morning in summer. They started pulling at six o'clock and rowed the entire distance home to arrive in the evening.[65]

David McLean unsentimentally recalled the obligation: 'It wis terrible on a Saturday mornin' if you were up aboot Cour or maybe further down, at the

Minister's Bay or along there, an' no' a breath o' win', an' maybe a torn net, tryin' tae mend the net, an' tryin' tae propel the boat along wi' the oars, an' the sun gettin' higher, an' you out all night without sleep.'[66]

Sailing

Sails were set whenever possible, for the obvious reason that they reduced the fishermen's labour. The area of canvas exposed was dependent on the vigour of the wind. For a light wind – a 'wee sar'[67] – the full spread of the lugsail would be allowed, and the big jib run out to the point of the bowsprit. As the wind increased beyond a good sailing breeze, the sail would be reefed repeatedly, and the big jib replaced by the second or middle jib, until finally, in a half-gale, the third or wee jib would be run out. Bowsprits were not generally carried on the earliest class of trawl-skiff (p. 59), but when the boats increased to the 25-30 feet (7.62-9.14 m.) range, a small bowsprit, about 4 feet (1.22 m.) long, was introduced. It carried a small sail, principally for steadying the boat and improving steerage on a tack. That bowsprit could be lifted in one hand and shipped on to the stem, and the cringle on the tack of the jib was simply slipped over a hook at the end of the bowsprit before the sail was hove up on the halyard and set smartly. No bobstay was used, but that hook was coiled so that the cringle, once positioned, could not be accidentally unshipped (dislodged).[68]

The later bowsprits would extend to almost 14 feet (4.27 m.) in length, about 4 feet (1.22 m.) of which was inboard,[69] but 10 feet (3.05 m.) was generally considered adequate. The bowsprit would be shipped only when making a passage, or when returning to port with a light wind.[70] Trawl-fishing being a two-boat operation, the frequency of the neighbouring boats' manoeuvring alongside one another, sometimes in rough weather, precluded the use of bowsprits in working conditions. Of 16 skiffs photographed leaving Tarbert harbour *circa* 1900,[71] bowsprits are visible on three, but only on one of these is a jib set. Ten skiffs are without bowsprits, and three others are too distant or obscure to discern details.

The jib was hauled out on a traveller, an iron ring which ran along the spar almost to its end, and on the upper side of which a hook projected to receive the cringle on the tack. The traveller was operated by a line fastened to it and leading through a metal sheave several inches from the end of the spar, and along its underside to the stringer where it was made fast. The second jib need not have been set out to the end of the bowsprit, but could be positioned midway along, and the third jib was often hooked on to the eye on the stem-head, thereby removing the necessity of the bowsprit.[72] When dowsing a jib the halyard was slackened and the sail simply collapsed, and could be grasped and hauled in, taking the traveller with it.

The bowsprit itself could be clamped into position at two points. The inboard end was always clamped to the 'palbitt' (mooring-stanchion) with a half-round clasp, pinned securely. On the starboard side of the stem-head, the top two 'strokes', or planks, were recessed to accommodate the bowsprit, and a clasp might be fitted there too.[73] On the very end of the bowsprit was a round belt of iron, termed the 'cranse', from the underside of which projected an eye. To that eye was shackled the bobstay, a wire rope which was always set when the bowsprit was in use. When the spar had been shipped over the bow and clamped, the wire was looped under the bobstay-cleat, which was bolted to the stem on the starboard side immediately above the water-line. The wire was then hauled tight from the foredeck, and made fast around the timber-head, on carvel-built boats a structural timber which projected 6 inches (0.15 m.) above the gunwale, and on clinker-built boats a specially fitted stanchion.[74] The end of the bobstay was spliced to form an eye, and a lanyard spliced to the eye. That lanyard would be doubled through the eye, hauled tight – so that the bowsprit actually dipped a little under the strain, countering the 'chug' of the mast[75] – and made fast on the timber-head. A man might go out on his stomach, almost to the end of the bowsprit, to give it a *cam* (bend)[76] while his shipmates tightened the bobstay.[77] Alternatively, either of the halyards could be used to tighten it.[78]

The jib-halyard was set on the starboard side and was rove through two single blocks, one of which was hooked to take the head of the sail. A sheet, made fast at the clew cringle of the sail, led aft on both sides, around the rigging and through a hole in the top strake forward of the break of the deck. The rope on the lee side, taking the strain of the jib, would be turned around the stringer or, more commonly, around a belaying-pin projecting from the underside of the break. The sheet would be passed around the pin and held by slipping a bight under the standing part, which, being constantly under strain, kept that loop secure.[79] Knots were avoided, because in the circumstance of a sudden squall the jib- and lug-sheets had to be freed instantaneously, lest the vessel capsize.

The original Lochfyne Skiffs were not rigged with forestays. Instead, the jib-halyard would be hooked to the eye-bolt on the stem-head. That arrangement was more convenient because these boats, as occasional drift-netters, lay head-on to their fleets of nets with masts half-lowered to reduce wind-resistance, and the daily procedure of raising and lowering the mast was simplified.[80]

Lugsail and Yard

The fishermen themselves rigged a newly purchased sail to the yard. The peak and throat of the sail were originally secured by tarry ropes, about 3 feet (0.91 m.) long, spliced to the cringles and passed through corresponding holes in the yard. The cringles would be hauled tight to the yard, bound repeatedly, and the rope knotted. At each end of the yard a small knuckle of wood projected, to prevent a slackened rope from slipping over the end of the spar. The rope used to lash up the throat was generally the heavier of the two,[81] but in the early 1900s an iron band, doubled over the end of the yard at the throat and bolted, came into use. For extra security, the cringle would be shackled to an eye on the underside of the band.[82] Between the ends of the yard, about 10 'roobans',[83] 2 feet (0.61 m.) in length, supported the sail. These lines were laced through a strand or two of the rope on the head of the sail and then bound around the yard, but not tightly, for their main function was simply to hold up the head of the sail.[84]

A yard with a bend on it might attract critical attention. Henry Martin recalled having heard 'oulder men' remarking, as they studied a skiff's rig: 'There's too much o' a kyam (*cam*) in it – he's loasin' a loat o' sell (losing a lot of sail).'[85]

A skiff's mast, from the step to the head, was approximately equal to the length of the vessel. The *Bonnie Jean* of Ardrishaig, at 35 feet 1 inch (10.70 m.) overall length, carried a mast 35 feet 2 inches (10.73 m.) long.[86] A noteworthy feature of the Lochfyne Skiff was the pronounced rake of the mast, varying between 15 and 20 degrees from the vertical. This merits comment. The small ring-net craft required maximum room amidships for the handling of the net and the reception of fish, and

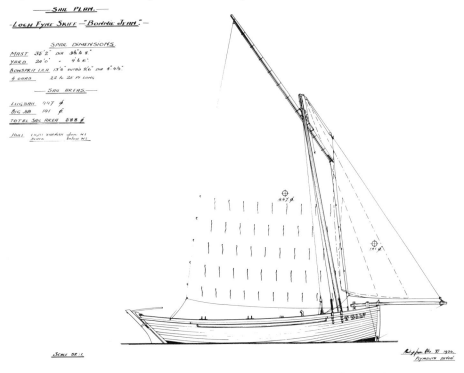

17. Sail plan of the Lochfyne Skiff *Bonnie Jean* of Ardrishaig, 1936. Photograph Science Museum, London

that room was obtained by positioning the mast well forward. But a mainsail rigged too far forward impairs speed, manoeuvrability, and good sailing to windward, and in the narrow waters where ring-net skiffs operated these qualities were indispensable. In the lochs, bays, and kyles, dangerous shores were constantly close, and a boat had to be good to windward and handy at answering the helm. The sail, therefore, required to come back towards the stern, and that was achieved by keeping the foot of the mast forward, but letting the head back – that is, raking it.

The extreme rake did, however, handicap a boat running before the wind. The head of the sail being well aft of the foot, the canvas presented an inclined surface to the wind and so could not trap the wind efficiently, which is the essence of running.

G

The head of the sail had therefore to be brought more into line with the foot, and so the mast was straightened up by sliding its heel along the grooved oaken step-block bolted to the keel. This 'stepping' of the mast when running fair was another distinctive feature of the Lochfyne Skiff.[87]

The mast was supported laterally by two pairs of 'stays', or shrouds. It was double-stayed on the starboard or 'back' side of the boat, and on the port side a single stay was rigged, with the lug-halyard doubling as a stay.[88]

At the head of the mast an iron cranse, to bear the stays, was fixed, and, immediately below that, a metal sheave, similar to that on the end of the bowsprit,

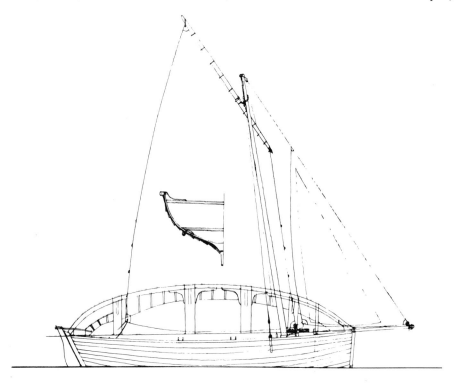

18. The Lochfyne Skiff — an arrangement of sectional end elevation, plan, elevation, and sail plan

but on a larger scale, was pinned inside the mast. The 4-inch (0.10 m.) deep notch into which the sheave was fitted was strengthened by the grip of the cranse, and the sheave itself was 'never seen tae nor touched'.[89] Through that sheave ran the pliable 'wire tie', which served to raise and lower the yard. On the port side the tie was made fast to a wire strop on the upper double block of the lug-halyard, and, on the starboard side, spliced to the traveller on the mast. The traveller incorporated an eye on to which the tie was spliced and seized, and on the underside a heavy hook. That hook caught up the eye of the metal strap fitted to the yard approximately a third of the way from the throat.

That strap, or 'batten',[90] would be manufactured by a local blacksmith. It was formed of a piece of iron, perhaps 2½ feet (0.76 m.) long, rounded to the shape of the yard, and buckled in the middle to create, when fitted, an eye. Four lashings of light wire – called bowsing- or seizing-wire – attached it to the yard.[91]

Setting the Sail

The procedure of hoisting the lugsail was this: the cringle in the clew of the sail (the after lower corner) and that in the tack (the forward lower corner) were attached to the sheet-block and the horse respectively. The sheet-block was shackled to a clip-hook ('clippax') with a spring tongue, which, when the cringle had been passed on to it, locked it in; and the horse was an upright heavy hook, its lower end formed as an eye, which ran on an iron bar, bolted horizontally to the deck on the fore side of the mast. The yard and sail lay stowed along the starboard side of the boat, with the fore end of the yard extending beyond the mast.

A man stood forward, caught the end of the yard and lifted it, assisted by one or two men aft of the mast. As another man, on the port side, slackened the halyard, the tension left the tie connecting with the traveller, which the man holding the fore end of the yard removed from the 'snotter' – a short becket attached to the heel of the mast – and hooked into the strap of the yard. When the man on the halyard began to heave, the upper block descended to the lower block, taking the tie, which in turn raised traveller and yard together. The sail thus mounted the mast, but the strength of one man was not sufficient to 'tent' (tighten) it. The entire crew had to put its weight on to the halyard to set the sail thoroughly, so the halyard was passed into a 'rickety', bolted upright on the foredeck, aft of the halyard blocks. The men ranged themselves fore and aft, each one, including the skipper, heaving on the halyard until the sail had been set as smartly as could be managed. The man furthest forward then secured the rope around the belaying-pin below the break of the deck (on the port side), as previously described. (The halyard, when hauled through the rickety, revolved a ratchet-wheel, and an engaging pawl ensured that the rope gained did not run back each time the crew paused.)

To lower the lugsail, the halyard was slackened and the traveller, carrying the yard, dropped down the mast until the tie could run no more. The yard was unhooked from the traveller, which was then hooked into the eye of the 'snotter', and a strain was pulled on the halyard until everything tightened up. The clew and the tack of the sail were then released, and yard and sail could be stowed to the side. If there was a 'weight' of wind to be reckoned with, a skiff would be hove to before the lugsail was lowered.[92]

On Lochfyne Skiffs the lugsail was always set on the starboard side, a rig which Kintyre fishermen compared favourably with the dipping lug of their East Coast counterparts who, when staying, had to lower the yard and shift it round to the lee side of the mast.[93]

If reefing – that is, shortening sail to a strengthening wind – the halyard was slackened to let the traveller down sufficiently to hook the second cringle on the tack of the sail on to the horse, and to pass the corresponding cringle on the clew into the clip-hook. The sail was then set again, and the slack of the foot tied up in reef-knots along its length using the 'points'. There were generally 10 points aligned across the sail, on each side, and the hanging foot, by the knotting together of the matching points, was gathered up 'as nate (neat) as ye can, an' that's no' very nate at times, wi' the canvas soakin' wet an' as hard as iron'.[94] There were five lines of reefing points on a sail, but four reefs would seldom be used.[95] After a night of hard wind, one crew might call to another, 'Whoot (what) kinna night had ye?' – 'Och, we had three reefs in.'[96]

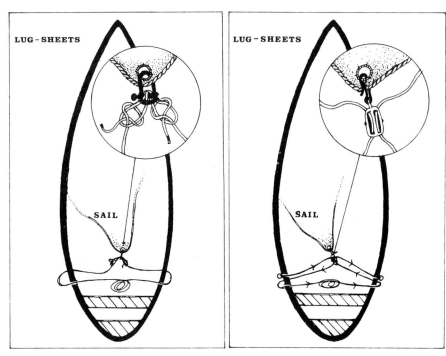

19. Lug-sheets. A. Shackle and rope-loops arrangement. B. Clip-hook and block arrangement.

The Lug-sheets

That the lugsails on the early trawl-skiffs were managed by single sheets only is a probability. The size alone of the craft would have dictated a simple arrangement. With the later class of skiffs – of 30 feet (9.14 m.) and more in length – double sheets were introduced. A 2-foot (0.61 m.) length of rope was spliced into a round, doubled through a shackle, and seized at the shackle with marline. Two loops were thus formed, and shackle and loops were attached to the cringle on the clew. Each end of the sheets – in fact, a continuous rope – was passed through a hole (bored into the top strake on both the port and starboard sides, just forward of the stern-beam) and made fast to the corresponding loop, with an ample allowance of rope free in the stern sheets (Fig. 19A).[97]

About the turn of the century the rope-bights were replaced on many boats by a double block, shackled to a clip-hook (already referred to). The ends of the sheets were made fast not at the sail, but at additional, forward, sheet-holes, one on each side, as illustrated (Fig. 19B). [98]

The sheet would be belayed only if the wind was dependably light. Otherwise, the skipper held the rope across his knees as he sat at the helm, [99] and perhaps placed, for safety's sake, a foot on the coiled running length of the sheet. [100] If a squall struck, the rope could be released instantly, allowing the 4 or 5 fathoms (7.32 or 9.14 m.), slack at his feet, to run free. When going about in a stiff breeze, a man would go aft to help the steersman to haul the sheet on the lee side. He usually stood on the stern beam and put a foot on the rail to haul, a dangerous stance should the sheet unexpectedly part. The son of a Campbeltown boat-owner, Duncan Black, almost drowned through just such an accident on the skiff *Speedwell:* 'It seems that Mr Black called to his son to come aft and assist him in hauling the sheet, and while he was doing so, as he was standing on the after beam the sheet gave way in his hand, and the skiff at the same time giving a lurch in the heavy sea, the young man was thrown overboard.' The crew of a nearby boat, the *Janet Mathieson,* was able to rescue him quickly, which was as well because he was fully clothed for fishing, [101] and when a man's seaboots fill with water, he sinks. An identical accident several years later cost Patrick Murray his life in Ardnacross Bay. A fisherman on the neighbouring boat remembered: 'We could dae naethin'. We had nae engines then. There wis nae rope on the sheet. The sail wis jeest lyin' flappin'. We went away clear o' 'im, but A don't think he came up again anyway.' [102]

In a light free wind, the sheet could be replaced with the sounding-line or a length of light buoy-string, one end made fast to the clew cringle, and the other to the sheet-hole. The sail was thus allowed to 'blow oot'. [103] Before beginning work, the sheet would be loosened on the port side, where the net was worked, so that if herring were located the net could be shot quickly and without obstruction. [104]

Sail-Care

To ensure that the sailing gear was maintained sound, the fishermen might devote an entire summer's day to opening the ropes on the lugsail and working Archangel tar into and between the strands. Particular attention was paid to the leech, which would be stretched with a block. That treatment obviated bagginess of the sail, which could be caused by successive wettings and dryings. A flat sail would 'shed the wind', whereas a baggy sail would not, and so slowed a skiff's passage to windward. [105]

On the pre-motor skiffs, a sail-repair outfit – needles, thread, and a sailmaker's palm – was carried, so that if a sail happened to tear during the working week, it could be temporarily patched. On returning to port the sail would be repaired by a sailmaker. Sets of sails were known to remain useful for 20 years and more. [106]

Sails, when new, were white, but many fishermen tanned them immediately with a solution of cutch (p. 129), a white sail being indiscernible at night, and therefore liable to put the skiff, invariably without navigation lights, in danger of being accidentally run down by another vessel.[107] Experienced fishermen could identify skiffs belonging to their home fleet by the cut of their lugsails alone. 'The shape wis that much imprinted on yer mind, seein' it that of'en,' said Hugh McFarlane, 'that if ye wid see the sail . . . "That's so an' so". Ye winna need tae see boat or men.'[108]

Sails would be barked several times annually to preserve them. The lugsail would be spread out on level ground which had first been swept clean, and several bucketsful of hot cutch solution would be carried from a nearby bark-house. To the solution would be added animal fat[109] or rancid butter (obtained cheaply from a grocery store),[110] depending on the preference of the skipper. When the additive had been melted, the mixture would be rubbed into the spread sail with brooms. When the 'barks' had been thoroughly applied to the upper side, the sail would be gradually folded in upon itself, and the exposed underside progressively scrubbed, until the canvas was reduced to a compact bundle.[111] It was allowed to lie, bundled up, for two or three days, to let the canvas absorb the solution, and then attached again to the yard and hauled up the mast to dry.[112]

The Regattas

The ultimate test of a skiff's performance under sail, and of her skipper and crew's co-operative skills, was undertaken in the annual regattas at Tarbert, Campbeltown, and Ayr. There were customarily two main competitive classes: the premier race, which was entered by skiffs provided with mast, bowsprit, and sails especially made for racing, and so rigged once only in the year; and the working class, for skiffs using standard gear.

Preparations were meticulous. The boats entered for a race would be beached several days in advance, their bottoms scraped and scrubbed, and a coat of black lead – normally used for cleaning household grates – brushed on, sometimes over two coats of black varnish.[113] The wives of crew members occasionally assisted by polishing the lead coating. Once polished, the bottoms would be 'lik' gless, an' she'd glide through the waater wi' the least wee bit air'.[114] To the ends of the long bowsprits would be attached 'whiskers', tightened to the chain-plates on the boat's shoulders.[115] These wire ropes served, on a tack, to strengthen the bowsprit against the additional strain it would be subjected to by the increased area of canvas carried.

John McCrindle of the Maidens, Ayrshire, remembered that a new tiller was made in Campbeltown especially for the *Seagull's* contest for the premier prize at the Ayr regatta several years before the First World War. As the *Seagull*, leading the other skiffs, thumped further ahead on the second round of the course, the tiller snapped at the neck, the boat lost steering way, and the clip-hook, after the sail had given 'two or three flaps, went up intae the leech, through the leech, an' tore right up'. The mishap was attributed to the carpenter's having cut the neck of the tiller too deeply.[116]

For a Tarbert regatta competition, remembered by Hugh McFarlane, the sand-bags from three other skiffs were loaded into the hold of the *Glen Garnock*, making a total of about 80 bags. During the race, the wind 'took off' and many of the crews began to throw out sand-bags to lighten their boats and so increase their speed, but the *Glen Garnock's* owner, Angus Livingstone, forbade his crew to dump a single bag, despite his boat's lead. 'We told him,' said Hugh McFarlane, '"Ye'll loss the race!" Says he: "We'll win the race fair."' The *Glen Garnock* did win the race, and her crew received a prize of £20 and a supply of groceries. Included in the supplementary crew were two other skippers, whose boats were not entered.[117]

The *Seagull's* crew was disqualified in another year's regatta at Ayr for reducing her ballast. 'We wir dumpin' thir san'bags,' remembered John McCrindle, 'an' if we'd tae dump them *in* the bag it widna have been bad, but we emptied them out, an' every crew wis seein' the patch o' san'.' The indiscretion was reported and an official disqualification was the result.[118]

In the regatta at Campbeltown in September, 1904, nine skiffs entered the premier race. The winning skiff, Charles Cameron's *Clemina*, completed the course in 4 hrs. 4 mins. 30 secs., 26 seconds ahead of the *Sweet Home* (Peter McMillan), and 2 mins. 12 secs., ahead of the *Angelus* (John McKay), in third place. Five minutes and one second, merely, separated the seven boats which completed the race, and 'as the boats closed in on opposite tacks on nearing the winning line, excitement was intense'. Six prizes were awarded: £15, with trophy, £12, £9, £6, £3, and a consolation prize to the last skiff to pass the marker-boat, so that only one crew received no reward. The only mishap of the race was the collision of the *Clemina* and the *Frigate Bird* at the outside marker-boat, which the reporter of the *Argyllshire Herald* described thus: 'It seems that at this stage *Clemina* was the overtaking boat, and either had to run down the marker-boat or foul *Frigate Bird*, unless the latter gave way to allow Mr Cameron's skiff to round. She did not do so, with the result that the skiffs became locked together.' The dialogue of the crews involved was not reported, but it may be assumed that it was of a less than delicate character. A protest lodged by the *Frigate Bird's* owner, John McIntyre, at the end of the race was not upheld.

In the race for skiffs with working sails, a time allowance of one minute per foot overall measurement was granted, so that shorter vessels would benefit proportionately, but the allowance was not of decisive importance that year, because the winning skiff, the *Polly*, finished 4 mins. 28 secs. ahead of the *Defender*. All four skiffs which completed the race received cash prizes, the highest award being £10 and the lowest £2. The fifth competing skiff, the *Maggie*, 'broke down'.

Side events occupied the attention of the regatta crowds while the racing skiffs were out of sight. Junior sailors, from H.M.S. *Calliope*, which was anchored that day in Campbeltown Loch, raced in 10-oared cutters; three crews of veteran fishermen competed in line-skiffs, the reporter noting that their style was 'a remarkable contrast to the rowing of the younger generation', though, disappointingly, he did not expand on the remark; local lumpers − quay labourers − raced with shovels, one crew disguised as negroes; open boats raced with lugsails; two classes of jolly boats − 23 and 18 feet (7.01 and 5.49 m.) long − competed in the loch; and on Dalintober quay a greasy pole competition was staged; to complete the entertainment, swimming and water-polo matches were organised in the harbour.[119]

Some boat-owners, perhaps goaded by sceptics, were not content with a regatta win. In October, 1876, for instance, Robert Brown of Torrisdale, whose skiff the *Zephyr* had taken first prize at the Campbeltown regatta, 'sent the bell through the town challenging any skiff in the harbour to run over the same course as on the previous day for £3'. Five Campbeltown skiffs accepted the challenge, but the *Zephyr* was again winner.[120]

With the advent of motor power, the tradition of racing declined and finally, by 1914, disappeared. Motor races were substituted briefly, but the very essence of regatta competition − the skill of boat management − had been removed, and the substitute proved uninspiring.

Caring for the Skiffs

In summer, when marine growth was strongest, the skiffs would be beached every two or three weeks, and their bottoms scrubbed. The routine was necessary to maintain the maximum motive capability of the vessels. A growth of *lìobhragach*,[121] the green stringy weed that sprouts on constantly submerged hulls, retarded the progress of a boat when under sail or oars. The skiffs would usually be moored well above low water-mark, and wooden legs fitted to each side of the hull to support her upright as the tide ebbed, and allow the men to get under each side and work. Originally the legs were secured by a length of buoy-string passed through a hole in the boat's side, but bolts replaced string on the bigger and heavier class of skiffs.[122] Some fishermen would beach their boats in shingly bays close to the fishing ground after a night's work, if the tide was suitable. They might 'ram her on' about four a.m., sleep for five hours, and then get to work when the tide had receded sufficiently.[123] Many of the Tarbert fishermen used bunches of old netting for rubbing the hull clean,[124] but long-handled brushes were generally used elsewhere.

The skiffs were beached annually in spring. 'They winna dae a turn ti' the cuckoo wid come,' said Hugh McFarlane.[125] In places such as Minard and Carradale, which lacked sheltered harbours, the boats were hauled ashore on rollers by teams of villagers. In Carradale, when a skipper intended to beach his vessel, he would send a messenger − usually a son − around the village to summon the men, and 30 or 40 helpers might assemble at Port na Cùile to put their weight on to ropes.[126] In Campbeltown and Tarbert the skiffs were simply moored along the harbour's edge on legs.

The stone or iron ballast was first thrown out and washed. Then the fishermen got to work with scrapers − Gaelic, *sgrìobanan*[127] − which were often files with the points cut off, turned up at each end conversely, and sharpened, all of which was done by a local blacksmith.[128] After scraping − which was not done annually, but every second or third year − the timbers would be rubbed over with sandpaper from stem to stern. A coat or two coats of boiled linseed oil − *ùilleamh bruich*[129] − and *roset* (rosin) mixed was then brushed on. The mixture, which was heated in a can over an open fire, was watched carefully to prevent its boiling over, because it required to be heated almost to boiling point. The tests employed were the dropping of a feather into it and the dipping in of the end of a brush − if the feather sizzled, or the hairs were singed, then the temperature was high enough.[130] Oil and roset mixed was originally the sole preservative coating on the upper hull, but, after

20. Nineteenth-century trawl-skiffs propped along Dalintober beach. The two vessels in the foreground have been scraped, evidenced by the light-coloured hull wood, but re-varnishing has begun. In the background is Campbeltown harbour, the masts of sailing-ships visible. Per E. Morrison, Campbeltown

the adoption of varnish, two or three coats of varnish would be applied. A second or third coat of varnish would be brushed on only when the previous coat had dried, therefore a spell of bright, dry weather was desirable.

The use of paint by fishermen in the Lochfyneside parish of Glassary – 'They change the colour of the paint frequently, the men of one place generally painting alike' – was noted in 1844,[131] but in Kintyre hull-painting was disapproved of until well into the present century, and paint would be used only for interior work and lettering. All paint was mixed, in paste form, with white lead, which such painters as Galbraith and Cochrane of Campbeltown purchased in five-hundred-weight drums, at £1 per hundredweight. The other ingredients in the mixture were raw linseed oil and driers. Browns, greens, blues, and all dark colours were mixed with boiled, rather than raw linseed oil. Boiled linseed oil cost, in 1910, 1/9d per gallon, and varnish 7/6d per gallon.[132]

Coal-tar or 'black varnish' (actually refined coal-tar) would be coated on a skiff's bottom, which would also be scraped periodically, though less frequently than the upper hull; but anti-fouling paint, mixed from red lead, first ground in a hand-mill,[133] eradicated the use of tar on a general scale about 1910. The hold of a vessel, from keel upwards, would be coated with coal-tar, after having been scrubbed with salt water and allowed to dry.[134] Before the introduction of coal-tar on a commercial scale, the fishermen of Kintyre and Lochfyneside probably themselves distilled varieties of tar. Osgood Mackenzie in his *A Hundred Years in the Highlands,* recounted that one day, after many native Scots firs had been felled for the building of Lord Elphinstone's lodge in Glen Torridon, Wester Ross, he and Elphinstone amused themselves by counting the rings on the trunks to discover the average age of the trees. Elphinstone pointed out to him that 'nearly every one of the trees had had a big auger-hole bored into it just above the ground level'. Elphinstone had been told by the old folk in the district that 'these holes had been bored by the Loch Carron people to produce tar for their boats'.[135]

The lockers and cupboards of the forecastles would be scraped and varnished too, and the practice of using a dried-out *mùrlach,* the spotted dog-fish, was common in all the Kintyre fishing communities. Line-fishermen would keep perhaps five or six, if hauled on the lines several days before beaching of the boat was due, and these they would behead and set aside.[136] The skin would 'grip any wud (wood) an' lee (leave) it as smooth as anythin''.[137] Some fishermen skinned the fish, but others used the carcase whole. The deck-head of the forecastle might also be varnished, but was ordinarily painted brightly in blue or green.[138]

The spring cleaning usually occupied the fishermen for three weeks, by the end of which period June, and the beginning of the regular herring season, would be impending. The pride of the crews in their boats, especially those which were family-owned, ensured that the utmost care and interest were exercised in their maintenance. The fishermen competed earnestly, but without rancour, in the tending of their craft, as in all other activities which engendered competition: fishing, sailing, and rowing. They did not care to be considered negligent, and perhaps believed that the integrity of a man could be judged by the appearance of his boat and the condition of his working gear. Such time-consuming pride has been sacrificed by most modern fishermen in the name of economics. As Hugh

McFarlane remarked: 'A thing they did, they did it substantial. In other words, time dinna coont (count). There wis no money in time. That's the whole thing. They liked the boats tidy.'[139]

REFERENCES AND NOTES

1. R.F.B.
2. J. McIntosh, Carradale, 3 June, 1958, recorded by E. R. Cregeen, and D. McSporran, 30 April, 1974.
3. Janet M. Black, Campbeltown, letter to author, 21 March, 1979, from McGeachy family tradition.
4. R.F.B. (Campbeltown).
5. J. T. McCrindle, 29 April, 1976.
6. 10 June, 1958, recorded by E. R. Cregeen.
7. R.F.B. (Tarbert).
8. 17 February, 1975.
9. *F.B.R.* (1896), 189.
10. *Ib.* (1902), lxi.
11. *Ib.* (1904), lvi.
12. *Ib.* (1905), 80-4 (App. A).
13. R.F.B.
14. 30 October, 1974.
15. R.F.B. (Tarbert).
16. H. McFarlane, 30 October, 1974.
17. D. C. Shedden, Saltcoats, Ayrshire (son-in-law of late J. Thomson), undated letter to author.
18. Obituary of I. Thomson, newspaper cutting.
19. D. C. Shedden, *op. cit.*
20. J. McIntosh, 3 June, 1958, recorded by E. R. Cregeen; D. McIntosh, 28 April, 1974; G. McKinlay, 23 February, 1977 (S.S.S.). That lugger was used for trading, by MacDougall himself, presumably after failure of the Kinsale fishery. She would be moored alongside a natural jetty – a long rock projecting into the sea close to Rudha nan Sgarbh – and loaded with potatoes from the Whitestone fields. The crop was stored in two potato-houses close to the shore. These still stand, roofless, but in otherwise fairly sound condition. The potatoes were sold at Howth, where herring would be bought and cured for the return trip. The business was conducted in association with the McKinlays, who farmed Whitestone. The families were connected by marriage. G. McKinlay, 23 February, 1977.
21. R.F.B. (Campbeltown).
22. R. Oman, 6 June, 1974.
23. G. McKinlay, retired farmer at Whitestone, 10 March, 1977 (S.S.S.).
24. D. McIntosh, 27 April, 1974, and R. Oman, 6 June, 1974.
25. *F.B.R.* (1903), 250.
26. R.F.B. (Campbeltown).
27. To J. Sinclair, Secretary of State for Scotland, on behalf of Campbeltown Fishermen's Association, 29 February, 1908, A.F. 65/1456.
28. D. MacVicar, 18 January, 1979 (H.D.S.G.).
29. J. McWhirter, 5 February, 1976.
30. 11 June, 1974.
31. H. Martin, 4 June, 1974; D. McIntosh, 9 December, 1975; H. McFarlane, 10 December, 1975.
32. J. McWhirter, 5 February, 1976.
33. As above, 5 March, 1975.
34. As above, 5 February, 1976.
35. J. Conley, 5 December, 1975.
36. H. McFarlane, 18 March, 1976.

37. J. McWhirter, 5 February, 1976.

38. H. McFarlane, 18 March, 1976.

39. J. McWhirter, 5 February, 1976, and D. McIntosh, 28 April, 1976.

40. H. McFarlane, 10 December, 1975.

41. As above, 18 March, 1976.

42. J. McWhirter, 5 February, 1976; H. McFarlane and J. Weir, both 18 March, 1976.

43. H. McFarlane, 18 March, 1976.

44. J. Weir, 18 March,1976.

45. J. Conley, 5 December, 1975.

46. J. McWhirter, 5 February, 1976, and A. Martin, 14 May, 1976.

47. H. McFarlane, 26 April, 1976.

48. 18 March, 1976, and D. McIntosh, 27 April, 1976.

49. 5 February, 1976.

50. 18 March, 1976.

51. H. McFarlane, 18 March, 1976.

52. J. McWhirter, 5 February, 1976.

53. D. McSporran, noted 25 June, 1979.

54. H. McFarlane, 18 March, 1976.

55. D. McSporran, letters to author, 27 June and 2 July, 1979.

56. J. McWhirter, 5 February, 1976, and H. McFarlane and J. Weir, both 18 March, 1976.

57. H. McFarlane, 30 October, 1974.

58. Inventory submitted to Fish. Bd. by J. Miller, 29 September, 1860, Leith, A.F. 37/20.

59. 4 May, 1974.

60. 5 February, 1976.

61. Gael. *putag*. H. McFarlane, 13 May, 1978 (H.D.S.G.).

62. J. McWhirter, 5 February, 1976, and J. Weir, 18 March, 1976.

63. 30 October, 1974.

64. 20 April, 1974.

65. 27 April, 1974.

66. 3 June, 1974.

67. H.D.S.G. (Kintyre generally). Language of its origin uncertain.

68. H. McFarlane, 5 April, 1976.

69. Sail-plan of Lochfyne Skiff, *Bonnie Jean,* T.T. 117, 1936. The Science Museum, London, negative no. 100/37.

70. D. McSporran, 11 June, 1974, and H. McFarlane, 11 June, 1976.

71. Photograph given to the author by the late John Weir, Tarbert.

72. J. McWhirter, 8 April, 1976.

73. As above, and H. McFarlane, 5 April, 1976. J. McWhirter qualified his information: 'For workin' purposes very few had them. I saw them puttin' them in for the regatta, when ye'd be puttin' in a longer bowsprit.'

74. H. McFarlane, 29 July, 1976.

75. As above, 5 April, 1976.

76. H.D.S.G. (Campbeltown). Gael., 'bend, curve'. Generally pronounced 'kyam'.

77. H. Martin, 10 June, 1976, and H. McFarlane, 29 July, 1976.

78. J. McWhirter, 8 April, 1976, and H. McFarlane, 29 July, 1976.

79. H. McFarlane, 11 June,1976. Technically, a slippery-hitch.

80. J. McWhirter, 8 April, 1976, and H. McFarlane, 11 June, 1976.

81. H. McFarlane, 29 July, 1976.

82. J. McWhirter, 8 April, 1976, and A. Martin, 14 July, 1976.

83. D. McSporran, letter to author, 4 July, 1979. A form of 'robands'.

84. A. Martin, 14 July, 1976, and H. McFarlane, 29 July, 1976.

85. 21 February, 1978 (H.D.S.G.).

86. Sail-plan of Lochfyne Skiff, *Bonnie Jean, op. cit.*

87. H. McFarlane, 5 April, 1976. Elucidated and expanded by Robert Smith, on whose explanatory account of 12 July, 1979 the preceding passages are based.

88. H. McFarlane, letter to author, 7 January, 1977, and D. McSporran, noted 25 June, 1979.

89. H. McFarlane, 5 April, 1976.

90. D. McSporran, letter to author, 4 July, 1979.

91. A. Martin, 14 July, 1976; H. McFarlane 29 July, 1976; D. McSporran, noted 25 June, 1979.

92. D. McSporran, noted 25 June, 1979.

93. H. McFarlane, 5 April, 1976, and J. McWhirter, 8 April, 1976.

94. H. McFarlane, 5 April, 1976.

95. J. McWhirter, 8 April, 1976.

96. D. McLean, 7 April, 1976.

97. H. McFarlane, 5 April, 1976.

98. As above, and D. McSporran, diagram and letter to author, 4 July, 1979.

99. H. McFarlane, 5 April, 1976, and D. McSporran, letter to author, 27 June, 1979, etc.

100. H. McFarlane, 5 April and 11 June, 1976.

101. *Argyllshire Herald*, 17 September, 1904.

102. J. McWhirter, 8 April, 1976.

103. D. Blair, 5 June, 1975, and H. McFarlane, 11 June, 1976.

104. D. McLean, 7 April, 1976, and H. McFarlane, letter to author, 1976.

105. H. McFarlane, 10 December, 1975 and 5 April, 1976.

106. As above, 10 December, 1975.

107. H. McFarlane, 20 April, 1974, and D. McLean, 8 December, 1975.

108. 20 April, 1974.

109. H. Martin, 31 November, 1974, and D. McLean, 5 December, 1974.

110. D. McSporran, 11 June, 1974, and J. McWhirter, 10 December, 1974.

111. A. Martin, 27 November, 1974; H. Martin, 31 November, 1974; D. McLean, 5 December, 1974.

112. H. McFarlane, 7 June, 1974, and J. Conley, 5 December, 1974.

113. J. McCrindle, 28 April, 1976.

114. J. McWhirter, 8 April, 1976.

115. J. McCrindle, 28 April, 1976, and H. McFarlane, 29 July, 1976. Regatta bowsprits could extend to 17 feet (5.18 m.) – H. McFarlane, 5 April, 1976.

116. 28 April, 1976.

117. 30 October, 1974. Assisting skippers were Peter Smith, a native of Gigha, and Archibald Paterson, a native of Carradale.

118. 28 April, 1976.

119. *Argyllshire Herald*, 17 September, 1904.

120. *Campbeltown Courier*, 7 October, 1876.

121. H. McFarlane, 10 December, 1976 (H.D.S.G.).

122. As above, 30 October, 1974.

123. D. McLean, 6 April, 1976.

124. J. Weir and H. McFarlane, both 5 April, 1976.

125. 30 October, 1974.

126. J. Conley, 5 December, 1974.

127. D. MacVicar, 22 October, 1978 (H.D.S.G.).

128. H. Martin, 31 November, 1974.

129. D. MacVicar, 22 October, 1978. Raw linseed oil – *ùilleamh amh*. H.D.S.G.

130. D. McIntosh, 27 April, 1976, and J. McWhirter, 10 December, 1974, respectively.

131. *N.S.A.*, Glassary, 691.

132. A. Cochrane, of Galbraith & Cochrane, painters, Campbeltown. Information prepared for author.

133. *Ib.*

134. J. Weir, 18 March, 1976.

135. London, 1949 edition, 46-7.

136. A. Martin, 14 May, 1976.

137. H. McFarlane, 30 October, 1974.

138. D. Blair, 21 January, 1975.

139. 30 October, 1974.

5
Daily Life Afloat

THE forecastle design of the skiffs from the earliest-built to the last remained fundamentally the same. In the later skiffs, however, increased space — for the forecastles reached 14 feet (4.27 m.) in length — allowed for the provision of roomier cupboards, larger stoves (therefore, improved cooking facilities), and more comfortable beds.

On both sides of the forecastle a 'bunker' or 'locker' extended forward from the bulkhead, built to the curve of the hull, but perhaps straight on the inner side. These bunkers, which usually ended short of the mast, were continuous boxed-in seats, raised about 1½ feet (0.46 m.) from the platform, and 2 feet (0.61 m.) or more at their broadest, for they also served as beds. Two or three covers were set in flush with the top of the bunkers, spaced along their length, with a finger-hole in the middle of each so that they could be lifted out easily. Inside the bunkers could be kept tools, etc.

Set into the very forepeak of the boat was a three-sided shelf, fixed midway between the platform and the deck. It was shaped to the curves of the bow, and its after edge raised several inches to prevent such articles as were kept on it from being thrown on to the platform when the boat was pitching around. On the earlier skiffs the weekly supply of a dozen or more loaves, and the main stock of sea-biscuits, were placed on the shelf,[1] but on the later skiffs, when an extra cupboard provided adequate room for all foodstuffs and crockery, the shelf served to contain the compass and foghorn and other odds and ends.[2] Below that shelf, on the platform, bundles of spare netting and the jibs were stowed, and, if the crews were fishing at a distance from home, spare clothing and towels would also be packed forward.[3]

Bunks and Hammocks

Each locker served as a bed for a man, as has already been mentioned. The other men — usually the nimblest — clambered into bunks or hammocks above the lockers. When Hugh McFarlane went to sea in the *Britannia*, in 1898, his bed was a wooden bunk, divided along its length, and the two parts hinged together so that the outer one could be folded up against the hull of the boat when not in use, and lashed out of the way.

The bunks — or *shelves*, as they were commonly called — were about 6 feet (1.83 m.) in length and 2½ feet (0.76 m.) broad, the inner board being some inches narrower than the outer. Two wooden supports jutted from the hull of the boat, and these, like the inner board, were fixtures; but, despite their length of 2½ feet they offered no hindrance to the men in the tiny den, one being secured immediately aft of the mast, and the other tight against the bulkhead. A board, 3 or 4 inches (0.08 or

0.10 m.) high, was affixed along the outer edge of the 'shelf' to keep the sleeper from rolling out on to the platform. Hugh McFarlane's pillow was a bundle of netting covered with his waistcoat.[4]

The most common type of bed was made of canvas, lashed either to a rectangular iron frame or to outer and inner longitudinal supports, and suspended from the deck, fore and aft, on hooked iron rods. These hammocks, which were 5 to 6 feet (1.52–1.83 m.) in length and 2 feet (0.61 m.) in breadth, were stowed compactly along the sides of the forecastle when not in use.

The canvas sheet of the former, framed type was perforated around its edges, the holes spaced perhaps 6 inches (0.15 m.) apart and strengthened by brass thimbles. It was lashed to its frame by a length of buoy-string passed through the holes and wound about the frame. The iron rods were about 2 feet (0.61 m.) in length. The upper end was fixed to the deck, and the lower end hooked under the frame of the hammock. The entire frame folded up on hinges against the hull. It was less commonly used than the other canvas type, possibly because the frame, after long use, began to buckle.[5]

The canvas of the latter type might not be perforated at all. The inner edge could be secured under a strip of wood of corresponding length, which was fastened to the hull by screws. In most cases, however, that edge was perforated and would be lashed to the strip or to a metal bar. But the outer edge of the canvas was invariably rolled about a metal pole and sewn slackly so that the pole could be removed, because the canvas was taken home each spring and boiled clean. For that reason, too, the canvas would never be fixed to the hull with anything other than screws.[6]

The breadth of the hammock could be kept extended by inserting sticks at the fore and after ends. These sticks, called 'stretchers', were notched at each end to accommodate the round of the pole. They were not entirely successful, by all accounts. An unsettled sleeper might knock out both sticks during his sleep, so many of the fishermen preferred to do without. The fore end of the hammock might, however, be lashed up to the mast.

When a crew had risen to prepare for the night's work, the rods would be unhooked from the frame and the ends of the outer pole slipped into beckets, so that the whole bed hung along the side of the forecastle. The bed of the young cook was the bare boards of the platform. He lay with his head below the mast and his body stretched aft, wrapped about in the big jib. His seaboots, covered with a jersey, served as a pillow.[7]

The Tarbert skiffs were normally manned by a crew of four, but in the Campbeltown and many of the Carradale skiffs a fifth crew-member was carried. He was usually a boy, employed direct from school. His tasks on board were menial, but quite as arduous as those of his experienced seniors. He generally began work on a quarter of a man's share of the profits (p. 214), and as his ability as a seaman increased he graduated to a half-share, and, finally, when he could competently mend nets, splice ropes, etc., he received a full share, and his duties as cook and drudge would sooner or later be assumed by a younger boy. His graduation might take from two to five years, depending on both his ability and desire to learn.[8]

The crew slept almost fully clothed. Some removed their jerseys and even fewer their trousers. The forecastles were warmed in winter by a fire which burned constantly in the small stove. Indeed, at times the heat was excessive and the scuttle above would be opened to let in fresh air. During the night, if fishing was slack, the skiffs might lie at anchor until the approach of dawn, when the herring would begin to strike offshore into deeper waters. The fishermen usually 'turned in' for several hours, so they preferred to be in a state of readiness to raise the anchor and get under way if other boats were heard shifting.[9] The competitiveness of the crews was keen, and there was perhaps no more shaming experience than to have been asleep while other fishermen were successfully at work. Many stories were told, with unconcealed delight, of large catches secured 'while the fleet slept'. The fishermen were also alert to the danger of their unlit vessel's being accidentally struck in darkness by another vessel.

Access to the den was by a small sliding-door – of variable dimensions, but, on average, 3 feet (0.91 m.) by 4½ feet (1.37 m.) – on the starboard side of the bulkhead, several inches out from the bunker. When the boat had been loaded to capacity with herring, the catch would be packed right to the bulkhead, thus preventing entry by the door. Instead, the men would drop through the scuttle into the forecastle. On some narrow-scuttled skiffs, hefty men would be precluded from that resort, and would have to remain on deck. A cup of tea and a 'piece' would be handed up through the scuttle to them, and undoubtedly they would meditate ruefully on the warmth of their fellows in the den below.[10] In some skiffs a block of wood was nailed to the forecastle bulkhead and served as a step for the men as they climbed through the scuttle.[11]

Cupboards and Stoves

Into each corner of the after end of the forecastle a cupboard was built, though in the earliest forecastles only one such cupboard was made, usually on the starboard side.[12] These cupboards extended from the bunker almost to the deck, and could be three-sided (that is, the outer side built over the corner) or in square form. They contained crockery and cutlery, and the weekly food supply.

In the middle of the bulkhead a small stove (*bogie*) was fitted. A section of the bulkhead, perhaps 2 feet (0.61 m.) broad and extending from the platform to the deck, would be cut out, and a three-walled box constructed. That *stove-box* was formed externally and thus was a feature of the hold rather than of the forecastle. The stove was set entirely into it, so that the front of the stove was almost flush with the interior of the bulkhead, conserving valuable space in the den.

The original stoves were crudely designed, utterly unsuitable for cooking on, and a fire hazard moreover. Hugh McFarlane remembered that the *Britannia's* stove regularly set the box alight, so close against the wood was the funnel. In that event, a bucketful of water would be poured down the pipe.[13] In later years, when the forecastles increased in size, the funnel could be fitted to the stove with a gap of several inches between it and the back of the box; and asbestos or sheet tin or zinc fixed to the interior of the box also diminished the possibility of fire.[14] With the original stoves, the kettle had to be placed inside the stove and on to the burning coals. The kettle was, said Hugh McFarlane, 'as black as the lum itsel'', and would become so hot that it could be removed from the stove only on the end of a poker. The kettle − or *fish-pan*, if a meal of herring and potatoes was to be cooked − was kept secure within the stove by jamming the door closed against it.[15]

About the turn of the century, 'Jack Tar' stoves became popular. These were of improved design, and incorporated a guard-rail on top for containing the pot or kettle. On some boats a small primus stove was carried in addition, and it simplified the cooking operation by enabling the boy to have both kettle and pot boiling at once.[16] Otherwise, the kettle was first boiled, the herring then cooked, and, finally, the kettle again brought 'through the boil' and tea added.[17] Fish and potatoes were usually cooked together in one large pot.

A second, larger 'Jack Tar' model later appeared, capable of holding both pot and kettle, but the first was of a more suitable size for the smaller skiffs, and remained in use on most until the final years of the skiffs. If the fishermen were working too far from home to return at the week-end, a pot of broth or mince might be cooked on Sunday, which relieved the monotony of their fish-based diet.[18]

The Diet of the Fishermen

The staple diet of the fishermen was fish, and especially herring, perhaps boiled for breakfast, boiled for dinner, and fried for tea. A single fisherman might consume in excess of 60 herring in a week.[19] 'When it wad come Thursday or Friday,' said Henry Martin of Dalintober, 'there wir yins that winna care, an' there wir other yins that wid tell ye tae put on a rasher o' herrin' on a Seturday mornin' comin' for the herbour. They wadna haiv dried breid, as they said.'[20] White fish − cod, flats, and whitings − whenever netted with herring, would be kept and eaten to vary the diet.[21]

In winter, a small barrel − kit or firkin − of 'salt herring' would be kept out in the hold. Some fishermen did not care to eat herring that had absorbed much pickle, and would dump the remaining content of the kit after a week or two of pickling, and salt down another batch. Pickled herring might continue to be eaten into the month of April, and even May.[22] When fishing for spawning herring on the Ballantrae Banks during February and March, large cod were often netted in great numbers, and some of these fish, which had congregated on the gravel banks to gorge themselves on the deposited spawn, would be filleted, cut into portions, and salted down.[23] A few cod could also be lifted from the herring-nets fixed out on the banks by local fishermen. As one Campbeltown fisherman remarked: 'Ye'd never want *kitchen* there if ye seen nets in the waater.'[24]

On a warm summer's day the fishermen might split open a couple of dozen herring, sprinkle them freely with salt and pepper, and spread them out on a board, perhaps with a canopy of netting raised on a brush and the boat-hook to keep gulls off,[25] but usually with the netting simply laid over the fish. These fish were termed by the Tarbert fishermen *sgadan gréine* (sunned herring)[26] and were so delicious that batches were occasionally prepared on a Friday for taking home next morning.[27] On

H

especially warm days, the fish would turn brown like kippers while the fishermen slept. 'Sgadan Gréine' survives as a name given to a rock, though the savoury fish are now seldom eaten. On the shore below Laggan Head, four miles south of Tarbert, that rock presents its face to the sea, and, approaching the shore by boat on a bright day, the marble forms resembling *sgadan gréine* can be seen in rows, 'pure white, a dozen or more, split on that rock'.[28]

Seabirds were regularly shot and eaten by the fishermen, particularly guillemots (called *dookers* by the Tarbert men and *marrotts* by the Campbeltown and Carradale men), but also shags (*scarts*), and even puffins. Shotguns were carried on all of the Tarbert, and many of the Campbeltown skiffs, and the birds might be shot and eaten once or twice a week, when available on the current fishing ground. The Tarbert men plucked and then singed the dookers, but many of the Campbeltown men simply skinned the birds, which was invariably the way of preparing scarts for cooking. In some communities, such as Carradale and Minard, birds shot for family consumption would be buried for several days, wrapped in cloth, to diminish their fishy flavour.[29]

The Tarbert fishermen, who acquired a reputation in all the Clyde fishing communities for their fondness of cooked seabirds, and hence earned the collective nick-name of 'the Dookers', went out from the village, two or three men in a small boat, and shot birds for the household, especially when meat was scarce.[30] If few dookers were about, the fishermen might make a *càbhach* by scattering on the water fish livers or broken herring or bread, etc. Gulls innumerable would descend on the spot to feed, attracting edible birds from afar.[31] If the fishermen were 'awfully hard pressed' they might make use of drowned dookers, brought back from Loch Striven, where they stuck in the cod-nets sunk there in late winter and spring.[32]

The birds could be stewed – one for each man – or several boiled to make a potato soup. When killed from the skiffs they would be lifted on board in a basket or lap-net. They were an especially valuable supplementary food in such fishing areas as Loch Ryan. 'If ye hanna been up in Stranraer,' said David McLean, '(and were) livin' on herrin', an' ye waanted a change, ye'd go an' shoot a few marrotts.'[33] The annual toll on these various birds must have run into thousands, as they were greatly esteemed by most fishermen. 'Tae us,' said John Weir, 'they wir a luxury, but A daresay that a person that wasn't used wi' them winna care for them.'[34] The practice of shooting seabirds declined as the skiffs were superseded by larger motor boats, and as the monotonous diet of the fishermen improved to include fresh meat.[35]

A Tarbert doctor, by name MacMillan, encouraged 'chesty' patients to eat dookers for relief, and the fishermen, some 70 years ago, brought in, at his request, quantities of the birds for such invalids.[36]

Much bread was eaten during the week. As many as a dozen loaves were carried to sea every Monday, but that stock had to last until Saturday morning, when the fleet returned to its home port. By then the stock might be almost depleted. John McWhirter repeated a maxim which was especially applicable on a Saturday morning: 'Big bites o' *stainlock,* an' wee bites o' loaf.'[37] Bread was eaten with meals of fish, and also during the night as *pieces* with bowls of tea. Several pounds of butter and a hunk of cheese were supplied with the stores each week, and perhaps a pot or two of jam, though it was considered by some crews as rather an extravagance, because it was consumed quickly. Treacle and syrup were undoubtedly more economical alternatives to jam.[38] A constant supply of vinegar was required to flavour the herring; pepper; salt; tea; sugar; tins of condensed milk (which young sons of fishermen used to surreptitiously guzzle in the den while their fathers were on deck); potatoes, and a necessary pack of matches. For the Silvercraigs fishermen, the preferred drink at sea was coffee.[39] The exchange of herring for eggs, baked produce, fresh milk, and butter-milk – a much relished addition to a meal of herring and potatoes – was carried on with farmers' wives on certain parts of the coast where the fishermen could land from the skiffs.[40]

A simple meal, which could be quickly prepared and cooked for breakfast or as a night-time snack, was 'meal an' *creesh*', onions and oatmeal fried in fat. The following story, from Dugald Campbell of Torrisdale, tells of a Tarbert skiff on which the dish was so frequently cooked for breakfast that it finally became, for one crew member, obnoxious. (There is an implicit reference to the practice of feeding oatmeal to young chickens as an introduction to solid food.) The cook rose late in the morning and, as usual, prepared a panful of meal and creesh. He succeeded in rousing two of his fellow crewmen, but the third was reluctant to leave his hammock, having smelt the too-familiar concoction as it fried. The cook, irate, summoned him for a third time: 'Right, Johnny, come on – yer breakfast's ready.' – 'Aye, aye,' replied the unhappy Johnny, 'A'll be up in a meenit – A'm jeest sherpenin' ma beak.'[41]

Prior to the advent of motor-power, great quantities of sea-biscuits – in Gaelic, *briosgaidean cruaidhe* ('hard biscuits')[42] – were carried on the skiffs. Quite a lore exists about these biscuits, undoubtedly because the fishermen relied upon them if becalmed or storm-bound remote from any port or accessible village. When the store of loaves was exhausted, the biscuits would be resorted to, though a few fishermen claimed to have preferred biscuits to bread, especially if the bread was becoming stale.

The biscuits – one type of which was rimmed all round – were commonly contained in galvanized metal 'tanks', which were attached to the bulkhead on the port side, close to the stove-box, so that, assured of regular warmth, the biscuits remained *crumpy* (crisp).[43] Close to the top of the tank would be a round hole large enough to allow a man's hand through. The hole would be covered by either a metal cap or a disc which could be swung to the side on a nail when biscuits were wanted.[44] Biscuits could also be contained in a bag of fine-meshed netting suspended from a nail on the bulkhead.[45]

Not every baker could produce acceptable biscuits, though the ingredients were simply flour, water, sugar, salt, and – optionally – lard.[46] Of inferior biscuits, Hugh McFarlane remarked: 'If ye wid lee (leave) them a month in that biscuit-box, ye wid hammer a nail intae them.'[47]

At meal times, the necessary articles would be laid out on the platform at the feet of the fishermen, there being no table. Cutlery was eschewed, and herring eaten with the fingers from dishes resting on lap or knee. Tea was made in the kettle, and, if the individual crew members' tastes for milk and sugar concurred, these might be added to the brew before it was poured into the deep bowls from which many fishermen drank. Bowls were more easily stowed than mugs in the cupboards, and if properly fitted one into the other and wedged securely would survive a gale.[48]

David McLean recalled a droll encounter between one of the Pankhurst sisters and an old Carradale fisherman who shall be identified only by his nick-name 'Perker', though he has been dead for more than half-a-century. Miss Pankhurst had been to sea for a night on board Mr McLean's father's skiff, the *Genesta,* and when she landed in the morning on Carradale pier, a group of fishermen gathered around her. She remarked on the novelty of drinking tea from a bowl, at which old 'Perker', in his gruff Gaelic voice, spoke up from the back of the crowd: 'Ach, the best tea ever I ate wis oot a pot.' Mr McLean, who was then a young boy, remembered the unrestrained mirth of the fishermen.[49]

Water

In the sailing skiffs water was usually contained in wooden casks, which would be topped up at every opportunity. Water was a precious asset, and the fishermen, before motor-power relieved their dependence on favourable weather, were constantly alert to the possibility of being storm-bound or becalmed, and observed all the precautions available to them. The importance of a heavy stock of sea-biscuits has already been mentioned. Water was equally vital, and if a skiff was moored close to a well or spring of good water, the cask would be carried ashore and refilled.

When fishing in 'strange places' – such as the Hebrides – the fishermen would ascertain, before watering at a burn, that the flow was clean and that no farms were in the locality.[50] Along the Kintyre coast, however, they had the advantage of knowing where the wells and springs were, and they could row their skiffs into the very shore-head at certain places. The Black Bay was a favourite spot of the Campbeltown fishermen. The water there emerged 'oot the solid rock', but was accessible only when the tide was well ebbed.[51] Another place where the water rose bright and untainted from the shore below high water-mark was An Sruthlag, north of Saddell.[52]

The Tarbert fishermen, when they ranged south along the shore from their village, could water at several big wells, such as that in Camus an Tobair (Well Bay), north of Skipness. These wells, which they faithfully cleaned and tended,[53] were once familiar in name and location to all fishermen, but have not been visited in almost half a century.

At most watering places there was no jetty, and so the skiff's bow would be put into the rock shelf, with an anchor out astern. The boat's boy would climb ashore over an oar, make fast around a rock the head-rope, by which the vessel could be hauled 'tight in'. The cask could then be put ashore.[54]

The type of cask with which Hugh McFarlane was familiar was constructed in a narrow barrel shape, tapering at each end and strengthened by three metal hoops. A metal handle was attached at each end, and the cask lay on its length in the hold, stabilised by its flattened form. Its capacity was about five gallons, but a smaller cask, of similar design, was also carried, and it could easily be lifted to the mouth.[55] The casks John McWhirter remembered lacked handles, but were otherwise similar. These were carried ashore in a herring-basket, a man to each handle, though 'a very few could put wan on his *shooder* (shoulder)'. A thirsty fisherman, by removing the bung and inserting an end of a rubber tube, could take a drink from a cask without having to lift it.[56] David McLean was not a 'great approver' of casks, because the water at times tasted 'woody'. He was more familiar with metal flasks, a large one, on to the neck of which a metal lid fitted, and a smaller one, perhaps stopped with a wooden bung.[57] Before basketing aboard a catch of herring, the casks or flasks would be removed from the hold to the foredeck, so to remain accessible.[58]

When skiffs began to increase in size, metal water-tanks were fitted in the stern sheets aboard some, but were later removed to the fore end of the hold, to accommodate the engines. It was the practice of some fishermen to deposit in the tank a small lump of limestone which, it was believed, gradually dissolved, purifying the water. The water was extracted from the tank using an elongated metal cup, perhaps 10 inches (0.25 m.) deep, with a short strip of lead soldered along the rim to cause it to capsize in the water and fill instantly. A thin wire or cord was attached to the 'dipper', as it was called, opposite the weight, and to the other end of the wire was bound a piece of wood, or, if cord, a knot was made, both of which prevented the dipper from dropping irretrievably into the tank. The tank could be drained of accumulated sediment by unscrewing a cock close to its base. Although the tank greatly increased a skiff's water-carrying capacity, some crews continued to carry a cask or flask, from which the kettle could more conveniently be filled.[59]

In spite of the greater quantities of water which could be carried on the bigger skiffs, the insistence on conservation was not much relaxed. David McLean managed to keep his face clean by surreptitiously dabbing it with a moistened handkerchief, smeared with a little soap. He was thus able to remove the daily grime, but his action would have been considered a 'crime' by his older shipmates had it been observed. 'Ye might as well put yer hand in their pocket an' take money oot,' he said.[60] The fishermen were able to wash during the week only if they happened to be moored near a spring or burn. They would queue with soap and towels at such places as Port Crannaig, Carradale, and Loch Ranza, Arran. Often, though, they remained unwashed throughout the working week. Dishes would be cleaned in a bucket of sea-water, which might first be boiled, but only if the food, such as fried herring, had been greasy.[61]

In his earliest days as a cook at fishing, John McWhirter's duty was to ensure that in summer-time meal and water was constantly available to the thirsty oarsmen. A small enamelled can was kept aft of the pump-beam, beside the water-cask, and would be filled nightly with fresh meal and water before fishing began. Said Mr McWhirter: 'Ye wir pullin' wi' the oars, an' ye wir workin', an' ye wir sweatin', an' everybody wis drinkin' meal an' water.' The herring-buyers, when they boarded the skiffs during the night, would immediately make for the can and put it to their mouths, even before examining the catch.[62]

David McLean remembered meal and water – known to Dalintober and Campbeltown fishermen as *mairach*[63] – as having been a 'soothing' drink. The containers which he used were tin pitchers, which hung on the bulkhead to a nail which was 'bashed up the wey, so that there wis nae question on a bad night o' the pitcher takin' charge'.[64] The drink was believed to relieve constipation; health salts, too, were drunk for that reason.[65]

Tobacco Habits

Prior to the First World War the custom prevailed in Campbeltown and Dalintober of pipes and tobacco being supplied with the provisions, and the cost similarly deducted from the boat's gross earnings at the end of the week – 'Off the whole heid', as the fishermen would say. It was suggested that the scarcity of tobacco during that war affected the generality of the custom,[66] but it did, however, survive on some skiffs until the 1920s.[67] A half-pound or pound roll of 'Thick Black' or 'Bogie Roll' tobacco, and perhaps a quarter-pound of the slenderly rolled chewing tobacco, would be provided. The tobacco was unrolled and cut into equal portions each Monday. The stock of clay pipes was available to everyone, and often, if there had been numerous breakages, some of the men would be reduced to smoking with a broken pipe. If almost stemless, it would be known as a 'jaw-warmer',[68] in Gaelic *cutag*.[69] A curious repair could be effected by sticking the broken ends together with blood. A finger would be pierced with a pin to release a trickle of blood, which was smeared on to the ends of the stem. The pipe would be left undisturbed for a few minutes to allow the blood to set, after which it was ready for use.[70]

Some crews were evidently more careful with their pipes than others. Duncan McSporran of Dalintober remembered that when he went to the fishing first, a dozen clay pipes were supplied each week, of which six or seven would be used, so that in time 'they'd be piled up till there wid be dozens in the boat'.[71] At that time cheese-cutter caps were still fashionable, and if the stem of the pipe was stuck into the cap-band, it was relatively safe.[72]

Many of the boys were introduced to the smoking habit by the obligatory duty of filling their skippers' pipes at night. They would be instructed in how he liked his tobacco packed in the bowl, and would fill it for him in the forecastle whenever he required a smoke, lighting it and puffing at it to get it going, before passing it to him at the tiller.[73] The general attitude at that time was, as Angus Martin expressed it, 'If ye dinna smoke, it wis yer ain fa'lt (fault) – the tobacco wis there for ye.' The few non-smokers often retained their portion of tobacco and gave it to a friend.[74]

Some of the old Tarbert fishermen were severely opposed to young smokers. Hugh McFarlane was 25 years old when he started, and yet would not smoke in their presence. He sat forward of the mast, when the sail was set, and puffed quietly there. 'Whether they kent ye or no', they'd catch the pipe an' throw it intae the loch,' he remembered. And they would growl at a suddenly dispossessed young man: 'Away, what are *you* smokin' for?'[75] The white moustaches of many of these old men were 'barked' (tanned) with tobacco smoke, through use of the 'jaw-warmers'. At times the broken shank of the pipe would be completely concealed by their whiskers, and only the smoking bowl would be showing.[76] New clay pipes might be steeped overnight in water to diminish the dry harshness concomitant with initial usage.[77] A clay pipe 'matured' with use, and was at its best when the bowl and connecting part of the stem had absorbed tobacco-juice and taken a deep brown stain. It was the custom in some fishing families that when visitors had assembled around the fire, a pipeful of tobacco would be lit and passed from smoker to smoker.[78]

Tobacco-chewing was not as prevalent as smoking, though some of the older fishermen favoured it, and would regularly fall asleep with a nugget of tobacco still in their mouths. Their teeth were thoroughly stained by the juice.[79] Chewing was often a substitute for a smoke when opportunity to refill the pipe was lacking, such as when hauling the net, or when the weather was too wet or windy to smoke. Cigarettes, being ready to light and not liable to blow out, would later solve these problems.[80] Some tobacco-chewers 'cerried it on ashore', and at the harbourside street-corner where the Tarbert fishermen congregated 'ye wir up tae the ankles in spittles', remembered Hugh McFarlane with explicit exaggeration. But the habit was undoubtedly a foul one. One Tarbert fisherman, who 'winna put the pipe in 'is mooth', chewed tobacco constantly, and the front of his jersey would be soaked by dribbled juice.[81]

Lighting the Forecastle

Illumination in the forecastles of the early skiffs was by candles and stable-lamps. A single broad candle would be stuck on the mantelpiece − or *brace* − that spanned the open face of the stove-box, a foot or two from the deck. Additional light could be provided by a stable-lamp hung on a bent nail to the bulkhead. Before the skiff got under way, the globe would be removed for safety. These lamps were also hung on the mast when lying to drift-nets, to avert collision in open waters.[82]

These were generally replaced about the turn of the century by a single paraffin lamp fixed to the bulkhead on the port side, between the biscuit-box and the cupboard. It was weighted at its base with lead and mounted on a swivel-pin, so that it rocked with the motion of the boat. A bright sheet metal reflector could be inserted at the back of the lamp to increase the light, but was not a general feature. The wick was usually screwed low when the boat was at sea, or else the flame extinguished.[83]

When the globe was cracked (often caused by an unevenly trimmed wick, which caused the *lowe*, or flame, to play on the glass, or by a drip of water − perhaps from a leaky deck − striking the hot glass) the pieces were covered over with a strip of newspaper moistened with condensed milk, which bound the whole when the heat took effect. The paper tended to turn brown with the heat, so the repaired side of the globe was usually turned aft, and thus the dimness was directed against the bulkhead.[84]

A fragmented globe could not, of course, be repaired, so, lacking a replacement globe, the following curious resort might be adopted. A specially conserved bottle, of a particularly suitable girth, would be brought out. A length of oakum (loose fibre obtained by picking old rope apart) would be soaked in paraffin and wound around the bottle several inches from the base, and the end tucked in. The oakum was then set alight and when fully aflame the bottle was dipped into a ready bucket of sea-water. The sudden immersion caused the bottle to cleave cleanly in two, and the upper part would be inserted on to the lamp.[85] (That practice must surely have been rare, and probably not much of a success, because, of the fishermen the author enquired of, just two were able to describe it, and some others ridiculed the very idea.)

The Coal Supply

On most of the later skiffs, a large cupboard was built up at the bulkhead on the starboard side of the hold, beside the forecastle door. One such cupboard, described by Angus Martin, was divided into two parts. The upper part was a shelf set in about 1 foot (0.31 m.) from the top and covered by two sliding doors, the after one of which could be pulled out completely if a large article was to be stored away or removed. The lower part was constructed as a box, with a rectangular opening near its top. A secondary supply of coal was kept in that box, being restocked when necessary from the main supply in the stern sheets. The proximity of the box to the forecastle door was a great advantage, particularly in wet or windy weather. Before a heavy catch of herring had been basketed aboard, the young cook would be ordered to 'Away an' get plenty o' coal intae the forecastle!', because the box would later be inaccessible beneath the fish. Indeed, the box itself would fill with fish, and his responsibility after the catch had been discharged was to remove these grimy fish.[86] (On other skiffs, a small coal-box, which could contain about a quarter of a hundred-weight, was kept outside the forecastle door.[87]) A common kindling fuel on the skiffs was wood-parings, often cut from old fish-boxes. These were known in all the Kintyre fishing communities as 'sleeshags'.[88]

On the top shelf of that cupboard might be stored a pot and a pan, but often these hung on the bulkhead outside the door, where they tended to *roost* (rust) if the boat lay idle for a spell. Henry Martin commented: 'Some o' the oulder fellas used tae say, "We'll need tae get ashore wherr there's a good san'y (sandy) beach, an' *skoor* (scour) them wi' san'"', but as often as no' they wir dumped an' new yins got. They had nae accommodation for them inside.'[89] If the shelf was not reserved for the pot and pan, the torch and wicks (p. 154) were kept there.

Seaboots and Oilskins

Leather seaboots, which were made to measure by cobblers in the towns and villages, were worn until the 1920s, when rubber boots generally replaced them. In the first decade of the present century, the standard price was 25/- per pair of 28 to

30-inch-high (0.71-0.76 m.) boots.[90] They were regularly treated with linseed oil and Archangel tar mixed, although some fishermen substituted grease for tar, which they considered 'left them hard'.[91] When the leather had stiffened with age, restricting the movements of the wearer, the boots would be replaced,[92] but boots carefully attended to would last many years. Fishermen who were 'caatious aboot themsel's' coated their boots weekly until they shone.[93] The men of the Campbell family of Silvercraigs rubbed home-produced cod-liver oil into their boots by hand.[94] A certain fisherman in the Loch Gair district would half-fill with linseed oil a new pair of seaboots – or *botainnean móra*, 'big boots'[95] – and hang them up for several weeks until the leather had absorbed all the oil that it could. His boots thereafter required but little attention.[96] Some Ayrshire fishermen applied 'giss grease' – the fat of the gannet or solan goose – to their boots. The birds were frequently hauled on board enmeshed in the drift- or set-nets, and the fat would be removed from the carcase and boiled.[97]

Oilskin suits were preserved by coating them with linseed oil. When the fishermen would 'feel themsel's gettin' damp' – perhaps after two or three weeks' work – they would take their suits home, and, if the weather was fair, wash them, coat them with oil to stiffen and water-proof them, and hang them out to dry. Less than 10/- would purchase a complete suit before the First World War. John McWhirter remembered that 'some o' the oulder fellas winna be bothered pullin' the oilskin ower thir heid – they had an oilskin jaicket wi' five or six buttons'. The trousers were held up either by 'ould *galluses*' (braces) or by a string tied at the waist. A 'sou'wester' hat completed the working rig.[98] On mild summer nights, jackets were discarded in favour of oilskin sleeves, which buttoned on to the woollen jersey, or were held by elastic or by a string which another member of the crew tied.[99]

Story-telling

Deep knowledge and appreciation of fishing lore continued unbroken to the generation of fishermen born in the final decade of the last century, and the first of the present century. The surviving members of that generation constitute its final living repository – thereafter silence. Folklorists researching all traditional industries are familiar with the prime cause of that break: industrial mechanisation. The divorce from awareness of preceding modes of existence, little altered from generation to generation in traditional industries, has been completed in the author's own generation. Sailing and rowing, manual hauling of nets, line-fishing and drift-netting, and natural lore have become incomprehensible, and even ridiculous, to a generation which understands (but imperfectly) only the mechanics of engines, winches, and electronic equipment. These machines exist now as necessities. All other resorts have proved dispensable, and with the denial of their practical value, denied also is the related lore.

That lore is present substantially throughout this book. When it occurs it serves to characterise a technical or a social observation, but a range of lore existed which was only marginally associated with practicalities. Much of it originated in the need to entertain and be entertained. Humour, that agent which temporarily dissolves anxiety, was an essential force in the lives of the fishermen. Their entertainment was self-created, as indeed it almost invariably had to be. Based from home for periods of months, at remote villages or anchorages on the coast of Ireland or the Hebrides, time free from toil was spent by gathering to discuss work and to relate the 'ould yarns'. And in the cramped and smoky dens the principal characters of the living and the dead and their epic or outrageous exploits would be gathered from memory and presented afresh. From such tales this brief selection has been made, but perhaps the best of them can only be told, and these have been excluded.

Oral tradition relating to the smack and lugger fisheries is scant indeed, considering its former importance. Two stories, however, merit inclusion here.

'It's a bit hot for you'

The first concerned a Campbeltown 'worthy' whose nick-name was 'Hullya': 'They had great laughs wi' him. He was makin' broth wan time they wir goin' tae Ireland. Dan (McKinlay) wis gettin' very hungry an' he wis listenin', an' he thought he heard the rattle o' a spoon, so he went for'ard an' looked down intae the cabin. "Is the dinner near ready, Hullya?" – "Yes," he says, "I'm takin' mine's – It's a bit hot for you."'

David McLean, Campbeltown[100]

The Irish Sea Compass

The other yarn describes a joke played upon a rather naive hired fisherman who was walking towards the harbour carrying a large lugger compass. One local fisherman asked him: 'Where are ye goin' wi' that?' – 'Oh, down to the boat.' – 'That compass is nae use. It's too big for these waters. That's an Irish Sea compass. Take that away back up.' The unsuspecting man returned to his employer's net-store and confronted him with the accusation: 'Ye're tryin' makin' a fool o' me, givin' me an Irish Sea compass.'

David McLean, Campbeltown[101]

The Watch and the Chain

Time, appropriately, has invested with a terrible poignancy the proud claim of a fisherman who owned a watch and chain when such possessions were rare among working people. He would produce it from his waistcoat pocket and, as he wound it lovingly, declare: 'The sun and the moon may go wrong, but my watch will never go wrong.'

Hugh McFarlane, Tarbert[102]

From the Far South

A skiff, crewed by men from Southend on the southern extremity of the Kintyre peninsula, came rowing into Tarbert harbour one day. Someone on the shore shouted to them as they rowed by: *'An tàinig sibh o dheas?'* (Have ye come from the south?). The reply shouted back was: *'Thàinig sinn o cho fada mu dheas 's a tha deas ann!'* (We have come from as far south as south there is).

George Campbell Hay, Tarbert[103]

REFERENCES AND NOTES

1. H. McFarlane, 18 March, 1976.
2. Angus Martin, author's father, from a sketch by.
3. D. McLean, 1 February, 1976.
4. H. McFarlane, 18 March, 1976.
5. J. Weir, 18 March, 1976.
6. A. Martin, 1 February, 1976, and J. McWhirter, 5 February, 1976.
7. D. Blair, 21 January, 1975; G. Newlands, 1 February, 1975; H. Martin, 8 February, 1975, etc.
8. D. Blair, 5 June, 1974, and J. McWhirter, 5 March, 1975.
9. H. Martin, 8 February, 1975.
10. A. Martin, 16 June, 1974.
11. A. Martin, 16 February, 1975, and G. Newlands, 1 February, 1975.
12. H. McFarlane, 18 March, 1976.
13. As above.
14. J. McWhirter, 5 February, 1976, and H. McFarlane, 18 March, 1976.
15. 18 March, 1976.
16. H. Martin, 8 February, 1975.
17. D. Blair, 21 January, 1975, and H. Martin, 8 February, 1975.
18. H. Martin, 8 February, 1975, and D. McSporran, noted 8 June, 1979.
19. D. McFarlane, 24 January, 1975.
20. 4 June, 1974.
21. A. Martin, 16 June, 1974.
22. H. McFarlane, 18 March, 1976.
23. A. Martin, 16 February, 1975.
24. J. McWhirter, 7 April, 1976.
25. H. McFarlane, 18 March, 1976.
26. H.D.S.G.
27. G. Newlands, 7 February, 1975.
28. H. McFarlane, 18 March, 1976, and J. Weir, noted from.
29. The burying of scarts was referred to in a letter from Mrs Nan (MacNab) Cowley − a native of Minard − to the author, 3 May, 1976. The practice was also known in Carradale, but was uncommon − D. McIntosh, 27 April, 1976.
30. H. McFarlane, 26 April, 1976.
31. H.D.S.G. (Tarbert). The form *càbhach* is very tentatively offered. The *-ach* ending is, with some informants, alternated with *-ag*, and in others replaced by *-aidh*.
32. H. McFarlane, 26 April, 1976.
33. 6 April, 1976.
34. 5 April, 1976.
35. In such communities as Gigha, seabirds continue to be shot for food.

36. H. McFarlane, 26 April, 1976. Dr MacMillan's prescription referred to by D. McIntosh, Carradale, 27 April, 1976.

37. 5 March, 1975.

38. H. Martin, 4 June, 1974.

39. Mrs J. (Campbell) MacBrayne, 22 February, 1977 (S.S.S.).

40. Principally along the 'West Shore', between Carradale and Skipness. J. McWhirter, 5 March, 1975, and H. Martin, 10 June, 1976.

41. 4 March, 1978 (H.D.S.G.).

42. D. MacVicar, 29 October, 1978 (H.D.S.G.).

43. D. Blair, 21 January, 1975, and G. Newlands, 7 February, 1975, etc.

44. D. McSporran, 11 June, 1974; H. McFarlane, 30 October, 1974; G. Newlands, 7 February, 1975; H. Martin, 8 February, 1975.

45. D. McLean, 17 February, 1975.

46. T. McCulloch, baker, Campbeltown, noted.

47. H. McFarlane, 30 October, 1974.

48. A. Martin, 16 February, 1975.

49. 17 February, 1975.

50. H. McFarlane, 26 April, 1976.

51. J. McWhirter, 7 April, 1976.

52. D. McLean, 5 December, 1974.

53. H. McFarlane, 26 April, 1976. Some fishermen translate *Camus an Tobair* 'Bend of the Well' – e.g. J. Weir, 5 April, 1976.

54. A. Martin, 14 May, 1976, and D. McLean, noted 9 June, 1979.

55. H. McFarlane, 18 March, 1976.

56. J. McWhirter, 7 April, 1976.

57. 6 April, 1976.

58. D. McFarlane, 24 January, 1975; H. McFarlane, 18 March, 1976; D. McLean, 6 April, 1976.

59. H. Martin, 8 February, 1975, and A. Martin, 16 February, 1975.

60. 17 February, 1975.

61. A. Martin, 16 February, 1975, and D. McLean, 17 February, 1975.

62. 5 February, 1976.

63. H. Martin, 10 June, 1976 and (H.D.S.G.) 21 February, 1978. Confirmed by A. Martin, 20 February, 1978 (H.D.S.G.) and J. McWhirter (noted).

64. 6 April, 1976.

65. A. Martin, 16 July, 1976.

66. As above, 16 June, 1974.

67. G. Newlands, 7 February, 1975.

68. A. Martin, 16 June, 1974, and H. McFarlane, 26 April, 1976, etc.

69. D. MacVicar, 29 October, 1978 (H.D.S.G.).

70. H. McFarlane, 26 April, 1976. He himself repaired clay pipes in that way. The repair was adequate until a replacement pipe could be obtained. (A resort in farming as well as fishing communities.)

71. 11 June, 1974.

72. H. McFarlane, 26 April, 1976. 'In latter years it was in the waistcoat poacket. Every time ye'd bend, that wis wan away. An' always it wid breck shoart – that's wherr the jaw-warmers came oot o'.'

73. James Wareham, Campbeltown, noted. Confirmed by H. McFarlane, 26 April, 1976.

74. A. Martin, 16 June, 1974.

75. H. McFarlane, 26 April, 1976.

76. H. McFarlane, noted.

77. R. Conley, 27 April, 1976, etc.

78. Mrs J. MacBrayne, noted.

79. A. Martin, 16 June, 1974.

80. A. Martin, 16 June, 1974, and H. McFarlane, 26 April, 1976.

81. H. McFarlane, 26 April, 1976.

82. As above, 5 April, 1976.

83. H. Martin, 8 February, 1975.

84. G. Newlands, 7 February, 1975, and H. Martin, 8 February, 1975.

85. A. Martin, 16 February, 1975. Practice known also to H. McFarlane, 18 March, 1976.

86. A. Martin, 16 June, 1974 and (diagrams) 1st February, 1976.

87. G. Newlands, 7 February, 1975.

88. Gael. *sliseag,* 'Spill or shaving of wood', etc. (H.D.S.G.).

89. 4 June, 1974.

90. D. Blair, 21 January, 1975, and H. McFarlane, 5 April, 1976.

91. J. Weir, 5 April, 1976. Boot-stitching could be dabbed with Archangel tar as a protective measure – G. Newlands, 7 February, 1975.

92. H. McFarlane, 5 April, 1976.

93. As above.

94. Mrs J. MacBrayne, 22 February, 1977 (S.S.S.).

95. D. MacVicar, 18 January, 1979 (H.D.S.G.).

96. As above, 16 February, 1977 (S.S.S.).

97. J. McCreath, 2 May, 1976.

98. J. McWhirter, 7 April, 1976.

99. J. McWhirter, 5 February, 1976, and J. Weir, 5 April, 1976.

100. 3 June, 1974.

101. 5 December, 1974.

102. Noted.

103. Letter to author, 22 March, 1978.

6

Making and Maintaining Ring-Nets

DRIFT-NETS were constructed of *deepens*, which were panels of netting 12 yards (10.97 m.) long and 4 feet (1.22 m.) deep. Five or six deepens joined in depth (see p. 190) were termed a *net*. Factory-produced deepens cost 2/- each in 1843, but weaving in the home continued long after these appeared on the market: 'The fishermen can seldom afford to purchase the whole. They and their children generally manage to net a considerable portion, as well as to repair whatever is torn or decayed, and for these purposes they prefer a twine twisted at a rope work to that spun at home.'[1]

The practice continued into the latter half of the nineteenth century. In December, 1851, with the enforcement of the first anti-trawling Act impending, an official of the Fishery Board was asked in Tarbert whether supplies of twine could be granted to some impoverished trawl-fishermen, for the manufacture of drift-nets.[2] Knowledge of the custom of knitting nets in the home survived in Tarbert. John Weir's grandmother 'when she rose in the mornin' did so much o' that knittin' afore she went tae school. Of course,' he added, 'there winna be much schoolin' in it at that time.'[3] David McFarlane remembered having spoken, in his youth, to an old woman who told him that she and her brothers and sisters 'winna get out tae play till they did so much nettin' every day'.[4] A wooden gauge would be used to ensure uniformity of mesh size.[5]

The first British net-factory had been established in 1820 by James Paterson, in his home town of Musselburgh,[6] 42 years after the first recorded patenting of a net-making machine.[7] The fishermen's prejudice against machine nets was, however, formidable, being based on the tendency of the knots to slip. It was not until about 1839, when the manager of Paterson's factory, James Low, discovered a method of mechanising the hand knot, that the business began to expand. In 1839 Paterson had 18 net-looms at work and employed more than 50 persons. Ten years later, after his death, Messrs. J. and W. Stuart, Musselburgh, acquired Paterson's factory and his patent rights to the invention. In 1854 a new and enlarged factory was built with sufficient space for 100 looms, and hemp-preparing and -spinning machinery to the extent of 3,500 spindles.[8]

Until about 1850, hemp twine was used in the manufacture of nets.[9] 'There's nothing like hemp for nets' was quoted by David Bremner in *The Industries of Scotland* (1869) as 'a maxim of the fishing fraternity from the earliest times'.[10] By 1860 cotton had superseded hemp and was, in factories, predominantly used, though in 1863 the factory at Campbeltown produced a few nets from flax.[11]

In 1863, at Campbeltown, seven net-machines were 'in constant operation night and day to supply nets'. An additional three machines were then 'fitting up'. Yet, two years before, only one machine had been in operation. Two shifts of workers were employed in winter, the first beginning at 6 a.m. and working for 14 hours until 8 p.m., when the second took over for a 10-hour stretch. These shifts would, presumably, have been alternated regularly. For a net 100 yards (91.44 m.) by 15 score (25 feet or 7.62 m.) deep, which took 30 hours to complete, a worker was paid 7/-; a net 18 score (30 feet or 9.14 m.) deep, which took three hours longer in the making, earned the worker 8/6d. The average weekly wage at the factory was 12/-. [12] By 1864, the net-factory at Campbeltown was selling its produce to fishermen far beyond Kintyre. Thomas Murtagh of Howth, for instance, paid £130 for nets at Campbeltown that year. [13]

An organised net-making industry had, however, existed at Campbeltown in the eighteenth century. David Loch reported in 1778 that £800 was paid annually to boys and girls in Campbeltown and district for dressing hemp, spinning twine, and knitting. [14] The rates were 7d for 1 lb. dressed hemp, 4d for that same amount spun and twisted, and 7d per yard of knitted nets. [15]

By 1869, 13 or 14 Scottish factories were engaged in net-making, employing more than 2,000 persons. About 600 looms were in use. The operation of these machines required great dexterity: 'In forming each row of meshes, the worker has to press upon half-a-dozen levers in succession, and pass from one end of the machine to the other.' Eight hundred workers were employed by J. and W. Stuart alone. In the spinning department, the average wage of men was 21/- per week, and of women 7/6d to 8/-; in the weaving department men earned 20/-, and women 10/- to 18/-. [16] In 1873 there were 30 looms in the Campbeltown factory, all operated by women. These looms are thought to have been manufactured locally, and one, bearing the date 1866, was dismantled sometime after 1950. [17]

The Design of Trawl-nets

The first account of an intentionally designed trawl-net, in 1840, reveals two features then absent from drift-nets. First, the *back-rope*, to which the upper edge of netting was attached, was kept afloat by corks closely strung along its entire length, compared with the arrangement on drift-nets of large buoys afloat above the train on strings of adjustable length, with smaller buoys, called 'duckers', attached at intervals between to the back-rope. Second, the *sole*, or bottom of the net, was attached to a rope weighted with 'lead sinks', whereas the sole of the drift-net was bare. The length of the trawl was given as 160 yards (146.30 m.) and its depth as 10 yards (9.14 m.). [18]

The design of the trawl had improved considerably by 1851. Its length had not, since 1840, increased appreciably, but a depth of 4 yards (3.66 m.) had been added in some cases. That depth of 14 yards (12.80 m.) obtained only along the length of the bag, where the herring accumulated as the *wings* of the net were drawn in, diminishing gradually from the bag outwards, to a mere 4½ feet (1.37 m.) at the ends. Each end was 'kept extended perpendicularly by a small pole attached to it'. That pole was termed the 'stretch-beam', and to it was tied the short 'bridle-rope', on to which the sweep-line (in Gaelic the *ceannair*[19] or *suidhinn*[20]), for hauling the net ashore, was made fast when required. [21]

Later evidence suggests that the end was 'crooped'[22] in its attachment to the beam; that is to say, a greater depth of netting than the 4½ feet suggests would have been gathered slackly to the beam.[23] A purpose of the beam may also have been to enable the fishermen to raise the end quickly to begin hauling on the back and sole ropes. Had the ends been free, then two men might be occupied for several valuable minutes 'running' the edges of netting inboard,[24] with the attendant risk of the netting's fouling.[25] In any case, longer beams in these boats which were less than 25 feet (7.62 m.) in length, would have caused inconvenience. Beams continued in use until the late 1860s,[26] but were replaced, on the end of each wing, with a 'bridle', a light rope, several fathoms deep, to which the netting was crooped.

Trawls were put together by the fishermen themselves on communal greens within or adjacent to their village. Towards the end of the century they were being assembled in six sections – two wings, two shoulders, and a lower and upper bag – but when that practice was established cannot be determined. What can be stated with certainty is that by the mid-nineteenth century trawls were being carefully constructed to a specific design, although the length and depth of the nets varied according to the requirements of individual crews. It may be supposed that most of the small nets seized were designed for occasional fishing, such as that of Hugh Turner and Ezikiel Caskey of Dunoon, who were detected hauling the net together in July, 1863.[27] A list of six nets, seizures by H.M.S. *Jackal* in the winter fishery of 1859/60, and placed in store at Greenock, illustrates the variability of dimensions. Three were 360 feet (109.73 m.) in length, one 320 feet (97.54 m.), one 310 feet (94.49 m.), and the smallest 132 feet (40.23 m.). The depths at the bag ranged from 42 to 24 feet (12.80-7.32 m.), and in the wings from 34 to 12 feet (10.36-3.66 m.).[2] The latter measurements were presumably not taken at the very ends, where the stretch-beams were secured, unless the beams had been removed and the free nets stretched out.

The bags of these nets varied in bulge from 'deep' to 'very slight'.[29] (The purpose of a capacious bag and the technique of creating it will be explained later in this chapter.) A narrow-meshed upper bag, which was by 1900 a standard feature in the design of trawls, is not referred to in any official accounts prior to 1907.[30] The object of the narrow-meshed bag was simply to eliminate a *rotach*[31] (meshed thickness) of fish in the centre of the net, and so reduce the weight in that part, easing the labour of the fishermen. The value of narrow netting had not, however, escaped the attention of the early trawl-fishermen, but the principle was applied to the entire net, as the Royal Commission report of 1863 explained: 'Their inducement to do this was, that nets of the legal size became obstructed by the herrings meshing themselves, so that it took a long time to clear the nets after a fishing operation; and during the time the men were thus engaged, the boats of the cruisers might surprise them in the act.'[32] Another device – which modern bottom-trawl fishermen would call a 'blinder' – was experimented with: 'They tried to put a couple of nets into the middle of the net so that there should be no fish meshed in them.'[33]

By 1875, buoys had been added to the back-rope of the bag,[34] a feature in the rigging of the net which continues to this day. The function of the buoys, which at that time were made of animal skins by the fishermen themselves, was to increase the buoyancy of the bag, especially in the case of an extraordinarily heavy catch.

An exact understanding of the assemblage of a trawl-net may be gained by examining the detailed description by assistant general inspector of fisheries, George Reiach, of a net seized in 1866: 'When this trawl net was spread out, I found its back rope consisted of 2 ¼-inch (0.006 m.) lines 316 feet (96.32 m.) in length, mounted with 174 pieces of cork as floats; the sole rope consisted of a ⅝-inch (0.016 m.) rope 317 feet (96.63 m.) in length and mounted with 37 iron and 3 lead rings as sinkers; the depth – at ends 4 feet 6 inches (1.37 m.), at about quarter length from ends 28 feet (8.53 m.), and at centre 38 feet (11.58 m.). There was a spar at each end, but these got broken while at the trial at Inveraray, and only pieces of them remain attached to the net.'

The separate lines forming the back-rope are a curious feature. Perhaps the doubled lighter ropes served simply as a substitute for a stouter single line. (It may be noted, however, that double lines, of different lay, were a feature of drift-nets, and are in use today on the back- and sole-ropes of purse seine-nets, the principle being that the opposing lays – one a right hand, and the other a left hand – counteract each other and prevent kinking.) The heavier sole-rope would be necessary to withstand hard ground. Reiach continued with a description of the formation of the bridle-rope: 'The back and sole ropes extend at each end beyond the net and form a junction at about 4 feet (1.22 m.), the sole rope extends about 10 or 11 feet (3.05 or 3.35 m.) further.'[35] The end of the bridle-rope would undoubtedly have been spliced back on itself to form an eye, into which the end of the sweep-line could be passed and knotted.

Between 1866 and 1875, a rapid increase in both length and depth of trawl-nets apparently occurred. From a length of 316 feet (96.32 m.) in 1866 – 44 feet (13.41 m.) shorter than the greatest length of the six nets detailed in 1860 – an extension occurred of between 114 and 284 feet (34.75 and 86.56 m.). The increase in depth at the bag was of the order of 22 to 40 feet (6.71-12.19 m.).[36] Little length would be added to the standard trawl-net until well into the twentieth century, though the depth would have doubled by 1900.

By the turn of the century, the practice of tapering nets had declined, and little trace of it remained except perhaps in the cutting out of a small triangular piece of netting at the very end. This was called, in Tarbert, a 'jib' – no doubt because its shape resembled that of the sail – and its removal reduced the amount of croop necessary at the end, as well as providing a spare section of netting, useful for repairs. The 'jibs' might be joined together to form a more conveniently worked rectangular piece.[37]

A part of the net, possibly not sufficiently distinctive to have attracted the attentions of the various nineteenth century Fishery Board officials who periodically examined the fishermen's gear, was the *ra* (row), also called the *deepen*, which is probably an application of the drift-net fishermen's term.

The *ra* was a strip of heavy-ply netting sewn along the entire length of the sole before the rope was attached. Its usual depth was 20 meshes in Campbeltown[38] and 40 in Tarbert,[39] and these – at 32 or 30 to the yard,[40] and even less – were wider than was customary in the main net. Its purpose was to protect the bottom of the net from damage on hard ground, hence its heavier construction. About the beginning of the present century, a *ra* was added to the back, 5 or 10 meshes deep,[41] but perhaps 20 at the bag, to strengthen that part of the net which was lashed to the side of the skiff when discharging herring from the net.[42]

The sections which constituted the completed trawl-net, of whatever dimensions, were mounted to the back and sole ropes with *monish*[43] or *steening-twine*[44] – the terms are synonymous – a thick woollen string suited to the purpose by the secure, yet gentle grip it held on the rope when knotted tightly. Monish, in 1859, cost the fishermen 1/7d per lb., and a tarred rope could be purchased at 27/6d for 66 lbs. in 1866.[45] The cost of a trawl-net, in 1864, was £20 or £25, including ropes, corks, leads, etc.[46] The bag of the net was evidently a distinct purchase, possibly because – as was the later custom, in the case of the upper bag – it was of a heavier twine, for greater durability, than the rest of the netting. In 1867 James Crawford, Auchagoil, and Alexander McGilp, each bought a 'troll bag' for £5 4/-.[47]

That the method of assembling drift-nets with *deepens* was also employed with the early trawl-nets is distinctly probable, judging by the practice which prevailed in Tarbert until the 1930s of measuring off the netting into *nets* of 12 yards (10.97 m.), which was exactly the length of the deepens. The measurements were, by then, merely theoretical, but that these sections may originally have comprised trawls is, as has been suggested, a likelihood. 'Every man made whoot they called a deepen,' said Hugh McFarlane, 'an' they sewed them all thegether an' made the trawl.'[48] Both John Weir and David McFarlane gave similar accounts.

The Techniques of Net-Assembly

In Tarbert the sections of the net were first laced together by stretching one edge against the other and picking up the two corresponding meshes with every dip through of the twine-filled needle. A knot would be made occasionally to secure the lacing and so strengthen the join. The bottom edge of the upper bag was first laced to the top edge of the lower bag, though after the introduction of narrow upper bags – usually 38 rows of meshes to the yard, compared with 36 or 34 rows throughout the rest of the net[49] – the meshes could not be lifted one for one. Instead, allowance was made for the extra 100 meshes along the edge of the upper bag (to instance a likely figure) by 'gathering' two meshes of the upper bag to one of the lower bag twice in every yard. In that way, the join 'came out' exactly at the finishing end, and a flap of slack overhanging netting there was avoided. To each vertical edge of the joined bags, a shoulder was laced – gathering again along the depth of the upper bag – and, finally, to the ends of the shoulders each wing was laced.

The centre of an assembled net was ascertained by measurement. Each half, top and bottom, would be measured with a yardstick into 12-yard (10.97 m.) 'nets', and a tag of monish tied at every terminal mesh. Netting 288 yards (263.35 m.) in length was thus measured into 24 nets. Meanwhile, the centre of both the back- and sole-rope had been ascertained, a simple procedure because each 120-fathom (219.46 m.) coil would be formed of two equal lengths spliced together, and the centre of the splice was therefore the centre of the rope. After the calculated number of lead rings and corks – 80 lbs. of the former and 42 lbs. of the latter, in 1877[50] – had been strung free on the doubled back and sole ropes, both centres would be hung up at chest height, perhaps to a trestle or horizontal pole. A length – perhaps 36 or 48 yards (32.92 or 43.89 m.), more if space permitted – would be stretched on trestles. The 'setting up' of the net could begin.

Four men worked along from the centre, one on each length of back-rope and one on each length of sole-rope, setting up netting to the rope with monish. The setting up of the back-rope was the more awkward job, because along the length of the 50

21. At the Ardrishaig 'stances', 1901. The fishermen are running leads and corks on to the sole and back ropes of a trawl-net, prior to setting it up. Per A. Campbell, Paisley

or 60-yard (45.72 or 54.86 m.) bag, the corks ('arcans'[51]) had to be closely strung to provide the necessary buoyancy in that part of the net which would contain the full weight of a catch. Four meshes of netting were lifted over the needle, and these 4 meshes were crooped, or gathered, into the length of 2½ meshes on the rope. For instance, setting up netting 36 rows to the yard (0.91 m.) – that is, each mesh forming a square inch (0.03 m.) – 4 inches (0.10 m.) would be knotted 2½ inches (0.07 m.) from the previous knot. That extra 1½ inches (0.04 m.) of monish hung and formed the 'drop', while the distance between each knot was termed a 'post'.[52] The net thus hung slackly from the ropes creating a bag.

Along the bag a cork was attached at every second or third post simply by looping the monish around it, and knotting. The cork was not tied to the rope, but was held in place by the bight of the 'drop'. Along the sole-rope, a half-pound lead ring would be enclosed within every fifth, sixth, or seventh post – the individual fishermen's theories on weight distribution differed – and later secured to the rope by a monish tie tucked through the strands of the rope on each side of it.

After a 'net' had been mounted to the ropes, each pair of men stretched sole-rope and back-rope together and compared the setting up. If a discrepancy was noticed, then the faulty setting up would be cut back and re-mounted, but such a necessity was rare because the sure eye of experience judged quickly and accurately the stretch of each post.[53] After each length of three or four 'nets' had been set up, it would be removed from the trestles and another length hung up. As the men worked outwards from the bag, the croop was usually diminished until at the ends they might be lifting 4 meshes and putting these into the length of 3 or 3¼. Less netting was required in the wings, which were hauled in first, but in the shoulders and especially the bags, a big 'flow' allowed the herring space to swim around in until the raising of the sole finally trapped them.[54] The completed net was thus crooped to the extent of about a third, with perhaps 300 yards (274.32 m.) hung to 200 yards (182.88 m.) of rope. From the extra length of bare rope in each coil would be formed the bridle-rope,[55] as described on p. 117.

A trawl-net could be completely set up in a day. A basket, containing beer and whisky and biscuits and cheese, was often set down beside the working men. 'Some o' them,' said Hugh McFarlane, 'wir *smok'd* (drunk), but damn the bit o't – they winna go an inch wrong.'[56] The completed net was slung backwards and forwards over the shoulders of the crew, and of any passers-by who cared to help, and they walked in line down to the harbour. 'Everythin' was light aboot the nets,' said David McFarlane, 'because they didn't want them too heavy for haulin','[57] a consideration which could be dispensed with when motor winches were introduced.

Two distinct technical differences in the setting up of trawls prevailed in the other Clyde fishing communities. The sections of the nets were not laced, but 'butted',[58] that is knitted knot to knot along the entire length and depth of the join. In that way the pieces were completely and unnoticeably joined. Each edge of netting could be 'side-meshed' or 'twine-meshed' (doubled along its length) initially, to provide extra strength at the joins.

The second technical difference was that beyond Tarbert the ropes, rather than the netting, were measured off before (or, by some crews, during[59]) setting up. The extent of the croop desired in the bags, shoulders, and wings was first decided, and then the posts – which in Campbeltown/Carradale and Ayrshire were termed 'bools' and 'keys' respectively – were chalked out on the doubled sole-rope, using a measuring-stick with the length of the intended post notched on it.[60]

The normal practice was for one man to begin setting up one half of the sole-rope, and after he had advanced some fathoms along, a second man would start on the corresponding length of back-rope, setting up exactly in accordance with the post-measurements on the sole-rope, against which the back was stretched. When that partnership had progressed sufficiently to allow a start to be made on the other half of the net, a second pair of men would begin, following along.[61] But a single pair of fishermen, invariably the most experienced of the crew, might set up the entire net, first the sole-rope, and then the back-rope.[62] Occasionally, a skipper or elderly fisherman, especially skilled in net-work, would remain ashore for a week and alone prepare a new net for the sea.[63]

Certain Ayrshire fishing families, notably the Sloans of the Maidens, simplified the process of setting up by contriving to engage two crews at the work. The idea was to dispense with the practice of one man's beginning at the middle of the net and, alone, setting up his half-coil right out to the very end. By subdividing the total calculated measurement of each rope, and measuring out the netting correspondingly, numerous 'starts' could be arranged. For instance, if the predetermined sub-division was 20 keys, then 80 meshes would be counted on the sole or back of the netting, and a tag of monish − or *oollen*, an alternative Ayrshire word − tied on the terminal mesh. A man could begin setting that section up to the rope, and 80 more meshes along another man could start, and so on, along each length of rope. The key measurements had first to be marked off (on the back-rope), as usual, but that swifter method required that the corks and leads be positioned appropriately on the ropes before the setting up commenced.[64]

That technique was evidently confined to Ayrshire. The great success of the families which practised it testifies to its efficacy, but there can be little doubt that the meticulous and tradition-conscious older generations of Tarbert fishermen would have disapproved of it. There is, admittedly, a superficial resemblance to the Tarbert method of setting up, in that it involved measurement of the netting, but there the uneasy relationship ends.

A rare theory pertaining to setting up deserves recording. A few skippers believed that the net was a more efficient fishing instrument with maximum crooping in the shoulders, rather than in the bag. One such skipper, James McCreath of Girvan, reasoned that the shoulders were 'where ye waanted the flow, an' where ye caught the herrin''. He set 4 meshes into the measurement of 2½ in the middle − the standard gather − but reduced it to 2 in the shoulders.[65]

Although trestles were commonly used in setting up, the sole and the back ropes could, where possible, be stretched out tight to net-poles, with a strop here and there along the length, keeping the net supported at a height at which men could comfortably work. Dalintober fishermen liked to set up a net in the spacious malt-barn of a distillery, an arrangement which was possible in summer when distilleries ceased production, and the malt-barns were not in use. The ropes would be hung to the conveniently spaced posts supporting the ceiling.[66]

Net-Mending

As trawl-nets, unlike drift-nets, were worked principally inshore, damage to netting occurred frequently, and thus trawl-fishermen generally acquired an expertise in net-mending. That which they made, they had also to repair. The necessity was fundamentally one of economics. There were two distinct forms of damage − distinct, that is, to an expert − a simple tear (meshes broken, but not extensively torn away) and a hole (part of the net missing).

A split, though relatively simple to repair, was often of great length. In working conditions the fishermen would 'grop it thegether'[67] – lace it quickly and without fuss – as soon as the net was aboard and the damage assessed. It would be repaired carefully at the first opportunity. If slight, it would be mended on board the skiff next morning, but, if serious, the entire net would be hauled ashore when time permitted, and stretched out on a quay or green. A point of the tear would be hung at a suitable height, or else held by one of the crew – usually the least experienced – so that the pattern of the netting was stretched 'fair', that is, in the diamond form in which it was knitted, row after row, and which it naturally assumes when under hauling strain. The opposite form was termed 'across the *mash*' (mesh), and netting so hung could not – and cannot – be repaired with correct judgement of the leg measurements, because the meshes are slack and mis-shapen.

22. Net-mending, at Carradale, on board the skiff *Perseverance* of Campbeltown. The skipper-owner, Archibald 'Try' Mathieson, is in the centre of the group

In the case of a long, simple split, an attempt would be made to 'meeter in' (Tarbert) or 'catch' (Campbeltown) broken meshes at intervals along its length. The edges of the split would be stretched together and meticulously checked by a fisherman, the appearance of broken legs being scrutinised for possible correspondence. If he discovered, on each edge, tails which seemed to belong together, he would pass that part of the net to an attendant fisherman and check along until he came upon another, obvious link. Men could begin mending at these points while he continued his examination of the tear. Thus, a 20-fathom-long (36.58 m.) split might be divided into four lengths, and four men could engage themselves in mending, instead of one man, working alone from start to finish.[68]

Mistakes in net-mending, though infrequent among experienced fishermen, occurred occasionally, and with deplorable consequences. An error was called a 'steelter', though the Tarbert fishermen had a peculiar term, *mogal-ghoid* ('stolen mesh').[69] The most common mending error was the knitting of a five-legged mesh, but a 'three-legger' – called by the fishermen of Ballantrae a 'cra's fit' (crow's foot)[70] – could also occur. Inexperience, haste, tiredness, and the inadvertent turning of a corner in a split were all factors potentially productive of an error,[71] but no matter the cause, the consequence was invariably that unnecessary netting was progressively knitted in, forming a swelling poke of slack meshes, not immediately noticeable. As Hugh McFarlane said: 'Ye wid never get oot. Them that kent wid say, "Are ye no' feenished wi' that. Stop. Put doon the needle till we hae a look." Haul back the net an' hae a look at wherr he started, run along an' ye'll come on the steelter, an' jeest cut it all right back. An' ye'd then a half-barrel o' nets that wisna needed, in ribbons.'[72]

A gaping hole could be more expeditiously repaired by inserting a replacement piece of netting. At night, a piece might be laced in quickly to render the net whole again and ready for use. 'It wid do for the time bein',' said Hugh McFarlane, 'but as soon as they wir idle they wid haul it up an' mend it in.'[73] Not all fishermen were 'good hands with nets', to use the common phrase, yet some were acknowledged masters of the intricacies of net-mending. Serious damage often attracted teams of sympathetic helpers from other crews, a spirit of co-operation which is manifestly rare now, as many elderly fishermen regretfully observed.

Carving Netting-Needles

The fishermen fashioned their own netting-needles – *snàthadan-leasachaidh*[74] – from a limited variety of woods or from animal bone. The most suitable of the available wood was 'boutrie' (elder), and it was extensively used by the fishermen of Kintyre and Ayrshire; but, during Hugh McFarlane's earliest years of fishing at Tarbert, at the turn of the century, the wood preferred by the older men was *earradhris* (dog-brier). A bush was selected on which the wood was growing old and heavy – 'As thick as yer wrist an' thicker' – and pieces of branch would be cut to the desired lengths. The bigger type of needle, for holding monish, would be 1 foot (0.31 m.) or more in length and would take two or three fathoms (3.66 or 5.49 m.) of the string; yet, it was slender, to pass through the meshes easily. A long needle was more suitable for setting up a net, because the more monish it held, the less knots there would be in the finished work. The smaller needles, for lacing, knitting, and mending nets, were about 6 inches (0.15 m.) long.[75]

The length of wood, after drying, was split into narrow sections, and one of these would be whittled 'until ye wid get the thickness, and as straight as ye could go'. The eye, or *crò*, of the needle would be shaped out and an internal tongue formed. A 'stringy' type of wood was necessary, so that the tongue was sufficiently pliable to bend, without breaking, when the twine was being wound around it. When the indentation on the bottom – an inverted 'U' – had been formed, the surface of the instrument would be scraped with a knife to smooth it.[76] As Angus Martin of Dalintober remarked, with some amusement: 'There wis no television then – ye'd tae dae somethin' tae pass the time.'[77]

The tough and flexible lancewood, which was imported into Britain from the West Indies and used for the shafts of horse-drawn carriages and pony-traps, was suitable for needle-making. When a shaft had broken and been discarded, pieces could be obtained by the fishermen.[78] If, during a meal at home, a suitably shaped bone was noticed in the beef, it would be kept until it dried and naturally split, and from the pieces would be carved needles.[79] The fishermen of Kames, Lochfyneside, would boil a suitable rib to soften it and then press it under a board weighted with stone or iron. When cold, the flattened bone could be shaped.[80] John McWhirter recalled that groups of fishermen used to sit at the Weigh House (now demolished) at the head of Campbeltown Old Quay, whittling needles on winter days.[81]

Making and Mending Buoys

Until the second decade of the present century, the fishermen of the Lochfyneside villages, including Tarbert, produced their own trawl-net buoys from the skins of sheep. There is no knowledge, in Kintyre or Lochfyneside, of the killing and skinning of dogs, a practice which, within living memory, prevailed in certain fishing communities on the East Coast of Scotland.[82] In Campbeltown, canvas buoys were introduced much earlier, probably through the influence of the local net-factory. The oldest surviving Carradale fishermen were unfamiliar with sheepskin buoys, but were aware of their manufacture by previous generations of fishermen there.[83]

Hugh McFarlane, the last surviving trawl-fisherman trained in the skills of buoy-making, produced his final set of four in 1911 or 1912. He purchased four sheepskins from Dugald Campbell, a Tarbert grocer, for 1/6d each. The skins were soaked in a tub of water until the fleece had been shed. Meanwhile, four wooden stocks had been cut from an ash or elder tree and shaped. These stocks were 5 or 6 inches (0.13 or 0.15 m.) in depth, carved in a round with a nub at the top, through which a hole had been bored for the string which would lash the finished buoy to the back-rope, and with a lipped base around which the skin would be bound. Through the stock, from the top of the dome to its base, an air-passage had been bored for inflating the buoy.

The maker first gathered the edges of the skin, then he would 'plett' (*plait* – 'fold') them about the base of the stock, and lash them temporarily with marline. He then inflated the skin to judge the roundness of it. One side might protrude irregularly, so he would slacken a fold or two of the skin and re-adjust its position until he was satisfied with the shape of the buoy. He then bound the skin round and round repeatedly with marline, tightening it firmly so that it could not come undone. The ring of skin below the binding could then be trimmed.

The finished set of four buoys – one for the centre of the bag, one for each end of the bag, where it joined with the shoulder, and one for the bridle-rope at the end of the net, to keep it afloat until the neighbouring crew could lift it – were hung to dry, having first been 'barked' to preserve the skin. When thoroughly dry, between a pint and a half-pint of Archangel tar and linseed oil mixed would be poured into each buoy through the air-passage, using a small metal filler. The liquid would then be '*rummelt* roon' inside the buoy so that it spread and soaked into the entire skin. The mixture served at least three purposes: it helped preserve the skin, water-proofed it, and also kept it soft and pliable so that on warm days it would not crack. The buoys were again dried in the sun, perhaps hung out at the back of the house for a day. They would then be inflated and plugged, and, finally, brushed over with tar.

The completed buoys would be thoroughly tested in the harbour by attaching 56-lb. weights to them and dropping them overboard from a skiff. In order to withstand the test, they had to remain afloat, and they invariably did. These buoys lasted for many years, and as successive nets deteriorated would be transferred from one to the other. When they did crack they were discarded, repair being impossible; but small quantities of tar and oil mixed would be poured in from time to time to maintain them.[84] The buoy – usually pronounced 'bow', as in 'now' – was known on Lochfyneside as *bolla*.[85]

The canvas buoys which gradually replaced the traditional skin type were factory-produced of canvas strips sewn together and similarly bound about a wooden stock. These could be repaired quite simply (if the fishermen could be bothered to do so, and many could not) by inserting a wooden plug, grooved around its girth, and by gathering the edges of the tear over the groove, and either lashing them with twine or tacking them into the wood.[86] Another device required a strip of lead in addition to the plug. The torn edges were pulled over, rather than around the plug, and the lead was placed over the top, and tacks hammered through lead, canvas, and wood to seal the hole.[87]

Drying the Nets

Much effort was expended by the fishermen to preserve their nets from rapid decay. Frequent drying and immersion in a preservative solution – originally the bark of native trees, but later cutch, imported from India – were the means employed. Until the introduction of synthetic fibres in the 1950s, fishermen had rarely a week-end free from the work of hauling their nets ashore to either dry or 'bark' them (in fishermen's speech the term was not replaced, after discontinuation of the use of the substance, by an equivalent and more appropriate term such as 'cutch').

Drying was originally done by spreading the nets out in fields,[88] but at some time during the first half of the nineteenth century poles were erected on which to drape the nets. A more specific period cannot be suggested, but there is the reference by a Fishery Board official, in 1851, to 'trawl nets hanging round the harbour as usual' at Tarbert.[89] The Tarbert fishermen even resorted to drying their trawl-nets before the fires in their homes, in 1860, to prevent their seizure.[90] Fishermen without net-stores were certainly in the majority until the present century, and when their gear was not in use they had no option but to store it in their homes. The fishery officer at Ardrishaig, George Thomson, visited Airds on Lochfyneside, in February, 1863, and reported that 'as their houses are generally very damp, they have to turn out their netting occasionally to prevent (it) from going to waste'.[91]

The net-drying poles or *stances*[92] were of differing design. Some, like those formerly at Minard, Lochfyneside, were simple erections, formed of single uprights. The bases of the uprights – which were latterly telegraph-poles, already coated with tar and creosote – would be sunk several feet into the shore about high tide-mark, and spaced, four or five of them, 20 feet (6.10 m.) apart. Along the tops of the uprights, slender tree trunks stripped of bark would be lashed with chain. A single loop of chain, which when tightened bit into the wood, was sufficient to secure the end of a spar. These were mounted end to end in 20-foot lengths, and, about 3 feet (0.91 m.) below, trunks of corresponding length were similarly secured. Along these foot-spars the fisherman walked, from end to end, spreading the net over the upper spar, against which he leaned to steady himself. The net would be hauled into a punt from the skiff, moored out in the bay, but could not be hung on the stance directly from the punt, an arrangement possible in some districts where the poles could be erected so that they stood out in the sea at high water. Instead, less conveniently, the net would be transferred from the punt on to hand-barrows supported by a man at each end. Each barrow would take about a third of a damp net on a Saturday morning, but on Monday morning two barrows might suffice to carry the dried net back down to the punt.[93]

Hugh McFarlane described the erection of a similar stance, or *crochan*,[94] the Gaelic word used in Tarbert, and meaning 'hanger'. Permission would first be granted from the laird of a nearby estate to select and fell the required number of trees, which were commonly spruce, and up to 60 feet (18.29 m.) in height. Once cut, they would be allowed to lie and *sùgh* (lose their sap) for a month or more in the summer-time. When the sap had mostly gone, the bark was stripped off and the trunks lashed to a cart, which was dragged by a horse out of the forest and down to the village. The location on the shore had already been selected, or else an earlier *crochan* was to be renewed.

A notch would be sawn off the foot of each upright, and a hole dug 'as far as the waater wid let ye', which was perhaps 5 or 6 feet (1.52 or 1.83 m.). To the top of the pole would be attached four ropes, with a man on the end of each to steady the pole while the men at its base secured it. The fishermen would try to wedge the notch against a stable rock, and might knock chunks off it 'tae get it intae this grip'. Once they had shaped it to their satisfaction, the hole would be packed with stones to stabilise the trunk further. Four more poles would be similarly embedded, and the cross-trunks (*péircean*[95]) – often the entire length of a tree, about 60 feet – lashed on with chain or rope.[96]

Two crews would be involved in the operation, and most pairs of crews had at least one set of 'crochans' between them, which they would contrive to use alternately. Visiting crews generally hung their nets on other crews' 'crochans'. They would normally secure permission first, but occasionally not, which might result in a dispute. 'I've seen the man that owned it lowerin' the net tae get his own up,' said David McFarlane.[97]

The poles at Campbeltown, Dalintober, and Carradale were usually designed with triple uprights – two angular trunks, crossed at the top and lashed to a central trunk. At Dalintober, nets could be 'powled' (poled) directly from a punt at high

23. Hanging a ring-net to dry. Donald McIntosh of Carradale on the poles at Port Crannaig, c. 1930. Per D. McIntosh

water, but the crews with poles along the face of the high sea-dyke hauled their nets from a punt on to the quay and poled from the road behind the dyke.[98]

Net-drying was not confined to week-ends. On a sunny morning during the week the fishermen might shoot and haul the net to wash it, and then drape the bag over an oar or spar in several *faiks* (folds). A sling was attached to each end of the bearer, and it was hoisted the height of the mast on the halyards.[99] Cotton nets – and especially the bags – tended to heat, and even to steam, if they had been worked among oily herring. Fishermen were especially careful in summer, and during the day the net might be periodically shifted forward, and then aft again, to keep it cool.[100] Coarse salt was traditionally used to prevent heating. The bag of the net, where most of the oil adhered, would be exposed, and salt scattered over it. Several stones of salt might be distributed between successive layers of netting hauled forward.[101]

Washing was occasionally resorted to, and even steeping. Crews putting in to Carradale on a summer's morning sometimes sank their nets close to the shore to keep them cool until evening,[102] and when Donald McIntosh of Carradale went to fishing first, in the harbour lay an old ship's lifeboat into which a net could be pulled, and the plug removed.[103] When engines were introduced, the same effect could be accomplished by the boat's being steered circularly to port, and while men aft shot the net overboard, others, forward, hauled it in.[104] Dalintober fishermen might, on a Saturday morning, put an end of the net on to the quay and haul it through the water from the skiff, lying offshore.[105]

On a Monday evening the upper wing of a thoroughly dried net might be soaked with buckets of water to reduce its buoyancy, otherwise, as Bob Conley of Carradale put it, 'If ye came on herrin' she winna go doon the same. Ye'd see the wing floatin' away, an' then ye wid maybe miss them.'[106] A floating wing also presented difficulties to a neighbouring crew approaching the end of the net to lift it – the netting might wrap up the propeller.[107]

The long sweep-lines used in the pre-1914 shore fishery were dried as regularly as nets. The young cook's job on a Saturday morning was to coil the rope – 100 fathoms (182.88 m.) or longer – from the stem to the stern of the vessel.[108]

Barking the Nets

The protection of nets was not restricted to regular drying and occasional washing. Additional proof against decay was required. W. Stuart, of the net-manufacturing firm of J. and W. Stuart, Musselburgh, wrote in 1956: 'Nets made from vegetable fibres have to be treated before and during use in order to prevent or reduce damage by bacteria, and also to protect the twine from mechanical wear.'[109] Until about 1850 the preservative substance used was the bark of trees, principally oak and birch.[110] The fishermen of Cumlodden on upper Lochfyneside, for instance, paid £20 in 1818 for a ton of bark.[111] The barking of nets was evidently a prolonged and tedious exercise, which had to be repeated with regularity throughout the fishing season. Fishermen would be occupied for several days barking a fleet of drift-nets.[112]

By 1840, native bark had begun to be replaced by cutch (also called *catechu,* both being from the Malay word *kachu*). Of the fishermen of Lochgilphead and Ardrishaig it was reported, in 1844, that 'all of them have abandoned bark for tanning their nets, and use catechu'.[113] The chief sources were the wood of *Acacia catechu* and *Acacia suma,* both trees indigenous to India. In the process of forming cutch, the bark of the trees was stripped off – to be used for tanning – and the trunk fragmented and boiled in water. The thickened extract was then decanted into iron pots and boiled again. Having attained a syrupy consistency, it was poured into moulds or on to mats and exposed to sun and air until it hardened. The resultant substance was dark-brown in colour, or, in mass, a lustrous black, and of brittle constitution. Cutch was imported into Britain in three different forms. The superior kind, preferred by fishermen, was known as *Pegu catechu,* and was marketed in blocks, covered over with large leaves, and boxed.[114]

Experiments were conducted in 1850 by N. Russell, fishery officer at Tarbert, to establish the relative shrinkage of meshes when immersed, new, in a solution of pure cutch, and, consecutively, in solutions of larch ooze (tanning liquor) and cutch/larch ooze. A piece of new netting had boiling cutch solution poured over it, and was left to steep for 12 hours. The second piece of new netting was first immersed in a cauldron of larch ooze for 12 hours, and then dried and steeped again for 15 hours in a mixture of cutch and larch ooze. The result of the experiments was that the shrinkage of both nets was about 1¾ inches to the yard (0.91 m.).[115]

The shrinkage of netting was a legal issue, of importance to both drift- and trawl-fishermen until 1868, when the law of 1819, prohibiting small-meshed nets,[116] was repealed. Between these years, the use of netting of less than 1 inch (0.03 m.) from knot to knot was prohibited in the herring fisheries. A reversion to the old mesh standard was, however, 'over and over again' recommended to the Royal Commissioners of 1877. The strongest reason advanced by the commissioners against these recommendations was that fishermen, because of the shrinkage of netting, 'might unwittingly be led to an infraction of the law'.

Heavier cotton, the commissioners stated, shrank more than finer material, and thus 'a net weighing 20 lbs. with 32 meshes to the yard, a net weighing 16 lbs. with 33 meshes to the yard, a net weighing 14 lbs. with 34 meshes to the yard, and a net weighing 13 lbs. with 35 meshes to the yard, would all shrink to . . . about 36 meshes to the yard'. With each fresh barking a net shrank a little more.[117]

The quantity of cutch used in the barking of a net would vary according to the age of the net and to the extent of the work to which it had been subjected. The initial barking of a new net, which would be naturally white, was the heaviest of all. The requirement was calculated in terms of a net's weight, the normal allowance being a quarter of the total weight of the net. For instance, a net weighing 1,600 lbs. would require 400 lbs. of cutch.[118]

Prior to the introduction of wide-meshed wings in 1928/29 (p. 227), a net might, when hauled, be a shimmer of hanging fish, except in the narrow upper bag. The oil and scum deposited on the wings and shoulders by such 'heavy mashings', and on the upper bag by the accumulation of the fish there, were detrimental to the net's durability, and it would be given a barking at the first opportune week-end. If a tub was available, the net might be allowed to steep throughout the week-end, and would be removed on Monday morning and hauled damp on board the skiff. If, however, other crews intended to bark their nets, or if the net had been little used since its last treatment, then a quick 'pull-through' would be settled for.

The quantities of cutch used ranged from 100 lbs. to 20 lbs. The wide discrepancy is attributable not only to the variable requirements of nets, but also to the conflicting theories of fishermen on what constituted a satisfactory barking. Angus Martin commented: 'Some believed in gettin' a lot o' cutch, an' some believed in gettin' less. A mind o' ma gran'fether sayin', "Mind an' don't make the bark too strong, now, becaas it'll no' go intae the net."'[119] Angus Martin's cousin, Henry, was also familiar with an old theory relating to excessive use of cutch: 'They aye maintained that ye lee (leave) her too heavy if ye bark often . . . They meant that when she got heavy she wanna fish as well, she wasna as *soople* (supple), an' winna form.'[120]

No record seems to exist of when specially built 'bark-houses' appeared, but it may be accepted that the facility was available about the middle of the nineteenth century, to judge by a report in 1852 of an 'excellent net barking house' built in Tarbert.[121] In smaller communities, however, the work was done in the open, either on the shore or in a field close to the shore.

At both ends of Minard, a copper cauldron was mounted on a pit dug into the face of the shore verge. The pit was open towards the sea, with an iron grate, about 1 foot (0.31 m.) from the ground, built into the three bricked sides. The cauldron, which held 60 or 70 gallons, was portable, with three handles fitted below its rim. After it had been partially filled with water from a nearby burn, a fire was stoked beneath it. The cutch was then added, and when that had melted the solution was transferred in the boat's draw-buckets to a wooden tub about 7 feet (2.13 m.) long, by 2 feet (0.61 m.) broad, by 2 feet deep. The tub was set down longways at one end of the 'barking-board', a wooden tray 12 feet (3.66 m.) by 8 feet (2.44 m.), and rimmed around by wooden strips 2 inches (0.05 m.) high. At the opposite end of the board, a basin formed by the lower half of a 40-gallon barrel, was embedded in the ground to collect the cutch solution which ran off the net as it was pulled through the tub and across the board. If the net required a steeping, the solution would be transferred from the cauldron into a barrel 6 feet (1.83 m.) high and specially made for the purpose. Staging was necessary to fill the barrel and to haul the net into and out of it. The nets were carried to and from the tub or barrel on hand-barrows.[122]

Facilities for the barking of nets were similar in the village of Carradale. Boilers were set up on the harbour shore, close to a water supply. About a hundredweight of coal was required to heat the water, and the fuel would often be laid on a Friday evening before the crew put to sea. 'All ye had tae do on a Seterday mornin',' said Donald McIntosh, 'wis put a match tae it.'[123] John Conley remembered that cutch was purchased in hundredweight boxes. A certain kind of cutch was brittle and could be broken quite easily, but another 'sticky' kind had to be cut with an axe, which was first sprinkled with water. The desired quantity, when detached, would be weighed on a spring balance machine. The broken cutch was usually held in a herring basket while it melted.[124] Some fishermen steeped the cutch overnight in cold water.[125] In a 'haul-through', a man stood with a pole pushing part of the net down into the tub for some minutes, before another pull was made and several more fathoms of netting immersed. When a net was to be steeped for a day or two, a few heavy stones would be placed on it to keep it deep in the solution.[126] The now-vanished fishing community at Torrisdale, near Carradale, maintained a single boiler, which was occasionally used by the Carradale fishermen.[127]

Two imaginative facilities existed at Tarbert at about the turn of the century. One, a dismasted hulk of a sailing-ship, the *Banton Packet*, lay ashore in the harbour. She had been laid up by her owner, John Forgie, after lucrative trading operations at the great Islay herring fishery which had run its course by 1895 (p. 198). Boilers were fitted inside her hull, and the fishermen would take their boats alongside her and haul their nets through a hole cut in her side.[128] Competitive facilities were provided in a wooden shed, also situated in the harbour. A set of rails ran from a window at the side of the shed out along the beach, and at low tide the fishermen would distribute a net between two or three wooden trolleys, which they trundled up to the side of the shed. The net was pulled through the window. At high tide, when the rails were submerged, the skiffs could be steered alongside the shed, which was then standing out in the water, and moored there below the window.[129]

In that shed, the cutch, once melted in the boiler – which sat on top of a square brick fireplace – would be baled into wooden buckets using a *spuidsear*, which was constructed of staves, fastened top and bottom by iron hoops. It had the size and appearance of a bucket, but a handle about 5 feet (1.52 m.) long was fitted through a hole cut near the rim, and nailed to the base on the opposite side. A man stood on the wall of the fireplace and filled the buckets at his feet with the scalding 'asso'.[130] The buckets were handed down in succession to the other crew members, who emptied them over the net heaped in a tub. The man who operated the *spuidsear* had, said David McFarlane, 'a wey o' copin' withoot spillin' a drop'.[131] The tubs were laid on stool-like supports about 2 feet (0.61 m.) high, and were arranged across one end of the bark-house. When the net had been barked, the 'asso' would be 'drawn off' by removing a wooden plug from the base of the tub. The liquid ran into a gutter in the cobbled floor, which directed it through the doorway and into a drain. Both balers and buckets would be tanned dark-brown by years of use.[132]

The iron-floored bark-house belonging to the author's great-grandfather, John Martin, was situated in Queen Street, Dalintober, near the family home. It was, in common with the bark-houses in Campbeltown about the turn of the century, relatively mechanised. The boiler, which was about 5 feet (1.52 m.) deep and rounded at the bottom, was seated on a brickwork fireplace. The cutch was contained in a basket suspended in the water from an eye-bolt in the ceiling. As the water began to boil, the basket would be shaken from time to time to spread the melting cutch. When the cutch had thoroughly dissolved, a cock on a pipe extending from the base of the boiler over a pit in the floor would be opened, discharging the solution into the pit. After the net had been barked, the solution would be baled out using buckets with ropes attached. Two men would tip the bark into a chute which carried the liquid out into a *sheugh* (gutter) in the street. Angus

Martin and his friends amused themselves in childhood by dropping pieces of wood into the chute and following their rapid progress until they disappeared down a *siver* (drain) further along the street.[133]

The sweep-line, anchor-cable, and other ropes were also barked regularly, which blackened them. Tar would later be used as a preservative, on nets as well as ropes.[134]

Tarred Nets

Tarred nets were apparently introduced in Campbeltown shortly after the turn of the century, but were not accorded general approval until about 1910. Acceptance was delayed even longer in the remote and traditionalist fishing communities on Lochfyneside. In Minard, for instance, tarred nets were not adopted until the late 1920s.[135] The main advantage of tarred nets was that they did not require to be barked as frequently as untarred nets.[136]

The fishermen themselves usually tarred a net. The sections of netting, if delivered white from the factory, were almost invariably barked first. They would then be dipped into a tank containing heated tar, passed through a mangle to expel excess tar, and hung on the poles to dry.[137] Tar, however, though a good preservative, gradually hardened the twine of the net, until 'she wid go intae yer bone' when hauling, and eventually burst apart all at once.[138]

The fishermen soon began thinning tar with creosote, which gave the netting a silkier texture. The ratio of tar to creosote varied according to individual preference, but Donald McIntosh of Carradale mixed one part creosote with two parts tar.[139]

After numerous barkings, and regular use, the tar 'worked off' a net. Some fishermen re-tarred nets, but the results tended to be unsatisfactory. 'I never saw one that wis a success,' said Robert McGown. 'They were always sticky, an' then they got brittle. They would tear easily.'[140]

A few fishermen believed in attaching hanks of tarry twine to the end of a new net, so that if mending later needed to be done, the twine had been given the 'same work' as the netting to be repaired, and was therefore compatible with it.[141]

REFERENCES AND NOTES

1. *N.S.A.* (Inveraray), 32.
2. L. Lamb, to Fish. Bd., 29 December, 1851, A.F. 7/85, 304.
3. 10 December, 1975.
4. 24 January, 1975.
5. H. McFarlane, 24 January, 1975.
6. D. Bremner, *The Industries of Scotland* (1869), Newton Abbot, 1968 (reprint), 314.
7. *Encyclopaedia Britannica*, 14th edition, Vol. 16, 247.
8. D. Bremner, *op. cit.*, 315.
9. *F.B.R.* (1901), 200.
10. *Op. cit.*, 312.
11. P. Wilson, fishery officer at Campbeltown, 4 March, 1863, A.F. 22/4, 12-13.
12. *Ib.* On 21 October, 1868, fishery officer A. Levack reported 31 machines operating in factory at Campbeltown. A.F. 22/5, 231.

13. *E.R.C.* (1864), 811.

14. *Essays on Trade, Commerce, Manufactures and Fisheries of Scotland,* Edinburgh, 1778-9, 159.

15. A. R. Bigwood, The Campbeltown Buss Fishery, 1750-1800 (unpublished M.Litt. thesis, University of Aberdeen, 1972), 84.

16. D. Bremner, *op. cit.,* 317-9.

17. *Third Statistical Account* (Argyll), 1961, 286.

18. A. Sutherland, fishery officer at Ardrishaig, to Fish. Bd., 12 February, 1840, A.F. 7/104, 220.

19. H.D.S.G. (Tarbert), and R. McLachlan, Ardrishaig, 9 June, 1975.

20. D. MacVicar, 17 September, 1978 (H.D.S.G.).

21. L. Lamb, rpt. on trawling to Fish. Bd., 1 May 1851, A.F. 7/85, 268.

22. From Gael. *crùb,* 'to contract or shrink' (H.D.S.G., Tarbert and Carradale). Also known in Campbeltown, but 'gather' was there the equivalent and regular term.

23. J. Weir, 10 December, 1975.

24. H. McFarlane, 26 April, 1976.

25. H. McFarlane, 26 April, 1976, and J. Weir, 10 December, 1975.

26. D. Mitchell, *Tarbert in Picture and Story,* Falkirk, 1908, 90.

27. Sgt. D. Fraser at Dunoon, to chief constable J. Fraser at Lochgilphead, 22 July, 1863, A.F. 37 151/3.

28. 8 May, 1860.

29. *Ib.*

30. R. Spink, fishery officer at Tarbert, 4 November, and F. Fraser, fishery officer at Rothesay, 1 November, A.F. 62/351.

31. H.D.S.G. (Tarbert).

32. *R.R.C.* (1863), 15.

33. J. McMillan, fisherman, Tarbert, *E.R.C.* (1864), 1181.

34. J. Murray, fishery officer at Ardrishaig, Remarks upon reading Mr Holdsworth's book titled *Deep Sea Fishing and Fishing Boats,* 16 February, 1875, A.F. 37/176, 8-9.

35. 20 January, 1866, A.F. 7/87, 243.

36. *Ib.,* and J. Murray, *op. cit.*

37. J. Weir, 26 April, 1976. An occasional practice only, but A. Alexander, the Maidens, Ayrshire (2 May, 1976) referred to the practice of 'goring', or tapering, in recent times. Six or seven fathoms (10.97 or 12.80 m.) would be tapered out to the 'gable', or end, for quick-hauling purposes and also to reduce the risk of the netting's fouling on the propeller in a strong tide.

38. J. McWhirter, 5 March, 1975; D. McLean, 8 December, 1975; R. McGown, 8 December, 1975.

39. J. Weir, 10 December, 1975, and H. McFarlane, 5 April, 1976.

40. H. McFarlane, 5 April, 1976, and J. Weir, 10 December, 1975, respectively. D. McLean, 8 December, 1975, quoted 24 rows.

41. D. Biair, 21 January, 1975; J. McWhirter, 5 March, 1975; D. McLean, 8 December, 1975; R. McGown, 8 December, 1975; H. McFarlane, 5 April, 1976.

42. R. McGown, 8 December, 1975.

43. The Eng. spelling. The Gaelic form, as recorded (H.D.S.G.) in Tarbert, Carradale, and from D. MacVicar, 18 January, 1979, is *monaist.* Etymology obscure.

44. Interchangeable with above, particularly in Campbeltown.

45. A. Fraser, *Lochfyneside,* Edinburgh, 1971, 16.

46. J. Martin, fisherman, Dalintober, *E.R.C.* (1864), 755, and D. Bruce, fisherman, Ardrishaig, *ib.,* 1147.

47. D. Fraser, *op. cit.,* 21.

48. 7 June, 1974.

49. J. McWhirter, 26 April, 1974; R. McGown, 8 December, 1975; J. Weir, 10 December, 1975. Forty-two row netting in top bag, or 'sling', also quoted.

50. D. McCallum, fisherman, Tarbert, *E.R.C.* (1877), 132. In 1938, 250 half-pound leads and 50 lbs. of cork, cut circularly, were required. Rpt. by then fishery officer at Tarbert, J. M. Steven, by whom copy was loaned to author.

51. H.D.S.G. (Tarbert) – Gael. *arcan*. Also D. MacVicar, 17 September, 1978 (H.D.S.G.) – plural *arcain*.

52. H.D.S.G. (Tarbert). 'O' pronounced as in Eng. 'lost'. Possibly from Eng. 'post'. The *Scottish National Dictionary*, 207-8, cites a quarrying usage, with the meanings 'seam' or 'section'.

53. H. McFarlane, 5 April, 1976.

54. Dunc. Blair, 1 May, 1974.

55. J. Weir, 28 November, 1975.

56. 5 April, 1976.

57. 24 January, 1975.

58. Recorded in Tarbert, etymology doubtful. D. McFarlane, 24 January, 1975; H. McFarlane, 5 April, 1976; J. Weir (H.D.S.G.), 10 December, 1976.

59. D. McIntosh, 27 April, 1976, and J. McCreath, 2 May, 1976.

60. R. McGown, 8 December, 1975; D. McIntosh, 27 April, 1976; J. T. McCrindle, 29 April, 1976; A. Alexander, 2 May, 1976.

61. J. McWhirter, 7 April, 1976.

62. R. McGown, 8 December, 1975, and D. McLean, 6 April, 1976.

63. J. McWhirter, 7 April, 1976.

64. M. Sloan, 29 April, 1976, and A. Alexander, 2 May, 1976.

65. 2 May, 1976. Also D. McLean, 6 April, 1976.

66. H. Martin, noted 23 June, 1979, and D. McSporran, noted 25 June, 1979.

67. H.D.S.G. (Tarbert) – Gael. *gròb*, 'join or sew together awkwardly'.

68. D. Blair, 21 January, 1975; H. McFarlane, 26 April, 1976; D. McIntosh, 27 April, 1976.

69. H.D.S.G. (Tarbert). Variant pronunciations recorded, but local interpretation of the term would seem to support the spelling. H. McFarlane: 'A stolen mash. Five legs. Ye see, ye stole wan on the wan side.' 10 December, 1976.

70. J. McCreath, 2 May, 1976.

71. H. McFarlane, 26 April, 1976, and D. McIntosh, 27 April, 1976.

72. H. McFarlane, 26 April and (H.D.S.G.) 10 December, 1976 (composite).

73. H. McFarlane, 26 April, 1976.

74. 'Mending-needle', D. MacVicar, 17 September, 1978 (H.D.S.G.).

75. H. McFarlane, 26 April, 1976.

76. As above.

77. 14 May, 1976.

78. D. McIntosh, 27 April, 1976.

79. J. McWhirter, 7 April, 1976.

80. D. MacVicar, 16 February, 1977 (S.S.S.) and 17 September, 1978 (H.D.S.G.).

81. 7 April, 1976.

82. Mrs Doreen Shepherd, letter to author, 1 May, 1977.

83. D. McIntosh, 9 December, 1975.

84. H. McFarlane, 10 December, 1975 and 26 April, 1976.

85. D. MacVicar, 17 September, 1978 (H.D.S.G.).

86. R. McGown, 8 December, 1975, and D. McIntosh, 9 December, 1975.

87. D. McIntosh, 9 December, 1975.

88. *F.B.R.* (1901), 201.

89. L. Lamb, to Fish. Bd., 27 December, 1851, A.F. 7/85, 302.

90. G. Thomson, fishery officer, 3 November, 1860, A.F. 37/22.

91. 21 February, 1863, Ardrishaig, A.F. 26/10, 34.

92. The general Lochfyneside term: R. MacNab, 25 September, 1975; D. MacVicar, 22 October, 1978 (H.D.S.G. – no Gael. equivalent known to him); Ms M. Campbell, 23 October, 1978 (H.D.S.G.). Also recorded from H. Martin, Dalintober, 31 November, 1974.

93. R. MacNab, 25 September, 1975.

94. H.D.S.G. Invariably used with Eng. plural, i.e. 'crochans'.

95. Singular *péirce*. Spelling represents D. MacVicar's pronunciation, 17 September, 1978 (H.D.S.G.), but similar forms recorded in Tarbert (H.D.S.G.).

96. H. McFarlane, 30 October, 1974. The operation of getting and erecting poles also described by H. Martin, 31 November, 1974.

97. 24 January, 1975.

98. A. Martin, 27 October, 1974, and H. Martin, 31 November, 1974. At ebb tide, nets had to be barrowed across beach to poles.

99. R. MacNab, 25 September, 1975; J. T. McCrindle, 29 April, 1976; T. Sloan, 29 April, 1976.

100. J. Weir, 10 December, 1975.

101. J. Weir, 10 December, 1975; T. Sloan, 29 April, 1976; K. MacRae, Portree, Skye, 19 February, 1974, recorded by W. Maclean.

102. J. Conley, 5 December, 1974.

103. D. McIntosh, 9 December, 1975.

104. D. McSporran, 11 June, 1974; J. McWhirter, 10 December, 1974; R. Conley, 11 June, 1975.

105. A. Martin, 16 June, 1974.

106. R. Conley, 11 June, 1975.

107. J. T. McCrindle, 29 April, 1976.

108. H. Martin, 8 February, 1975.

109. The Manufacture of Nets, article in *The Scotsman Survey of Fishing Industry,* 23 November, 1956, 1.

110. *F.B.R.* (1901), 201.

111. A. Fraser, *op. cit.,* 12.

112. *F.B.R.* (1901), 201.

113. *N.S.A.* (Glassary), 692.

114. *Encyclopaedia Britannica, op. cit.,* Vol. 5, 27.

115. 24 May, 1850, A.F. 7/105, 184. He refers to the fishermen's practice of treating nets with larch bark and catechu combined, which suggests a transitional compromise.

116. *R.R.C.* (1878), xxv-vi.

117. *Ib.,* xxvi.

118. J. R. Stuart, managing director, J. & W. Stuart Ltd., Musselburgh, letter to author, 4 May, 1976.

119. 16 June, 1974.

120. 31 November, 1974.

121. L. Lamb, 11 June, 1852, A.F. 7/105, 285.

122. R. MacNab, 25 September, 1975.

123. 27 April, 1974.

124. 5 December, 1974.

125. D. McIntosh, 27 April, 1974.

126. R. Conley, 11 June, 1975.

127. J. Conley, 5 December, 1974.

128. H. McFarlane, 30 October, 1974, and J. Weir, 27 June, 1975.

129. H. McFarlane, 10 December, 1975.

130. H.D.S.G. (Tarbert). Gael. form of 'ooze', tanning liquor?

131. 24 January, 1975.

132. H. McFarlane, 26 April, 1976.

133. A. Martin, 16 June, 1974. Also H. Martin, 31 November, 1974.

134. H. Martin, 8 February, 1975.

135. R. MacNab, 25 September, 1975.

136. D. McIntosh, 27 April, 1974; D. McLean, 17 February, 1975; R. McGown, 8 December, 1975.

137. J. McWhirter, 10 December, 1974, and R. McGown, 8 December, 1975.

138. J. McWhirter, 10 December, 1974.

139. 27 April, 1974.

140. 8 December, 1975.

141. H. Martin, 8 February, 1975, and A. Martin, 16 February, 1975.

7

The Evolution of Ring-Netting

THE simple origin of ring-netting – that is, the attachment together of several pieces of drift-nets to create a crude beach-seine for surrounding herring in shallow bays – has been documented in Chapter 1, and need not again be referred to except briefly, to clarify the subsequent developments in design and technique.

The evolution of ring-netting can be considered in four stages:

1. It was, originally, an exclusively shore-based operation, using improvised drift-nets.

2. The net was still being dragged to the shore, but the fishermen had also begun to set it, when necessary, in deeper water, and the net itself had been designed specifically for *trawling*.

3. Contact with the shore had virtually ceased, but the fishermen still preferred to haul the nets, which were of greatly improved design, into shallow water.

4. With the introduction of motor-power, dependence on inshore fishing was entirely removed, and the later adoption of winches enabled deeper nets to be used, thereby increasing the efficiency of the operation in offshore waters.

The first account of an intentionally designed trawl-net, in 1840, described only one method of using it. By that method, the fishermen surrounded the shoal of herring with the net in shallow water, and 'if the quantity is large they allow the net and fish to ebb, but if the net can bare (*sic*) the weight, it is drawn on shore'.[1]

The clamour struck up against trawling induced the Fishery Board to examine in detail both the methods of fishing and the design of the nets used. The resultant reports – the first in 1851, and the second (published) in 1852 – revealed that since 1840 three other methods had been developed, by any of which the net could be set in offshore waters.

Undoubtedly the fishermen had experienced the absence of herring in the shallow bays – 'low in', to use their own terminology – and were forced to devise means of surrounding the herring 'high off'. David McFarlane of Tarbert reasoned: 'In a year or so the herrin' got wary; it wisna comin' in just as low – it shows ye how it adapts itself – an' they had tae have a neighbour tae haul in the end o' the net.'[2]

The Methods

Of the three methods, two involved the use of *sweep-lines,* attached to the ends of the net, with which the fishermen could set it further offshore and yet still haul it into the beach. As many as 150 fathoms (300 yards or 274.32 m.) of rope could be used on each end,[3] depending on how far offshore the herring had been located, or were supposed to be.

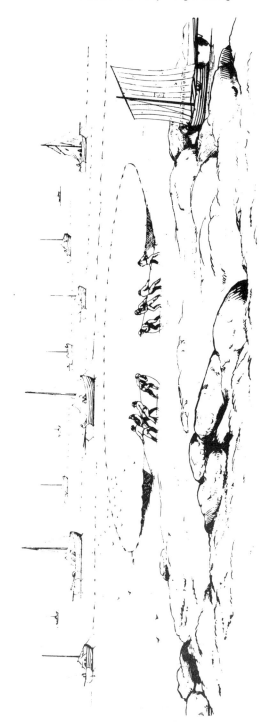

24. Trawl-crews about to meet as they close the circle of the net. The sweep-lines are almost entirely hauled, and on the right one of the fishermen has grasped the beam on that end of the net. Moored inshore is a trawl-skiff, while drift-net boats offshore lie to their nets

In hauling the net from the shore, one of the crew of four men was landed on the beach with the sweep-line attached to the first wing of the net, and there remained holding that end while the skiff was rowed offshore and the rope paid out. As the boat curved on her course towards the shore again, the net itself was set and the second sweep-line streamed out until the boat touched on the beach. The crew landed, taking with them the end of the second rope. The hauling then began, with two men on each rope, 'the parties gradually approaching each other as the bag or centre part of the net approaches the shore'. When the net had been hauled to the shore, the enclosed catch, if heavy, was discharged from the net into the boat with baskets, but, if light, was hauled on board the boat at once and emptied from the bag of the net.[4] That method was restricted to clean stretches of shore, as rocks within the sweep of the net would cause tearing and loss of catch.

To increase their operational range, trawl-crews paired and hauled the net to their boats, anchored out clear of the stretches of rocky shore. The net and sweep-lines were similarly set − or *shot,* which was, and still is the fishermen's term − but the end of the first rope was secured on board one of the skiffs, which had dropped anchor. The crew in the net-boat rowed out, and, when net and sweep-lines had been shot, they anchored in line with the neighbour-boat, but at a distance from her, to keep the mouth of the net spread.[5]

The crews then began to haul on the ropes. 'When this has been partly done,' wrote Laurence Lamb in 1851, 'they raise their anchors and row the skiffs a short distance towards each other, and both anchors are then let go again. This is repeated several times when hauling the net, until the bag . . . is pulled in. The skiffs are rowed alongside of each other, and the crews join their strength to raise the net with the herrings into one of the skiffs. If they are unable to do this, they either get the assistance of more skiffs and crews, or they endeavour to pull the net to the shore. Sometimes 150 crans of herrings are taken at once by one of these trawl-nets.'[6]

These two methods evidence the inclination of the fishermen still to haul their nets into the shore, and, when possible, *on to* the shore. But yet another method had come into existence, the most significant of all, though its potential would not be fulfilled until the advent of motor-power almost 60 years later. It was a mid-channel operation which did not employ sweep-lines, and throughout which no contact was made with the shore.

When worked from a pair of boats, one end of the net was passed into the boat which was to remain stationary, while the other boat was rowed around the shoal making a ring of the net (hence the term *ring-netting*) and returning to her neighbour. The net was then hauled by both crews into the boat from which it had been shot. Using a single boat, the end of the net was made fast to a buoy; the net was then shot, and, when the boat returned to the buoy, hauled in.[7]

The discomforts attached to hauling the net directly on to the shore have already been noted − 'The fishermen having to wade into the sea are constantly wet to the neck . . .' (p. 40) − but the practice was, by 1861, still preferred to hauling the net into a boat, if the evidence of Ardrishaig fisherman Duncan McBrayne may be relied upon. He stated, probably in less orderly language than credited with: 'It is desirable, when convenient, to haul the net on the shore, but if the herrings happen

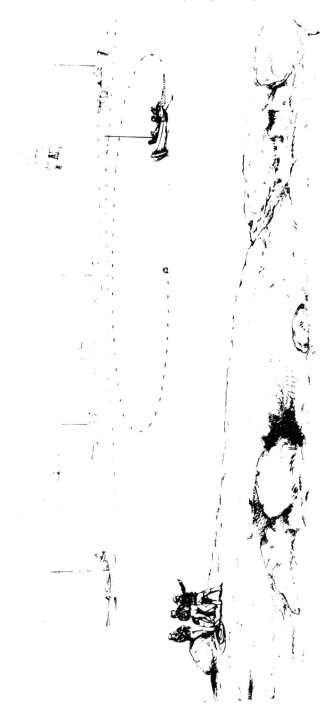

25. Ring-netting, the earliest-known illustration (1852). As a single fisherman at the stern shoots the remaining few fathoms of net, the skiff is rowed around towards the end-buoy

to be in deep water, or should the shore be rough, then both ends of the net are brought together (i.e. ring-netting – author). When the net is hauled on shore the men land from their boats which they heave on to the beach.'[8] McBrayne's account is particularly valuable, being the statement of an actual fisherman, brief though it is by comparison with the wordy products of some officers of the Fishery Board.

A rock was named after one of the Tarbert shore fishermen. Johnny McQuilkan's Rock (Creag Eonaidh) is exposed in Camus na Ban-tighearna when the tide ebbs. The small nets used to be hauled to it, and the story goes that one evening Johnny and his crew set the net and were just beginning to haul it, when Johnny saw that it was full of herring from end to end. He immediately began to jump with excitement, crying *'Tha mi beairteach gu bràth!'* (I am wealthy forever!), and jumped so much that he flattened the rock. John Weir concluded his version of the story on an unhappy event – the net burst with the weight of the fish – but the other two versions allowed Johnny his wealth, and the author would have it that way.[9]

Knowledge of the Shore

The transition from drift-netting to trawling forced the fishermen to acquire a new range of skills and knowledge. Shoals of herring had ordinarily to be located before the trawl-net was shot (the contrary and less frequent practice will be explained on p. 141), whereas at drift-netting the fleets of nets were set arbitrarily, more or less. Trawl-fishermen therefore became adept at recognising signs of herring, and in the next chapter these signs and how they were interpreted will be described.

Furthermore, because the net was usually worked in shallow coastal waters, the fisherman had to find out by trial and error where the rocky ground – *heckles*[10] – lay, if he was to avoid damaging or even losing nets. The fisherman of long experience knew the exact location of every rock and reef along the shores which he customarily fished, and was able to set his net on patches of clean ground barely greater in area than the net itself. But he was always wary of wind and tide, which could sweep his net, or even his boat, quickly on to rocks. The following anecdote, though hardly in a serious vein, suggests the rather daring character of the inshore fishery. A Tarbert crew was seeking herring along the shore on a black night, when suddenly the man on the bow shouted aft to his skipper: 'Keep 'er off – I can smell the *roideagach* (bog-myrtle)!'[11]

As herring fishing is done mainly at night, the fisherman had to be able to recognise his 'marks' in darkness. His marks were physical features on the land or in the sea: headlands, prominent rocks, islands, houses, trees – in short, anything recognisable. These marks enabled him to know the exact position of his boat, and where, in relation to the boat, the foul ground lay. He would judge the set of the net accordingly, allowing for the direction and force of wind and tide.

All these features were given names, and the process of naming was probably advanced, as Henry Martin suggested,[12] by the fishermen's unfailing curiosity to know where a successful crew had hauled a catch. Thus: 'Where did ye get them last night?' – 'Oh, MacNair's Tree', or 'The Flat Rock', or 'South o' the Deer Shed'.

A great part of that body of names has probably not survived the transition from trawling to ring-netting in deep water, a transition which emerged rapidly from the beginnings of motor-power. It would be wrong, however, to suggest that shore work ceased. In fact, it simply became less prevalent. Knowledge of the more obvious landmarks was transmitted from the generation of camper-fishermen to their sons, but these men, of whom it has been said, with more truth than is now possible to credit, that they 'knew every stone and tree on the shore', took most of their knowledge with them into extinction.

The Tarbert fishermen were, by 1864, ranging far from their Lochfyne haunts. John McMillan remarked in that year: 'We could work anywhere if the wind and weather permitted. We could work on any shore if there was clean ground, but in general we never pass what they call Carradale Point. It is too far for us.'[13]

The Sunk Nets

The existence of an *eye* (shoal) of large, deep-swimming herring off the shores of Kintyre and Lochfyne in the summer months was known before the trawl came into general use. B. F. Primrose referred in 1852 to a practice common among drift-net fishermen of removing from their trains two or three 'barrels' of nets, usually of the widest mesh, and 'securing one end to the shore, carry the other end out into the loch, and make it also fast there by an anchor or stake'. Such set-nets were sunk to the seabed and were 'very often successful in taking herrings of large and fine size'.[14] These large herring were later known as 'ground-keepers' because they seldom rose from the seabed.[15]

The trawl-net was more suited for the capture of that class of herring, for, as John Bruce of Ardrishaig commented in 1864, 'They get very fat, and they would not swim about, so as to mesh readily.'[16] So reliable was the fishery that the fishermen were accustomed to setting their nets without even an attempt to locate fish. The practice was explained by John McMillan of Tarbert in that same year: 'In the years 1846 and 1847 and from that up to 1852 or so we used to draw the bays without seeing any appearance of herrings at all. I have seen the time when we have been taking half the full of this room without seeing anything either jumping or playing about. That was just trying our chance.'[17]

There is no evidence that nets were then being weighted and sunk to the seabed to reach these slow-swimming herring, but the possibility must be accommodated. The above-quoted John McMillan stated in 1862 that the trawl-fishermen 'usually went out from the shore about 80 fathoms, sometimes 150, but rarely'.[18] And in that year a Tarbert merchant and ship-chandler, James McLarty, declared that he had 'sold ropes to the trawlers 150 fathoms long for each end (of the net)'.[19] Judging by the localities of seized nets during the period 1852-1867, the Tarbert trawl-fishermen regularly frequented the coast between their village and Skipness. The bays and bights along that coast deepen quickly, and to effectively work a net set 150 fathoms (900 feet or 274.32 m.) from the shore, weight would have to be added to its sole-rope to keep it on the bottom, a considered necessity at that time.

The first proper evidence of fishermen sinking their nets appears in a report by fishery officer John Murray in 1875. 'These nets,' he wrote, 'are always mounted so as to sink to the bottom when used as seans (*sic*).' Further description reveals that the nets were actually sunk beneath the surface: 'Besides the corks on the upper rope, three buoys are generally used, one in the centre of the bag, and one on either side about mid-way from the end of the net, which keeps the net perpendicular, and marks the centre of the net, *but which sink with the net* (the author's emphasis).' The sinking was effected by sheet lead wrapped around the sole, or iron rings 'slipped on'.[20]

As one would have expected, oral tradition has something to tell. The brothers John and Bob Conley of Carradale related versions of a story which attributes the innovation of a sunk net to a Claonaig or Grogport fisherman, and the idea of a deepened net to a Tarbert fisherman. The versions are substantially the same, excepting details of place and persons. This, in part, is one: Two crews from Grogport had been fishing, without success, for most of the night. The herring were *dooking* below the shallow nets, and the fishermen were exasperated. 'This man the name o' McKinven,' said John Conley, 'jeest all of a sudden took an invention intae his heid. He says: "A'll tell ye whoot we'll do. D'ye see that water-jar there? – Let me see it doon here." He filled it full o' water an' tied it tae the middle o' the net, an' then took off the buoys that wis on the net, an' let the net sink doon. An' he wis gettin' maybe fifteen cran or that each haul, an' he filled the two boats. He went intae Tarbert, an' there wis another man came doon: "Ye wir lucky men last night. How did ye get them?" – "Oh," he says, "A'm no' goin' tae tell ye . . ." But at last he told him, an' the other man says, "Well, A'm goin' tae invent a better method than that". So, he got a deeper net, put maybe fifteen score (25 feet or 7.62 m.) to her, an' that wis the first o' the ringin'.'[21]

The other version of the story has the innovator tie heavy stones along the sole-rope,[22] which was in fact what the Tarbert fishermen did at the 'long ropes' fishing along the bays between there and Skipness, as will be described at a later stage of the chapter.

The methods of trawling changed little from the earliest years until the last. The principal development, which has been noted but not yet explained, was that the crews no longer hauled their nets from the beach. The explanation will be speculative, lacking any firm evidence to go on, but not, the author hopes, unconvincing. The anchoring of the boats to haul the sweep-lines began as a means of avoiding hard ground at the beach-head, but as the practice was perfected the fishermen probably became less willing to put themselves to the work of dragging their boats on and offshore, particularly as that exercise combined with the hauling of the nets often resulted in their getting 'very wet and draggled' (p. 40). They may be imagined, soaked and chilled to the flesh, raising an empty bag to the shore and cursing loudly the circumstances that brought them to the job.

No longer dependent on beach fishing, the boats were increased in size, and the transition from beach- to anchor-trawling was completed. Boats of 23 feet (7.01 m.) in length, however lightly built, could not have been easy craft to drag on to a beach, but the bigger class of boats, at 30 feet (9.14 m.) or more, would have been impossible so to manage, except with ropes, rollers, and much manpower, and the extensive range of fishing operations precluded such provisions.

Contact with the shore was not, however, lost. If the inshore ground happened to be too coarse to give an anchor grip, the skiffs could, along certain parts of the coast which shelved steeply to the sea, be rowed right up to the very rock ledges, and a man would clamber ashore, usually over an oar, and wedge the anchor into a 'skleef' (cleft) in the rocks. Once back on board, the skiff would be slackened back on the anchor-cable, and the hauling of the sweep-line could begin.[23] (The occasional practice of landing a man on the shore continued until about 1914, when hauling to anchors was finally abandoned.)

Shooting the Net

The net was almost invariably set with the tidal current, either partly or wholly, so that a wing, or the net's entire spread, carried on to the shoal and not away from it. This was termed 'taking a weather gage'.[24] When shooting to the shore, the wing on the tideward side would be the longer.[25]

As the net was laid aft on the port side, it was better shot directly over that quarter. That was the preferred practice and was termed a 'lee shot'. In that way, the net streamed out unobstructed by the rudder as the boat curved on her course. For example, if she was sailing into the tidal current and a *tom* or *tòrr*[26] of herring 'rose to play' on the port bow, the skipper would simply hold his course until he had got 'above' the shoal, and then sweep around to port shooting the net to encircle it.

But he might decide to try a 'weather' or 'backside' shot[27] if he considered it expedient. In that case, he would make his ring on the starboard circuit, which meant that the net would be streaming out across the boat's stern, and that the man shooting the after part of the net would be constantly occupied throwing bights of the back-rope and netting overboard to prevent its fouling on the rudder-head during the shot.[28] There was an additional hazard of a loop of the sole-rope's falling over one of the buoys and so hanging up the middle of the net, an occurrence which invariably caused loss of the shot.[29]

The decision to 'shoot the wrong way', as the fishermen expressed it, was usually taken to avoid turning the boat around or perhaps even through a shoal of herring and so disturbing it,[30] or to get the net shot ahead of another nearby crew which was aware of the same 'spot' of fish and was 'getting into position' for a lee shot.[31]

A backside shot quite often resulted in the net's becoming 'rolled' or 'twined' (Gaelic *toinneamh*[32]), an occurrence which was costly in terms of both loss of catch and of time. Rolling happened as the net was being hauled over a shelly or corally sea-bed, where hard fragments snagged on the walls of the net, which were being swept by tide ahead of the sole-rope. As the rope was hauled, it rolled up the netting, collecting the shell or coral fragments, so that the men were heaving up a weighty, thickened sole, 'jeest lik' a big bale o' straw'.[33] The unravelling of such a net might take many hours, and when hard coral was involved mallets might be used to break it up before the meshes could be disentangled.[34] Hugh McFarlane recalled spending an entire winter's night in a wee *port* (bay) near Otter Spit after hauling a badly twined net. The net was 'absolutely a mass o' clam an' *spoot* (razor-fish) shells', and the fishermen cleared it by breaking the shells with their leather boots, *'sweirin'* an *champin''* (swearing and crushing), as Mr McFarlane remembered with laughter. [35]

Occasionally, strong cross-tides – an unpredictable factor – caused the net to 'over-bang' (the Campbeltown term) or 'back-fill' (the less obscure Tarbert term), so that instead of the walls of netting flowing outwards, and thus allowing the shoal within space to swim around in before the raising of the sole finally trapped it, they were collapsing inwards, a circumstance detrimental to the efficiency of the net. [36]

When hauling a net in an area of 'coorse gr'un'' (coarse ground), the fishermen risked a bad tearing. If the net caught on the ground, and a strong tide was running,

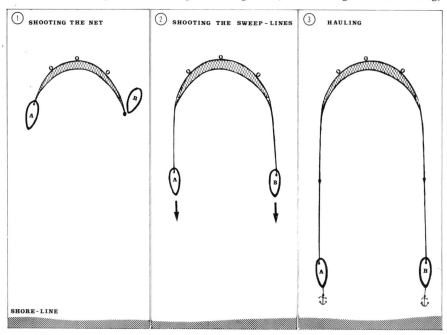

26. Shooting and hauling a net to anchored skiffs

then the fishing operation might be abandoned and the net hauled 'end on' – that is, one wing let go altogether – until the grip had been cleared. In extreme conditions, the fishermen might 'hang on' until the tide eased or changed. But the sole might 'jump' clear – 'Well, the thing then, if the sole jumps maybe the net is caught . . . an' when the net catches it generally lees (leaves) a bit behind.' The retrieval, from a bad 'fast', of little more than the bare ropes of a net was a calamity not unknown to fishermen. [37]

Tidal direction could not always be accurately determined, and a ring might inadvertently be made against the flood or ebb, resulting in the net's being swept away from the shoal. Such an occurrence was explained thus: 'The tide wis goin' the opposite wey.' [38] At 'corners', such as the Cock of Arran, an accurate gage on the tide might not be possible; and after heavy rainfall, a stream of fresh water flowing down narrow upper Lochfyne in opposition to a flood tide invariably ruined a fishing.[39]

The actual fishing operation began soon after a shoal had been located. Immediately the skipper announced, 'We'll take a "corrom" (chance) at this',[40] the man on the bow moved aft and on to the end of the bridle-rope bent a buoy, or perhaps even an oar or a hold-board.[41] The two oarsmen then gave their full strength to the oars, and, once upstream of the tide, the skipper removed the tiller, inverted it and replaced it in the socket so that the downward curve was cocked upwards, allowing greater room aft for the shooting of the net. (Not all skiffs were, however, fitted with a curved tiller.)

Once in position, the end-buoy was dropped overboard for the crew of the neighbour-boat, or *comannach*,[42] to collect, and the bridle-rope began to twist out after it. As the *Fame* (to name, for the sake of simpler description, the first of a pair of boats), curved around, the net began to flow out over the quarter, the skipper aft throwing out bights of the back-rope, and the man at the forward end of the heaped net similarly 'helping out' the sole-rope. Before the bottom coil of bridle-rope had begun to run out, he bent an end of the sweep-line on to the eye of the bridle-rope with a single or double sheet-bend knot, which could easily be *lowsed* (loosened) later. The 120 fathom (219.46 m.)-long sweep-line was coiled in a specially constructed wooden box below the break of the deck on the port side.[43]

Meanwhile, the buoy had been picked up by the crew of the neighbour-boat, the *Renown*, using an ordinary boat-hook (Tarbert) or a 'skidoag' (Campbeltown and Carradale),[44] a 9-foot (2.74 m.) wooden pole with an angular cross-piece nailed close to its end for catching up the rope. The buoy was quickly untied and an end of the *Renown's* sweep-line made fast to that bridle-rope. They then proceeded to the shore, paying out sweep-line, and dropped anchor as close inshore as possible, slackening back on the cable until the anchor was gripping tightly. The net had by then been shot from the *Fame,* and she too was approaching the shore, her sweep-line streaming out until exactly the same length as on the other end had been shot. Her anchor was then dropped. The length of sweep-line used was variable, depending on the extent of haul necessary. The distance between the anchored boats would also vary, but was always rather less than the spread of the net.

Hauling

The hauling of the sweep-lines began as soon as the boats were secured at anchor. The line was marked at perhaps 10-fathom (18.29 m.) intervals along its length by appropriately notched strips of leather worked into the strands of the rope. These marks enabled the crews to haul in the net evenly, so that the curve was not mis-shapen. The shape of the net was always a concern of the fishermen, a poor ring almost invariably yielding a poor catch (pp. 156-7). Each mark would be 'sung out' by the crew gaining it first – 'We've fifty fadom yet tae haal!'[45] – and that crew would pause until the corresponding mark had been reached by their counterparts, and then resume hauling.

In the hauling of the ropes, the men were lined fore and aft, facing aft, the skipper seated on the stern-beam, a man on the pump-beam, a man on the forward-beam, and a man on the break of the deck, who coiled the slack of the rope into the sweep-line box after each pull. The Campbeltown, and many of the Carradale boats, carried an additional crew-member, a young cook, and his job was to pull the slack through a rickety, which was a small detachable block on the port side of the foredeck, either hooked into an eye-bolt in the deck, or attached by a strop around the mast. The rickety − which operated on the same principle as the sail-rickety (p. 87), but was distinct from it − had been tried by the Tarbert fishermen, but was evidently dispensed with before the end of the nineteenth century.[46] The reason for its disappearance from the Tarbert boats was probably that it necessitated the deployment of one of the four hauling men, which would have reduced the speed of

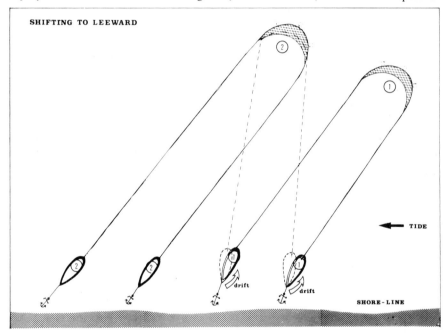

27. Skiffs shifting to leeward

the operation. The majority of the other Kintyre crews, however, with an extra member could employ the device and increase their efficiency. Indeed, it was considered an indispensable aid to shore fishing until about 1914.

After both boats had gone to anchor and the hauling of the ropes had begun, wind and (especially) tide conditions were still preoccupying considerations. Along the shores of Lochfyne, if, during a haul, the strength of the tide was such that it was likely to carry the net past the boats and thereby increase the strain of the pull, the anchors would be lifted and the boats rowed downstream. When sufficient ground had been covered and equal lengths of sweep-line streamed out again from each boat, the anchors would be dropped again and hauling resumed, with the net flowing down on the tide towards the boats, a naturally lighter pull.[47]

That practice was intentionally employed on certain stretches of shore between Tarbert and Skipness. If herring were supposed to be lying to windward (or upstream) of a known rock, or had been located there, the net would be set and the ropes shot away to leeward. Having anchored the boats, the crews would haul the ropes until the net was judged to be close enough in to drift down clear of the rock. The fishermen might even rest and light their pipes to 'have a smok' till the tide wid take the net by the rock'.[48] Then, if necessary, they would shift downstream and drop anchors again to take the final haul. The shifting was essential to the operation because, as Hugh McFarlane explained, 'They cou'na haul in against the tide'. The distance they would shift to leeward would depend on the rate of the tide – the stronger the current, the longer would be the shift. Their sure knowledge of the distance offshore and depth of such rocks as Grianan Rock and Laggan Rock, and their ability to judge with exactitude the strength and course of the tide, ensured that the operation was almost invariably completed without damage to gear.[49]

During the nights of brief darkness in June, the fishermen would be fortunate to manage two hauls of the net. Said Hugh McFarlane: 'We'll start maybe between eleven an' twelve. We'll go oot an' shot an' by the time we had the herrin' boated an' the net at her stern, it wis wan or half-past wan, startin' tae the break o' day.'[50]

These nets were sunk with about 15 long stones, called in Tarbert 'snigs'.[51] These stones – each weighing perhaps 7 lbs. – would be lashed to the sole-rope in the middle and at the ends, in addition to the regular lead rings. The strings attaching the three buoys to the back-rope could be lengthened according to the depth to which the net was to be sunk.[52]

There is little evidence that sunk nets were much used by the Campbeltown fishermen about that time, but Angus Martin remembered that when he went to fishing first in 1910, many of the Campbeltown and Dalintober boats carried stocks of metal sinkers with lashings still attached. These were sections of old fire-bars, each weighing 3 or 4 lbs., and he was told that they had been used for sinking nets. He also heard reference to the use of stones.[53]

As the bays which the Campbeltown, Dalintober, and Carradale fishermen customarily frequented, along both shores of the Kilbrannan Sound, are relatively shallow, the sinking of nets was not ordinarily necessary. Instead, the practice of *fleeting* developed. That involved shifting towards, rather than along the shore. The net would be set in perhaps 20 or 25 fathoms (36.58 or 45.72 m.) of water, and both boats would stream out from 100 to 120 fathoms (182.88-219.46 m.) of sweep-line before dropping anchors. When the net had been hauled almost to the boats, the anchors would be lifted and the crews shift once more towards the shore, shooting the sweep-lines. They were anchored again in shallow water, where the net would be 'taking top and bottom'.[54]

Boating the Net

When the sweep-lines were almost entirely on board, the *Renown's* crew lifted their anchor, or, if conditions demanded quick action, tied a buoy to the cable and temporarily abandoned it, then 'put on the oars' and rowed down to the boat from which the net had been shot, the *Fame*. (The crew to windward – that is, upstream of the current – invariably shifted along to the neighbour-boat, an obvious rule which eliminated laborious rowing against the current.) As the *Renown* nudged alongside the still-anchored *Fame*, a man on the bow of the *Renown* handed a coil of the sweep-line to the bow-man on the *Fame*, who, passing it around the forestay, hurried aft to begin hauling on it. On the stern of the *Fame* the final fathoms of her sweep-line were being hauled, and the bridle-rope about to be gained.

The next stage of the operation was to ensure that the net could be managed aboard quickly, which necessitated countering wind and tide, no less considerations than in the earlier stages of the operation. The anchor-cable was usually transferred from the *Fame's* 'palbitt', on the bow, to amidships on the starboard side, so that

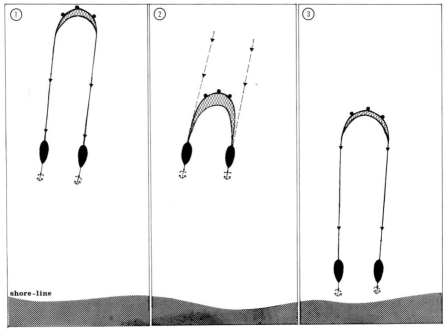

28. Fleeting a net

she swung around broadside to the weight of her anchor. On some skiffs a cleat was fitted, on to which the cable could be wound in several figure-eights, or else a bight was passed twice around the stringer and secured by a half-hitch. The *Renown* simply lay moored to the starboard bow of the *Fame*, clear of the cable, while the net was hauled. If wind or tide was putting a heavy strain on the net, and the combined crews were having difficulty getting the net aboard, the cable would be slackened out a fathom or two to lighten the strain.

On breezy nights, however, that procedure would be abandoned to avoid possible damage to the boats by their knocking together. Instead, the *Fame's* anchor-cable would be passed to the *Renown's* skipper, remaining alone on board his boat, and he would secure it forward. The end of his sweep-line, which had been transferred to the *Fame* when the boats met, would be made fast amidships on the *Fame* where the cable would normally be, and again on the stern of the *Renown*, lying about 15 yards (13.72 m.) off. The *Fame* was thus lying broadside to the weight of the neighbouring boat and anchor combined.

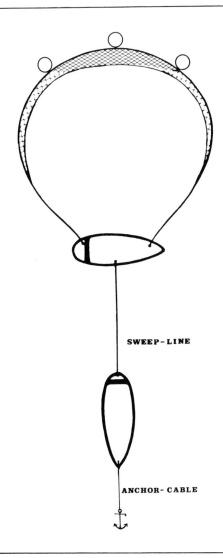

SWEEP-LINE

ANCHOR-CABLE

29. Boating a net

The linking rope would be slackened when necessary by the *Renown's* skipper. If a strong tide was running, the rope might be eased out and made fast, again and again, allowing the *Fame* to drift downstream or downwind fathom by fathom. (When working hard ground, the rope would likewise have to be slackened, to allow

L

the net to lift back off any grip in its way.) If too little end remained on the sweep-line to release any more, the anchor-cable would then be slackened.

The latter system, involving the transfer of the anchor-cable from one boat to the other, was, within living memory, the only one employed by Campbeltown fishermen. The other system, employing only the net-boat throughout the entire hauling operation, was preferred by Tarbert fishermen.

On the *Fame* two men were employed hauling each end of the sole-rope until the bottom of the net was entirely aboard, and the encircled shoal trapped. These men would join the others who, fore and aft, had been hauling the back-rope and 'flow' (netting). The entire crew could then gather in the walls of the net until the catch was contained within the top bag. That accomplished, the linking rope would be slackened on the *Renown*, and that slack passed around the *Fame's* rudder-head to the port side. The *Renown's* skipper then hauled her, by that rope, across to the stern of the *Fame*, where he would be rejoined by his crew. They, using stout poles, would push their boat around the back of the net until the two boats lay together, with the bag of herring between them. This was called 'squaring '. They were roped together to prevent their drifting apart, and a man with a fending-pole on each boat − forward on one, aft on the other − ensured that the boats did not come together, particularly in a strong wind. The protection of the net was also an essential consideration. If the keel of either boat were to catch and tear the bag even slightly, the weight of the catch might split the bag open in an instant. In the squaring of the boats, the risk of similarly tearing the net and losing a catch was even more pronounced, and the anxious admonition from the crew of the net-boat to the neighbouring crew would be, 'Keep 'er heel clear!' [55]

The skiffs were protected in all their manoeuvres together by fenders slung over their starboard side, particularly on the shoulder. These fenders were made from old netting and ropes bound together, and were 3 or 4 feet (0.91 or 1.22 m.) long. They could be transferred to the port side, if necessary, when going alongside a buying-steamer or a quay. Old motor car tyres were later substituted. [56]

Discharging Herring from the Net

After the boats had been secured together, the back-rope of the bag was passed across to the *Renown*, and her crew lashed it fore and aft to the stringer. The netting on the *Fame* was also lashed fore and aft. The bag was thus drawn tight between the two boats, and the herring 'dried', that is raised to the very surface so that they flurried almost literally in a dry state. As the bag was emptied of the catch, the slack netting would be gathered in and lashed down, keeping the herring 'dry'. The short manilla ropes used for tying down the bag were known as 'soogans'. [57]

The catch would be removed from the bag using quarter-cran baskets. It was usually baled into the tow-boat − in this case the *Renown* − unless sufficient to fill both boats, in which case both crews would get to work with baskets. (The liquid Gaelic word for a big catch was *steall*, [58] which is still generally used. The Campbeltown Scots terms were *sleever, glam,* and *tappie.* [59]) The work was extremely arduous. One man had to go down on his knees, and, taking one handle, lean over the rail of the boat and dip the basket into the mass of the herring. A second man would stoop to catch the other handle, and together they would raise the basket. Each man's weight would be borne by his stomach.[60] To lessen the strain of bending, 10 or 12 sand-bags would be stacked along her port side, weighting the boat over until her mouth was almost lipping the water.[61] When the basket had been raised, the fishermen would pause momentarily with the basket resting on the rail to allow water to drain out before tipping the fish into the hold. They preferred a catch to be quite dry in the hold, because if much water had been taken on board with the herring and a motion was on the sea, the catch might be softened and de-scaled as it

30. Basketing herring from the bag of the net. Taking a catch on board the *Busy Bee* of Campbeltown, off Heisgeir, south Minch, early 1930s. The catch is a heavy one, and two baskets are in operation. Stooping over the rail and pushing a basket into the bag is 'Big' Dennis McKay, behind whom – holding the hauling line – is Frank McKay. Raising the second basket is Sweeney Wareham, while skipper-owner Jackie Lavery helps tip it. On the *Blue Bird* is John Durnin, who holds the 'back-string' of the after basket. Fending off on the bow of the *Busy Bee* is Tommy Maguire. Photograph by A. Mathieson, Campbeltown

tumbled around within the compartments. Such 'washed' herring, as they were called, would likely fetch a low market price.[62] On rainy nights or when heavy seas were breaking on the boat and sending spray across her, 'washing' could not, however, be prevented.[63]

When, in the early 1900s, the boats became bigger and their freeboard was increased, a pole was added to the basket. It was passed diagonally through a handle and lashed tightly to it, and the end lashed to the bottom of the basket on the opposite side. To the other handle was attached a short rope, on which another fisherman pulled to assist with the raising of the basket. But a third man was necessary, on the neighbour-boat, to quicken the operation. Each time the basket was lifted, he pulled it out clear of the rubbing-strake with a lanyard attached to its rim.[64]

That simple improvement spared the fishermen much labour and discomfort. Previously, they would, on a breezy night, be drenched to the armpits repeatedly as they stooped down to the net with the basket. The invention was attributed to various Campbeltown fishermen, and to the McBrides of Pirnmill, Arran,[65] but one account, which credits it to Charles Durnin of Campbeltown, will suffice: 'Wan o' oor pairs went oot on a Saturday mornin', which was illegal, an' they got a big shot o' herrin' in Prestwick Bay. Well, they couldna get them dried up. The herrin' were aboot six feet or that fae the surface, an' they were standin' lookin' at them. "Oh, boys, this is loast herrin'. We canna dae better as this." An' ould Charlie Durnin, the "Duke" they caaled him, he says tae the boy: "Gie me that skiddoe." And he bent the skiddoe on tae the basket, an' they stood there an' filled the boats.'[66]

A Lochfyneside alternative to the basket was a lap-net with a handle 4 or 5 feet (1.22 or 1.52 m.) long. Called in Gaelic a *slat-àibh*, it could lift the equivalent of half-a-basket of fish from the net. It was carried on board Tarbert trawl-boats, in the final years soleiy for the discharging of small catches, but the introduction of the *stick-basket* eliminated it from that fleet about the turn of the century, so that by the 1920s only Ardrishaig fishermen continued to employ it. In its making, a naturally forked stick was cut from an ash or hazel tree to form the handle, and to the ends of the fork was spliced a short length of supple wood, thus forming a round. The joins were secured by lashings of twine, with a nail tapped in at each side perhaps. A bag of specially woven netting bound to the mouth completed the instrument. The lap-net was undoubtedly less uncomfortable to work than the original stick-less baskets, but the slowness of the operation rather diminished that benefit.[67]

On breezy nights, when the squaring of the boats could not be attempted, problems would result from the absence of a neighbour-boat to which the back-rope could be lashed up, opening the mouth of the bag and allowing the basket to be dipped in freely. In such conditions, an end of an oar or pole could be lashed to a gathering of the back-rope at the centre of the bag, and a halyard block hooked into a loop of the lashing. A man would operate the halyard, lowering and raising the back-rope in accordance with the swell on the sea, while another on the inboard end of the oar or pole kept the back pushed clear of the boat's side, thus enabling basketing to proceed efficiently. Alternatively, a pole or skidoag alone could be used, with a single man handling the entire operation.[68]

The 'stirrup' was an imaginative aid to fishermen in the basketing of herring in rough weather, and, though evidently little used, deserves description. Of the two accounts received, only one was the product of personal experience. James Reid of Kyleakin, Skye, was familiar with the stirrup and remembered it simply as a length of rope with an eye spliced on one end. The eye was lowered over the port side almost to the water-line, and the other end tied to the stringer on the starboard side. One of the fishermen, straddling the gunwale, placed a foot in the eye and steadied himself with his other leg on the deck. He caught one handle of the basket and dipped it into the bag. A second man, pulling on a rope attached to the other

31. A post-war use of the stick-basket. William Munro and Tommy Lawrence (right) of the *Summer Rose* of Dunure, Ayrshire, are lifting aboard, before brailing, a trial basketful of Lochfyne-caught herring which will be counted out to establish their quality. 1951. Photograph by courtesy of the *Glasgow Herald* and *Evening Times*. Per J. Gemmell, Ayr

handle, helped him to raise it. The device was 'handier' in a single-boat operation than the stick-basket, which tended to catch on the rubbing-strake as it was lifted, but less than comfortable for the straddling man, who was subjected to 'water runnin' everywhere, down yer neck, yer legs . . .'[69]

(With a pair of skiffs loaded to capacity, and a bulk of herring remaining in the net, the Tarbert fishermen of the nineteenth century might, when working in shoal water, cut out the bag, gather together and lash the edges, and leave it lying buoyed. Having discharged their loads into carrying-steamers, they would return and refill the boats. As Donald MacDougall remarked: 'They never lost a herrin' nor brok' a mash.' The practice was revived in Loch Ryan and Ayr Bay in the early years of the present century.[70])

Finally, after the bag had been emptied, it was hauled on board, and the boats separated. The net had then to be *redd* aft for shooting again. The after end had been carefully laid in the stern sheets by the fishermen as they hauled it, back-rope and sole-rope coiled, with the netting between well spread in layer upon layer. The forward end of the net, which lay strewn along the footboards, was quickly but skilfully gathered aft on to the mound of the after net.

The Flambeau

When searching for herring, the inviolable rule was that the skiff remained in darkness; but after the net had been shot and the sole raised, the work would be completed in the flickering light of the *flambeau*.

It was a crude, kettle-shaped paraffin can with a straight spout or neck 8 or 10 inches (0.20 or 0.25 m.) long (which prompted some Campbeltown fishermen to refer to the can as 'the goose'). Into the spout was pushed a tight wad of cotton fibres, bought at a ship-chandler's store, cut into lengths of about 2 feet (0.61 m.), and twisted together to form a thick wick. It was soaked in paraffin and when lit issued a naked flame. When it was beginning to dry out, the boy might be ordered to 'Gie the flambeau a dip'.[71] Unless attention was paid to the refilling of the can, the wick, being dry, would burn down quickly and be useless.[72]

Maintenance of the flambeau was the sole responsibility of the young cook. He had to ensure, before fishing operations began, that the wick was of sufficient length. When it had begun to burn down the spout, he would ease it over the top again with a little metal poker attached to the side of the can on a chain.[73] John Weir, however, always ignored the poker and quickly plucked the wick upwards with his hand.[74] The fuel was replenished from a bottle, or perhaps a beer-jar, handily placed. The can held about two pints of paraffin, a gallon or a half-gallon of which would be ordered when necessary.

The can was fashioned (by a local blacksmith) with a round handle on top, and hung on a standard. The standard – an iron rod hooked on one end – could be inserted into holes bored along the boat's rail, and was thus transferable from one part of the vessel to another, according to where work was proceeding. On a breezy night, when the boat was pitching about and the flambeau swaying with the motion, the boy might be instructed to take it and stand with it, a despised task which might result in a soot-blackened face.[75]

A type of torch similar to those which the rebellious trawl-fishermen of Tarbert carried with them at nights to warn one another of the approach of patrolling naval crews, was still in use on board the old drift-net boats in the final years of the nineteenth century. Wads of tow would be stuffed into the necks of paraffin-filled bottles, and when light was required the fishermen simply set fire to the wads.[76]

A single flambeau seen at night signified that a crew was hauling its net, but, when a catch had been secured, the recognised signal to the buyers on the steamers was two torches.[77]

Ringing

The first reference to ringing appeared in 1852 (p. 138), but that the net was more efficiently worked offshore when set with sweep-lines attached is certain. Ringing was seldom resorted to, as John McLean of Tarbert told the Royal Commissioners in 1864. To the question, 'You agree that going out to trawl in deep water would be an exceptional thing?', he replied: 'Yes, it is for our advantage to keep near the shore, and when we can get herrings there we always capture more than we could get by going out into the channel.'[78]

The Royal Commissioners in 1877 heard similar accounts in Tarbert. Archibald Campbell argued that 'ringing is more injurious than shore trawling, because the fish are less easily caught and therefore more liable to be lost'. He contended that 'strange fishermen from Ardrishaig and Campbeltown' were more 'given to' ringing than the Tarbert men.[79] Perhaps the Campbeltown fishermen, being, in general, relative newcomers to the recently legalised methods, were initially less prejudiced against ringing. Certainly, Dugald Robertson of Campbeltown remarked, without qualification, that he used a trawl-net from August until December, 'both for inshore fishing and ringing'.[80]

The judgement on the issue reached by the Royal Commissioners in 1865 may relevantly be quoted: 'When the fish are collected together in shoals near the shore, or in water of such depth that the circle net (trawl-net – author) can be properly worked, there can be no doubt that it is, by far, the most expeditious and cheapest mode of taking the fish. On the other hand, in deep water . . . when the seine cannot be worked, there is equally little doubt that the drift-net is the better engine.'[81]

Towards the end of the century, the aversion to ringing was breaking down, possibly facilitated by the introduction of the Lochfyne Skiff, a heavier class of boat than had hitherto been employed, and therefore capable of lending more weight for the hauling of the net in mid-channel, though in severe tide and wind conditions that advantage would certainly have been inconsequential. All of the surviving men who went to fishing between 1900 and 1910 remembered ringing in their earliest years at the job – particularly the daylight 'gannet fishings' in summer (p. 166) – but that there remained a strong opinion hostile to ring-netting, among the older men, is apparent.

The principal reasons for the trawl-fishermen's reluctance to fish in deep water lay, first, with the limitations of sail and oar, and, second, with the shallowness of their nets, and ringing remained an unsatisfactory alternative until the introduction of motor-power.

As to the second consideration, Henry Martin remarked: 'They always wanted tae haul in (the net) till the sole-rope got the ground, otherwise − they made out − they wanna get the herrin'.'[82] That condition could not, of course, be satisfied in deep water. John McWhirter recalled: 'It taen (took) ye a long time shottin' roon a ring wi' the oars, becaas if ye wirna well placed at the breck o' day, the herrin' wir gone afore ye'd time tae get intae position an' shot. It wis different on the shore. When ye shot the net ye had the herrin' *gegged* (jammed) then. They wir comin' off, an' they jeest came off intae the net.'[83] And, finally, Duncan McSporran: 'It wis mostly a

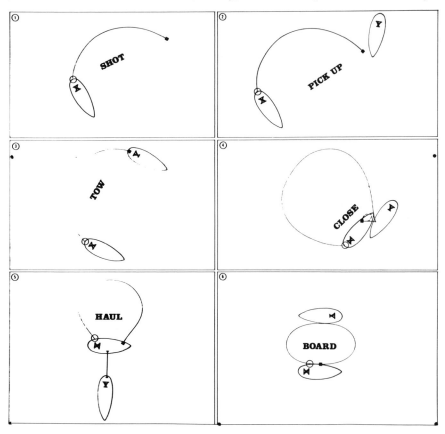

32. Ring-netting − the method

shore fishin', but ye'd tae go out intae deep waater when the herrin' wir in deep waater. As long as the herrin' kept up ye wir all right, but when they sank down ye wir oot. Ye had tae chance that, of course.'[84]

As ringing was practised in mid-channel − to deal now with the first consideration − without the advantage of a stable hauling base such as the shore or anchors provided, the operation of hauling the net so that it retained its circular shape was rendered uncertain. The net's shape was a primary concern, for if the ring became elongated by the force of wind or tide, then the net's effectiveness was lost. The area within which the herring shoal was surrounded being reduced, there was a corresponding reduction in the likelihood of the shoal's failing to make its escape before the sole closed beneath it. The only device which the fishermen could employ to counter these natural forces was, in effect, to substitute one of the boats for the shore, and that was accomplished by the tow-boat's keeping a strain, by oars or by sail, on the net-boat (linked together, as in trawling, by a length of sweep-line).

If the tide was putting a heavy strain on the net, and the hauling crew having difficulty getting the net aboard, the order 'Less wight!' (weight), or an equivalent, would be shouted to the two men remaining on the *Renown*. With that, the sail would be lowered a bit or – if a windless night – the rowing ceased. With the tension removed from the tow-rope, the net would be lightened and hauling could be resumed, the determined purpose being to raise the sole-rope quickly and so close the net.

If there was insufficient strain on the net, and the ring was collapsing, the cry would be 'More wight', 'A *kennin* (little) more', 'Keep her at ye's', or 'Give 'er a good tip, me boys'. The sail would then be set higher, or rowing resumed, and the *Fame* would slowly be pulled back off the net, and the net itself gradually come under strain until its shape had been restored. The weight would then be removed so that hauling could continue. So, gradually, by alternately creating and removing strain on the *Fame*, her net could be hauled in. On calm nights and with a slack tide, the operation would be much simpler and the net might be boated with relative ease. [85]

An undoubtedly apocryphal story which belongs to the period of transition from trawling to ring-netting describes a pair of crews hauling a net to anchored boats. While they were at work, another pair of crews appeared and shot their net in a ring within the sweep of the first pair's gear. They made a catch, and while they were basketing the herring into their boats, one of the still-toiling fishermen shouted to them from inshore: 'Did ye lee (leave) anything in her at all?' [86]

With the introduction of motor-power, ringing replaced trawling as the more efficient method. The lengthy and tedious hauling of the sweep-lines was eliminated (though short sweep-lines were still 'given' to the net in ringing), and motor-power ensured that the towing boat was seldom defeated by the force of wind or tide. The dependence on human strength had been removed.

Invention on the Brink of Change

A few years after the turn of the century, the candle *winky* was introduced to light the end of the net and guide the neighbouring crew to it. Its benefit would be fully realised when the shore fishery declined and the fleets began to work closely together in open waters. In such conditions, the contact which had been possible between neighbouring crews along the shores was less easily maintained, and the boats would often be quite distant from each other when one crew decided to shoot its net. A reliable mark on the bridle-rope was therefore essential. (The usual shooting signal to a neighbouring crew was several lit matches tossed overboard. [87])

The invention of the winky was popularly attributed to 'Daingle' (Daniel) Morans of Campbeltown, who, it was said, borrowed the idea from the type of lamp used on horse-drawn carriages.[88] The candle was lodged inside a can, which was mounted on a hollow metal float, circularly shaped, with a projecting lead-weighted base. Winkies, which were manufactured locally by plumbers, had often attached to them a large metal handle by which they could be lifted aboard. Three or four small windows, into which glass could be slipped, were cut around the can, along with a series of tiny holes, close to the top, to let in air. The candle's height in the can was maintained by spring action – as it was burning away, the spring would be easing it upwards gradually.[89]

The winky was kept constantly lit, in preparation for a quick shot. If a 'good tightener o' win'' was blowing, the fishermen would try to shelter the flame, but not always with success. Henry Martin remembered: 'Many's the time there wid be a row aboard the boats when they wid go tae put the cannel winky away. There wis nae light in't. There wis that much air blowin' in, it kept the light goin' strong, an' the cannel wasted away.'[90]

A novel, but altogether too brief experiment in quick closure of a trawl-net was undertaken in October, 1905, by a pair of Tarbert trawl-crews. The Tarbert fishery officer at that time, Robert Spink, considered the idea 'feasible', but explained that 'as the herrings remained in from 60 to 80 fathoms (109.73-146.30 m.) of water when experiments were carried on, results were very unsatisfactory'.[91] The net, which was designed and rigged by the McFarlane family, and operated from the *Britannia* and the *Mary*, was kept a secret from the other fishermen of the village, but Spink, who lodged at the family house, Rockfield, may have acquired, in conversation with the brothers, the details for his inadequately explained account.

The idea was, basically, to deepen the net. The additional netting was an old shoulder, 43 score of meshes deep (about 12 fathoms or 21.95 m.) and 34 rows to the yard. One edge of the netting was lashed securely to the sole-rope of the bag, and to the centre of the lower edge an end of a 50-fathom (91.44 m.) length of buoy-string was made fast. The string was 'stopped' to one half of the netting by three or four pieces of monish, lightly knotted, and the other end made fast to the after bridle-rope. As soon as the boats had met to complete the ring, and the towing boat's crew had been transferred, the after bridle-rope was hauled in and the string loosened off. That end was taken amidships and hauled on so that the monish stoppers parted from the bottom edge of the additional netting, which could then be pulled to the boat in a V-shape. The edges of netting on each side – the 'arms' of the V – were then gathered in, thus quickly and simply closing the net. One good catch was secured using the method, but it was discarded after its trial and was not re-adopted.[92]

Several years after that experiment, a Tarbert fisherman, Archibald Campbell, fitted a manual winch on board his skiff, the *Princess May*. It was bolted to the foredeck on the port side, and its purpose was to dispense with manual hauling of the long sweep-lines. Two men turned the handle, and a third coiled the rope into the sweep-line box. It was rejected after a single summer's trial, because manual hauling was considered the quicker method.[93]

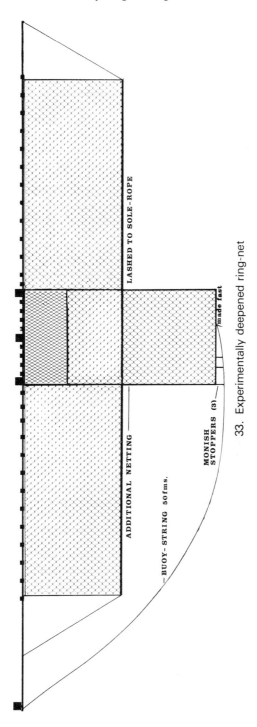

EXPERIMENTAL RING NET
TARBERT 1905. 80–90 fms.

LASHED TO SOLE–ROPE

ADDITIONAL NETTING

BUOY–STRING 50 fms.

MONISH STOPPERS (3)

made fast

33. Experimentally deepened ring-net

The experiment, though unsuccessful, was a precursor of the massive and rapid change which would begin in 1907 with the installation of the first motor engine in a West Coast fishing boat. Within 25 years, the familiar working ways would have vanished.

REFERENCES AND NOTES

1. A. Sutherland, fishery officer at Ardrishaig, to Fish. Bd., 12 February, 1840, A.F. 7/104, 220-1.

2. 4 May, 1974.

3. J. McMillan, fisherman, Tarbert, *E.R.C.* (1862), 13, and J. E. McLarty, merchant and ship-chandler, Tarbert, *ib.*, 12.

4. L. Lamb, rpt. to Fish. Bd., 1 May, 1851, Tarbert, A.F. 7/85, 269.

5. *Ib.*, 269-70.

6. *Ib.*

7. B. F. Primrose, *Paper* (1852), 3.

8. Precognitions, Hawton/Parker case, A.D. 14 61/275, 14/2-3 (marginal annotations).

9. D. McFarlane, 4 May, 1974; H. McFarlane and J. Weir, both 28 November, 1975. Creag Eonaidh = Johnny's Rock, noted 9 May, 1979 from G. C. Hay, whose version was that Johnny slid down the shelving rock and into the net, which burst. An alternative name was Creag Mhic Cuilgean − McQuilkan's Rock.

10. H.D.S.G. (Tarbert). Probably a figurative usage, a *heckle* − or hackle − being a comb used in the dressing of flax or hemp. The broad-bladed tangles found growing on such ground were termed, likewise by Tarbert fishermen, *ruffles*.

11. R. Ross, 25 May, 1978 (H.D.S.G.).

12. 4 June, 1974.

13. *E.R.C.* (1864), 1179.

14. *Paper* (1852), *op. cit.*, 2.

15. H. Martin, 4 June, 1974, and R. Conley, 11 June, 1975. Expression still in use among Kintyre fishermen.

16. *E.R.C.* (1864), 1140.

17. *Ib.*, 1175.

18. *E.R.C.* (1862), 13.

19. *Ib.*, 12.

20. Remarks upon reading Mr Holdsworth's book titled *Deep Sea Fishing and Fishing Boats*, 16 February, 1875, A.F. 37/176, 8-9.

21. 1 March, 1975.

22. R. Conley, 11 June, 1975.

23. H. McFarlane, 7 June, 1974; D. McLean, 17 February, 1975; J. Conley, 1 March, 1975; R. Morans, 26 September, 1975.

24. H. McFarlane, 5 April, 1976.

25. D. McSporran, letter to author, 31 July, 1979.

26. H.D.S.G. (Tarbert and Carradale). Both words have the meaning 'heap', *tòrr* being the greater. An equivalent Lochfyneside term was *sùil*, 'eye' (D. MacVicar, 22 October, 1978, H.D.S.G.). In English, however, an 'eye' of herring was usually 'a huge mass . . . that more or less filled the channel', as Robert Ross of Tarbert put it (24 March, 1978, H.D.S.G.). Kintyre equivalents of *tom*: 'ball', 'spot', and 'lump'.

27. The Campbeltown and Tarbert terms respectively. Shooting 'over the weather' − R. McGown, 8 December, 1975, and D. McLean, 6 April, 1976; 'off the backside' − J. Weir and H. McFarlane, both 5 April, 1976. J. McCreath, Girvan, gave 'shooting with the sun', 2 May, 1976.

28. H. McFarlane, 5 April, 1976.

29. D. McLean, 6 April, 1976, and J. McWhirter, 7 April, 1976.

30. R. McGown, 8 December, 1975; J. Weir, 5 April, 1976; J. McWhirter, 7 April, 1976.

31. H. McFarlane, 5 April, 1976.

32. H.D.S.G. (Tarbert). 'Rolled' was the Campbeltown term and 'twined' the Tarbert term.

33. H. McFarlane, 5 April, 1976.

34. D. McLean, 3 June, 1974.

35. H. McFarlane, 5 April, 1976.

36. D. McLean, 8 December, 1975; H. McFarlane, 18 March, 1976; J. McWhirter, 7 April, 1976.

37. J. McWhirter, 7 April, 1976 (quoted), and D. McSporran, letter to author, 31 July, 1979.

38. J. McWhirter, as above.

39. As above, and J. T. McCrindle, 29 April, 1976 – Lochfyne, 'The spate overflowin' the tide'.

40. Gael. *cothrom*. H.D.S.G. (Tarbert and Carradale).

41. H. McFarlane, 20 April, 1975. If the neighbour-boat had been close by, the end of the bridle-rope would simply have been thrown across – As above, 7 June, 1974.

42. H.D.S.G. (Tarbert). Also D. MacVicar, 17 September, 1978 (H.D.S.G.).

43. The box – in Tarbert known simply as the 'rope-box' – was built into the forward port corner of the hold, and was from 3 to 4 feet (0.91-1.22 m.) deep. It could be constructed either rectangularly or with the after side sloping down to the bulkhead, in a wedge-like shape, a design which effectively reduced fouling of the coiled rope. All boxes were built of widely-spaced slats to promote quick drying of the rope, a precaution against premature rotting.

D. McSporran, 11 June, 1974; J. Conley, 1 March, 1975; R. Conley, 11 June, 1975; J. Weir, 27 June, 1975.

44. Also 'skiddoe'. Both forms recorded (H.D.S.G.) in Campbeltown and Carradale. In Tarbert the pronunciation 'skiddaw' was recorded, but there, however, 'boatuck' (boat-hook) was the regular term (and, indeed, instrument). Etymology obscure.

45. J. McWhirter, 5 March 1975.

46. J. Weir, 27 June, 1975, and H. McFarlane, 4 May, 1976.

47. H. McFarlane and J. Weir, both 28 November, 1975.

48. H. McFarlane, 7 June, 1974.

49. As above, 28 November, 1975.

50. As above, 7 June, 1974.

51. D. McFarlane, 24 January, 1975, and J. Weir, 27 June, 1975 (H. McFarlane, 10 December, 1976, gave 'sling-stones' instead).

52. H. McFarlane, 7 June, 1974.

53. 16 June, 1974.

54. H. Martin, 3 May, 1974; D. McSporran, 11 June, 1974; J. McWhirter, 10 December, 1974.

55. R. Conley, 11 June, 1975 (quotation).

56. H. McFarlane and J. Weir, both 5 April, 1976; D. McLean, 6 April, 1976.

57. Gael. *sùgan*, 'A straw or heath rope'. H.D.S.G. (Tarbert, Carradale, and Campbeltown). Also applied to a short rope used for other purposes, e.g. to lace together a tear in netting, or to gather, for additional strength, a section of back-rope. A Campbeltown and Carradale equivalent of *sùgan* in its textual sense was 'torrigan'. Variants include 'torrikan' (H.D.S.G.).

58. H.D.S.G. (Tarbert, Carradale, and Campbeltown).

59. H.D.S.G.

60. D. McFarlane, 4 May, 1974; D. McLean, 17 February, 1975; D. McSporran, noted 15 June, 1979.

61. J. Weir, 27 June, 1975, and J. McWhirter, 5 February, 1976. Sand-bags were also used, when basketing herring aboard, for keeping the bag of the net taut, and the catch 'dry'. Hugh McFarlane, 30 October, 1974: 'Jeest take a bunch o' the net oot . . . a san'bag on't, an' that kept it there withoot a man hadin' (holding) on. Ye put wan on each corner, an' then men wir free tae fill the boats.'

62. H. Martin, 8 February, 1975.

63. D. McLean, 5 December, 1974.

64. J. McMillan, 2 March, 1975.

65. D. McIntosh, 27 April, 1974.

66. D. McLean, 5 December, 1974.

67. R. MacNab, 25 September, 1975; H. McFarlane and J.Weir, both 28 November, 1975; D. MacVicar, 16 February, 1977 (S.S.S.) and, discussing pronunciation and meaning, 17 September and 22 December, 1978 (H.D.S.G.). The spelling *slat-àibh* represents D. MacVicar's pronunciation. *Slat* applies to the handle. The second element may derive from the Norse *háfr*, which gives *tàbh*, 'a hand-net', in other dialects, but some doubt remains. (As with pole-basket, could be worked with a hauling-lanyard on neighbour-boat.).

68. H. McFarlane, 18 March and 26 April, 1976, and D. McSporran, noted 25 June, 1979.

69. 2 February, 1975, recorded by W. Maclean. Also described, but in less detail, by R. McGown, 7 December, 1975. He had heard his father, John McGown, discuss it.

70. H. McFarlane, 11 June, 1976, and J. McWhirter, who himself, in Loch Ryan, was present at the buoying of a quantity of herring in the bag of a ring-net. The catch subsequently had to be discharged from the bag using a basket lashed to the loom of an oar – 10 December, 1974.

71. A. Martin, 10 February, 1975.

72. H. McFarlane, 18 March, 1976.

73. A. Martin, 10 February, 1975, and J. Weir and H. McFarlane, both 18 March, 1976.

74. J. Weir, 18 March, 1976.

75. H. Martin, 8 February, 1975.

76. H. McFarlane, 18 March, 1976. A. Henderson, commander of H.M.S. *Jackdaw,* reporting to the Fish Bd., 13 September, 1861, on the seizure of three Tarbert trawl-skiffs, the *Catherine,* the *Solan,* and the *Mary:* 'They had all the usual implements of Trawling . . . (including) a machine for signalizing with, which is a bottle of some sort of spirits, and a stick with cotton on the end, attached to the cork, this they burn as a signal whenever any of our boats are seen approaching.' A.F. 37/157.

77. J. McMillan, 2 March, 1975, and R. Morans, 26 September, 1975.

78. *R.R.C.* (1865), lvii.

79. *E.R.C.* (1877), 130.

80. *Ib.,* 157.

81. *R.R.C.* (1865), lviii.

82. 3 May, 1974.

83. 10 December, 1974 and 5 March, 1975 (composite).

84. 30 April, 1974.

85. H. McFarlane, 30 October, 1974 and 5 April, 1976; J. Weir, 5 April, 1976: D. McSporran, noted 25 June, 1979, etc.

86. H. Martin, 3 May, 1974.

87. R. Conley, 11 June, 1975, and J. Weir, 10 December, 1975, etc.

88. G. Newlands, 7 February, 1975; D. McLean, 17 February, 1975; J. McMillan, 2 March, 1975.

89. H. Martin, 8 February, 1975. Also D. McSporran and D. McLean, both noted 25 June, 1979.

90. H. Martin, 8 February, 1975.

91. *F.B.R.* (1905), 222-3.

92. H. McFarlane, 24 January, 1975.

93. D. McFarlane, 24 January, 1975, and H. McFarlane, 28 November, 1975.

8

Appearances: The Natural Lore

THE means of detecting herring were varied. Some were more likely to succeed than others, but no matter the season of the year or weather conditions, the fisherman had ways of finding out if herring were present nearby. His sense of the natural world − whether consciously appreciated or not − developed with his skills as both seaman and fisherman. To work his boat safely and exploit her finest qualities, he had to understand the vagaries of wind and sea; and, equally, to fulfil his function as hunter, his predatory drive had to be conditioned by an accumulated knowledge of the behaviour of the herring and of the various fish, birds, and animals which also preyed upon it. His existence was *of* the natural world simply because he depended upon the signs it yielded to him in order to stay alive upon the sea, and, being alive, to hunt the herring to its death. These signs were known as 'appearances', in Gaelic *coltais*.[1]

Listening for Herring

On a calm, dark night the main-sheet would be tightened on the 'back' − or starboard − side of the skiff, so that, with sail taut, the skiff would be 'more driftin'' than sailin' . . . Ye could go along,' said Hugh McFarlane, 'an' the boat wid take steerin' wey, an' that wis aa' they waanted'.[2] The men would listen, with hands cupped about their ears, for the jump of a herring. One plop was sufficient to justify the shooting of the net, for, as David McLean remarked, 'That wan herrin' jumpin' gies away the millions below him.'[3]

A range of terms existed − principally among Tarbert fishermen − to express the different degrees of sound. The slightest sound of all was termed a 'bleeng' or 'bling', which has been variously explained by Tarbert fishermen: 'Breckin' the waater wi' their tails.' − 'A very light motion.' − 'A fish maybe puttin' its nose above the surface.'[4]

A complete jump could be described as a *plout* (Scots) or, in Gaelic, a *plub*[5] or *plubartaich*.[6] The fishermen could often, but not always, distinguish the sound of one kind of fish jumping, from another. The herring liked to leap completely from the water, making a 'clean plout'.[7] Larger fish caused a stronger sound, of course, but certain other fish similar in size to the herring behaved differently. The mackerel, said John Weir, is a more 'rash' fish,[8] and the young saithe would make a *spleeter,* or splash, with its tail on the surface.[9]

A small flurry of herring was known, to Carradale as well as to Tarbert fishermen, as a *fliuchan,* described by Denis McIntosh of Carradale as 'a wee watery spot o' herrin''.[10] The description is apt etymologically as well as phenomenally, the literal meaning of *fliuchan* being 'a wet spot'. The Tarbert term *bratan* is now virtually

163

interchangeable with *fliuchan,* and any edge of difference which may originally have existed has been blunted. The etymology of *bratan* is difficult, but is most likely to be 'small covering'. A third term – likewise confined to Tarbert – is *fras* or, in its oblique case, *frois. Fras* is 'shower', which quite nicely describes the sound of a bunch of herring flurrying on the surface.[11] The word was also current, in that sense, among the upper Lochfyne fishermen, as in: *Tha e a' froiseadh an deoachainn* (It's showering here).[12]

There was a real division of opinion among fishermen on the degree of accuracy attainable when listening for herring, but an unmistakable sound was a big *play* of herring 'renting the hills'. Such a forceful play of herring or, oftener, mackerel – a noisier fish when rushing on the surface in bulk – could be expressed by any of four Gaelic terms: *raibheic, gàir, toirm,* and *eas.*[13]

Before daybreak, herring might be heard – often from a distance of several miles away – 'playing off the shore', that is striking out from the shallow water towards the deep. Decided skill was required to successfully surround playing herring at daybreak, so quickly would they leave the coast. The mouth of the net would almost invariably be set across the shore to block their passage.

The fishermen preferred to shoot their nets on a single herring that had jumped, believing that a shoal so signified was 'steady'. A big play, especially far into night, suggested that the shoal was shifting away, or was about to sound to the deeps. 'On dark nights when we'd hear a big play we'd say – "That's them, they're shiftin' away clear o' that, or they're goin' tae lie doon." They might go an' lie doon for weeks, away doon in the deep, an' they winna rise at all,' said Donald McIntosh of Carradale.[14]

When listening for herring a strict silence was insisted upon, and young boys especially would be castigated repeatedly until silence became, for them too, habitual.[15] The slap of the sea on the overlapping planks of clinker-built boats reduced the likelihood of hearing a jump, and for that reason carvel boats were considered more satisfactory.[16]

The early trawl-fishermen were accustomed to listening for herring. 'There is nothing that you can perceive them by at the beginning of the season except as they jump and play about,' Tarbert fisherman John McMillan told the Royal Commissioners in 1864.[17]

Fire in the Water

In the late summer and autumn nights, when phosphorescence lit the sea, herring were located principally by sight. A man or two men would lie watching over the bow of the boat, and one would intermittently strike the gunwale with the anchor-shank. The sound-wave transmitted would cause a shoal of herring to start, producing an incandescent flash. That reaction, which was peculiar to the herring, was termed 'answering the anchor'. The Gaelic-speaking fishermen called phosphorescence *losgadh,*[18] the meaning of which is represented exactly in the English term which has totally replaced it, *burning.*[19]

The organism most likely to produce 'fire in the water' is the dinoflagellate *Noctiluca*, a very abundant member of the *phytoplankton* (minute drifting plants) in western Scottish waters. It emits a phosphorescent glow when disturbed, for an as yet obscure reason or reasons. [20]

In late evening, before darkness thickened and the search for herring could begin, a fisherman might stir the sea with an oar or boat-hook to test the strength of the burning. Sparks of phosphorescence in the swirl were termed a 'greegerty'. [21]

An individual herring sighted in the burning was known by Tarbert fishermen as a *leus*, [22] Gaelic for a light or glimmer. For a small bunch of herring, shining suddenly close to the surface, an extensive terminology existed. In Kintyre generally the English 'scatter' was commonly used, but less so its Gaelic progeny, *sgadarach*, which in Carradale was understood to represent a thinner spread of herring. [23] *Sràbhail* – a strewing – was confined to the vocabulary of Tarbert fishermen. [24] *Breac* – a speckling of fish – was used by upper Lochfyne fishermen, [25] and *dràbhag* can be considered a vigorous survivor. [26] Quantitatively, these appearances were all insubstantial and would not merit a haul. As evidence of the possible presence of worthwhile quantities of herring in the vicinity they merited, however, the attention of the fishermen.

For a thicker spread of herring the Tarbert term was the etymologically difficult 'yemarach'. 'That's not solid,' said John Weir, 'but very good.' [27] A still-greater phosphorescent flash was called, by Kintyre fishermen generally, a *stroke*, the potential yield of which Dugald McFarlane of Tarbert estimated at 60 or 70 baskets, [28] or about three tons. The greatest manifestation of all, however, was known by Tarbert fishermen as a 'meck'. [29] In English, the terms for a broad, solid mass of herring sighted in the burning were 'field' and 'park'. It was with profound nostalgia that some elderly fishermen recalled those 'fields of herring', suddenly luminous-green in a dark reach of sea, those immense 'fields' which may never again startle the sight of fishermen.

The early trawl-fishermen had trained themselves well in watching as well as listening for herring, as Tarbert fisherman John McLean's statement to the Royal Commissioners in 1864 indicates: 'We know the distinction between large and small fish. The one is a solid light flash, and the other is a simple flash.' [30] The Commissioners' interest in the trawl-fishermen's ability to discern large herring from small had its source in the claim by opponents of trawling that massive destruction of immature fish was caused by trawl-fishermen. John McMillan refuted the claim. 'As experienced men,' he said, 'we know the size mostly of the herrings when we look at them in the water . . . and we do not shoot at all among them if they are only small herrings.' [31]

A 'trained eye' was also required to distinguish between a shoal of herring and a shoal of other, unwanted fish; and, especially, to discern a glow deep down. The response of deep-swimming herring was often noted as a mere 'change in the water'. Robert McGown of Campbeltown remembered: 'Wi' the big herrin' ye used tae say there wir a blue flash off them. Ye always thought they wir good herrin', whether it wis the solidness o' them that made them dim.' [32] David McLean remarked: 'Sometimes the big herrin' show just lik' a dim wisp o' smoke.' [33]

M

The ability to detect the evanescent glow of an 'answering' shoal deep down was, as Robert McGown remarked, 'really expert work'. Expertise was, however, the product of many years of observation, both of the night waters and of the skills of an experienced man on the bow, and young boys fresh at fishing were invariably awed by the mysterious abilities of their seniors to locate and catch fish of which they themselves had seen not a sign. Said Robert McGown: 'When I went tae the fishin' first, I'd be lyin' for'ard on the bow wi' my father, an' my father wid be shoutin', "Right round wi' her − here's a fine spot!". I'd be lookin' an' couldna see a thing. Ye marvelled at them until ye got intae the way o' it yerself.'[34] In time, the instructors, their senses dulled, became dependent on their sons, and yet another generation lay forward and looked conscientiously but in vain upon the glittering waters ahead.

On still nights, the knock, knock, knock of anchors would be heard incessantly as the fleet ranged about, with not a light visible except if a fortunate pair were getting up the bag of the net or summoning, with two torches, the herring-buying steamers stationed in readiness here and there amid the fleet. John McWhirter remembered that during exceptionally heavy fishings in the East Sound (between Arran and the Ayrshire coast), 'Ye wid sweir there wis two or three shipbuildin' yerds workin', anchors goin' the whole night, *chappin'*, *chappin'* the whole night. It wis a wonder there wis a herrin' left above at aa'. Quietness tae (till) the burnin' came in the waater, an' as soon as the burnin' came in the waater, a man furrid wi' his anchor.'[35]

The Maidens, Ayrshire, fishermen when they adopted ring-netting on a general scale after the First World War, indulged less in opportunist striking of the anchor, and would generally *chap* only when they judged that herring might be present.[36] Many did not use the anchor at all, but struck the foredeck with the side of their leather boots, or knocked the gunwale with a marline-spike, bolt, or mallet.[37]

The introduction of motor-power caused a problem which several Ayrshire fishermen referred to. The reflection of a bright nocturnal sky on patches of surface waste oil occasionally deceived some fishermen, who mistook it for a flash of herring; but men with a keen sense of smell would not be so mistaken.[38] On a night of 'heavy burning', shoals of herring behaved untypically by 'flying' away from the boats in bright sheets, rather as mackerel would do. In such conditions, the best times for fishing were evening and morning, when the burning was 'slack'.[39]

The Solan Goose

That imperial fish-hunter, the gannet, unwittingly led generations of fishermen to great catches. His presence was always investigated if his behaviour suggested that he might be working on herring. Indeed, summer daylight fisheries were periodically based on the presence of great concentrations of the birds. The warm summer sun would bring to the surface banks of the copepod *calanus,* to which the herring would rise and 'feed in the sun'.[40] Over the shoals ranged the gannets, and constantly watchful of these industrious birds the fishermen ranged too, on the bright sea. Daylight fishing was uncertain work, however, because the herring were able to see the nets set around them and might evade capture by *dooking* below the nets before their closure. 'Ye missed them of'ner than ye got them,' John McWhirter remarked ruefully.[41]

A gannet prowling in the sky with a peculiar persistence was an almost sure indication that a shoal was below it. 'When ye see them hingin' thon way, cockin' thir nebs,' said Donald McIntosh, 'that's when the herrin' wir right thick.'[42] Explanations of the phenomenon of the uncertain bird were offered by Robert McGown and James Reid. The former remarked: 'We always knew when a gannet circles round and round (that) it's on a spot. It must just be waiting to get the edge o' the herrin'.'[43] The latter remembered: 'We used tae say about a gannet comin' along an' seein' herrin', if he made a dive an' then turned back: "Oh, they're too thick." And we'd ring on that sign.'[44]

A high vertical plummet – an almost infallible indication of herring – was termed a 'stroke'. The height from which a gannet struck was considered proportionate to the depth of penetration necessary. Gannets were especially valuable to fishermen in late evening, when the herring were beginning to seek the shore, and it was during that dusk period that the fishermen might 'ring on a gannet'.

If a bird was working on 'sile' (*sìol,* herring fry), *brathas* (the fry of other fish),[45] or mackerel, his dive would be shallow and angled, though gannets fishing with that technique on evening herring close to the surface were not absolutely unknown.[46] No Gaelic terms descriptive of that angled dive seem to have survived, but a more than ample range of Scots terms remains, including *skiting, skaving,* and *sketching.*[47]

The winter 'spawny herring' on the Ballantrae Banks attracted gannets in their thousands from the nearby colony of Ailsa Craig. John McWhirter recalled the spectacle: 'A could safely say that wis a solid mass o' herrin' on these banks for aboot five miles. Ye could hardly see the sky wi' gannets. Ye wir almost afraid tae sail on it wi' gannets. No' an odd gannet hittin' here an' there – they wir comin' pourin' oot the sky lik' shrapnel the whole blessed day fae mornin' until night. A skyfu' o' gannets pourin' down . . .'[48] The Ayrshire fishermen called the formations of fishing gannets 'rallies'. 'They'd gather, just a few,' said John Turner McCrindle, 'an' they went round and round till the sky wis white. Ye could hear the noise a mile away, them goin' down plop. An' an odd time ye saw the unfortunate one that had been *het* (hit) by some o' the others.'[49]

When fishermen came into an area where gannets were resting full-bellied on the surface after intensive feeding, upon approaching one it would invariably regurgitate the contents of its stomach to lighten itself for take-off. In that way, the fishermen discovered whether or not the unlucky bird had been feeding on herring. Eight or nine herring might be *boaked* up by a bird so disturbed.[50]

'Putting Up'

In daylight or at evening, a man or two men might lie forward watching the sea ahead for strings of bubbles rising to the surface. The phenomenon has been scientifically investigated, notably by T. H. Huxley in the 1880s. It is caused by the expulsion of gas from the swim-bladders of the fish, which enables them to maintain their natural buoyancy. The bubbles are small at the depth of release, but because they fuse together and expand in volume as they rise through the water (Boyle's Law), they eventually reach a greater size. The gas is thought to be expelled through the anal vents of the fish, but how the fish restore the gas when they go deeper is not known. [51]

The Tarbert fishermen of the nineteenth century called the appearance of these bubbles *frying*. [52] The general later term was 'putting up' – in Gaelic, *Tha e cur an uachdar* [53] with *belling* (from the Scots *bell*, bubble) also sometimes used, mainly by Campbeltown fishermen. The Tarbert *cobharach* described the patches of froth which concentrated 'putting up' might cause. [54] 'Old putting up', which lay on the surface as a dead froth, was a less valuable indication of herring, but the fishermen might linger near it and locate herring with darkness. As David McLean remarked, 'Any man but a trained fisherman wid never take the slightest notice o' that.' [55] In 1877, William Bruce of Tarbert spoke of the trawl-fishermen's watching for the 'bubbles made in the water by the fish breathing'. [56] He was undoubtedly a better fisherman than a biologist.

When a dense shoal of herring was swimming and 'putting up', the fizz of the bursting bubbles might be heard even through a skiff's hull. Robert Morans of Campbeltown remembered hearing, from within a forecastle, a mass of herring 'belling' as it departed Loch Broom. [57] An eerie mist, hanging over the surface of the sea for miles, was sometimes discernible on a bright winter's day. Robert MacNab witnessed such a sight off Wemyss Bay. The day was calm and sunny, 'an' ye wid see this – it was a mist just about a foot high on the water, as far as ye could see, an' there were five whales among it'. Periodically, the whales, suffocating in the dense concentrations of fish, would break the surface and shower herring from their backs 20 and 30 feet into the air. [58]

Whales and Porpoises

Common rorquals, lesser rorquals, and sei whales, collectively known to fishermen as 'herring whales', were once abundant visitors to the West Coast, but come no more in significant numbers – indeed, do not exist in significant numbers.

Both the whale (known to the Gaelic fishermen as the *muc-mhara*, or 'sea-pig') and the porpoise herded the herring and fed on the outer fish. Archibald Stewart of Campbeltown expressed delightfully their strategy: 'They jeest swept round and round an' nibbled a wee yin now an' agane.' [59] The fishermen knew when a whale was working around a shoal of herring by the infrequency of its blowing. [60]

A whale bursting to the surface clear of a dense shoal of herring was naturally a valued sign. Hugh McFarlane recalled: 'The old men used tae say, The whale wis throwin' it off 'er back at such a place . . . It wis jeest lik' a bing (heap) o' silver up in the air, an' then it wis all spread away.' [61] (Crashing noisily through the water was apparently a practice important to the hunting strategy of the porpoises. Said Robert Morans: 'They wid do a bit o' beltin' durin' the night tae lash the herrin' down so as they'd get a go at them.') [62]

Occasionally a whale would inadvertently be surrounded by a net. John Weir recalled that one, realising itself trapped, had burst its way through the bag of a net being hauled off Holy Isle, Arran, resulting in loss of the catch;[63] but a whale, similarly enclosed in the Kilbrannan Sound, was credited with cunning by one fisherman observer. Henry Martin, who was then a young cook, recounted: 'They shot roon a gannet or two that came an' struck. A'll aalways remember it. It wis jeest in the evenin'. There wis a whale in the ring, an' somebody aboord the boat kept harpin' that we'd lose everythin', the whale wid go through (the net). An' A mind there wis wan fisherman there who said, The whale wid dae *nothin' o' the sort*. He could keep there tae (till) the last, but he wid get oot through the openin' . . . and so he did.' The haul was complicated by a very heavy meshing, which floated the net up. 'A can aye mind them sayin' that it wis through the whale that the herrin' wir *breengin'* (dashing) away aa' roon an' mashin'.'[64]

Red Water

Explanation of that beautiful winter phenomenon, 'red water', which may never again be seen in western waters, is best left to the fishermen themselves, who observed it in the years of the herrings' great abundance. Said Hugh McFarlane: 'It's a sheen off the herrin', an' it throws a rid (red) skin on the waater, if there's any sun at all. That must be caused by the rays o' the sun strickin' (striking) the silvery ridness o' the scales. Ye'll see it fae a distance, a bloody colour.'[65] And John Weir: 'It's a common occurrence where there are herrin' in a big shoal. Probably it's accordin' to the light. It's only a sort o' sheen, a reddish sheen.'[66] James Reid recalled that, 'The old boys used tae say in the yarns that the herrin' were packed so close that it was the red in the back o' the herrin's head that was shinin' in the waater, makin' a show o' red.' He spoke of the potently beautiful appearance of low-lying Gott Bay, Tiree: 'An' Gott Bay's that flat, ye'd think the houses were standin' on the top o' a red sea.'[67]

Hugh McFarlane noted a characteristic in the behaviour of herring located by that sheen, hazardous to the fishermen. Such herring, especially when frost had come and the shoals congregated more, tended to plummet to the bottom when surrounded. He said: 'When they're seein' they're caught, there's somethin' pullin' against them. They've this instinct tae lie on the bottom, an' ye'll no' move them. Now, that's in the nature o' the fish. As soon as they wir feelin' they wir hemmed in, every herrin' wis away for the bottom, an' ye had tae stan' clear . . . I've seen them near away wi' net an' men. Stand clear till it wid get the ground. Ye cou'na stop it. Go away jeest lik' a stone. They'd be off − there winna be any herrin' in the net.'[68]

To attempt to rouse 'heavy' herring to the surface some fishermen would cut open sand-bags and *tim* the sand overboard and on to the shoal below. The practice was evidently not much approved of other than by Carradale fishermen, who attested to its occasional success.[69]

In dull weather, shoals of herring might be observed close to the surface swimming in 'black lumps'.[70]

Smelling Herring

A *fèath* was a calm on the surface of the sea caused by a patch of fish oil, and a *samh* (pronounced 'sev') the fishy odour that might accompany it. Among ring-net fishermen the use of these Gaelic words was confined to Tarbert. The equivalent English terms for the former were 'smelts', 'smeltings', 'smelty water', and 'greasy water', and in Carradale a fishy smell was called a 'soll', an etymologically obscure word.

Neither indication was considered greatly reliable of itself, but would often encourage fishermen to remain in the area until darkness, when herring might be heard or located in the 'burning'. Rings were occasionally made at night on the strength of an odour of fish, as James Reid humorously tells: 'I says to the Old Man, "There are plenty in here the night. Are ye smellin' them?" So we had a ring, an' got three-quarters o' a basket, an' he says, "Ye've a damn good nose". But they were there all right.'[71] The account also serves to illustrate the uncertainty of any judgement involved.

'Croy' and the 'Herring Cleaning'

Two related natural appearances may be mentioned. The rather prosaic English term 'red feeding' described concentrations of copepods, in particular *calanus*. This the Tarbert fishermen knew as 'croy' (*cró*, blood).[72] The red colouration is caused by oil contained in a sac which lies along the gut. It is ordinarily colourless, or at most faint pink, but there are times when the colour is much intensified and swarms of *calanus* may appear red.[73]

When it drifted close to the surface in banks, the herring would rise to feed on it in the warm summer days. In breezy weather, the fishermen usually searched for herring along lee shores, where the patches of feeding would be driven.[74] Herring so feeding carried a 'poke' of semi-digested copepods, and the condition of these fish, when caught, deteriorated in warm weather, their bellies softening and bursting, which drastically reduced their market value. Carradale fishermen called such fish *sgadan-brochan* ('porridge-herring').[75]

Come autumn, in advance of spawning – when the herring were *brògaidh*[76] – that feeding was all excreted, a phase termed by fishermen 'the herring cleaning'. That was a sign that the shoals were preparing to leave Lochfyne and the Kilbrannan Sound on their annual migration south to spawn on the Ballantrae Banks. For that very reason, the 'herring dirt', or *glar*,[77] was not a valuable sign in the location of herring. It was, as Hugh McFarlane remarked, 'leavings behind'. It was most evident when the tide gathered it, and resembled 'croy', being 'lik' wee rid worms'.[78]

When the herring did leave Lochfyne, their departure might transiently be marked by the wash they caused. Referred to as a 'herring wave', it was most noticeable on a calm sea, and such would be the speed of the shoal that any attempt to surround it would almost certainly fail.[79]

A winter daylight fishery occurred irregularly when a plankton organism the euphausiid *Meganyctiphanes Norvegica* – known throughout Kintyre as 'seel doo' (*siol dubh*, black seed) − brought feeding herring to the surface. A sunny day was good for the fishery, but 'as soon as the sun went low, they disappeared'. Flocks of gulls marked the areas where the feeding was, and herring might be seen leaping among it, which earned them the name 'hoppers'.[80]

Dog-Fish, Gulls, and Basking-Sharks

Among infrequent appearances signifying the presence of herring can be included the piked dog-fish (*gòbag*), the sea-gull, and the basking-shark. When dog-fish were rampaging among herring and biting, here and there, chunks from the fish, the oil so released would rise and form patches on the surface, causing a distinctive and detectable odour.[81]

John Turner McCrindle recalled a queer sight by moonlight on a calm night: 'When we looked ar'und us, everywhere wis doags' eyes. They wir prinklin' stars in the sea . . . They had just risen up wi' the moon.' Fishermen tried to avoid netting dog-fish because the damage caused by a thick meshing of these heavy, powerful fish − which can attain a length of four feet (1.22 m.) − could virtually spoil a net. No attempt would be made to shake out meshed dog-fish. They were hauled on board as they hung and would lie in a 'heavin' mass' on the deck. After the unhappy creatures had died, the net would be cleared and each one would leave a hole in the netting through which a man's hand could pass. Neighbouring Girvan skippers, John 'Lowe' Forbes and Willie McKenzie, had, in later years, an electric searchlight trained at times over the bows of their boats, and if dog-fish were seen swimming in shoals close to a 'spot' of herring, they would try to ring clear of them.[82] The fishermen handled dog-fish carefully, as the spikes immediately in front of both dorsal fins have associated poison glands which can cause an unpleasant sting.[83]

The gull was watched by daylight, and when seen pecking on the surface of the sea with a motion which some Ayrshire fishermen described as 'walking on the water',[84] sometimes put the fishermen on to herring. The old Campbeltown and Dalintober fishermen might comment negatively on both the absence *and* presence of gulls on the Gull Rock, two miles north of Campbeltown Loch. If absent, fishing prospects were judged to be poor: 'There's no' nothin' in it − there's no' a gull on the Gull Rock at aal.' If in evidence, however, the conclusion was equally pessimistic: 'That's fermers' gulls, lookin' for a worm.'[85]

Fishermen noted that herring were often present in the vicinity of basking-sharks,[86] which likewise feed on copepods. They, like dog-fish, caused damage when hauled in a net, tending to curl into the 'flow' netting and roll it about their prodigious bodies. The foul-smelling slime which they deposited on the net would rot the affected part quickly, and the fishermen would usually cut that piece away and replace it with new netting.[87] The regular Gaelic name for the basking-shark is *carbhan*, but *seòldair* ('sailor')[88] is the Tarbert name, which can still be heard. The name originated, obviously, in the resemblance to a sail of the shark's tall dorsal fin,[89] an image which is also represented in 'sail-fish'.

Feeling for Herring

In the very early years, a long pole was used for feeling herring on breezy nights when there would be too much turbulence of wind and water for herring to be heard. [90] While four men rowed the skiff along offshore, one man stood on the bow sounding with the pole. [91] The presence of herring would be felt on the pole as they struck against it, and the density of the shoal could be judged by the frequency of the taps felt on the wood. Fishermen probably also used the poles for 'feeling their way' about untried shores.

The practice probably disappeared with the decline of beach fishing. The estimates of the pole's length offered by the few Tarbert fishermen who remembered having heard references to it were of the order of 7 to 10 feet (2.13-3.05 m.) . . . 'aboot the length o' an oar'. [92] The smallness of the boats would have imposed a limit on the length of pole carried, and when the nets began to be used offshore − either as ring-nets, or with sweep-lines attached − the pole would have proved insufficiently long for sounding purposes.

The elderly Tarbert fishermen were of the opinion that the once-current expression, to 'feel herring', when the obvious meaning was 'hear', 'smell', or 'see', suggested a legacy of that practice. As Hugh McFarlane commented: 'That's the kinna language they had at wan time. They *felt* wan jump − they winna hear it.' [93] The author, however, mindful of the expansiveness of the Scots 'feel', reserves his judgement.

The garvie trawl-fishermen of the Firth of Forth (p. 2) sounded the estuary waters with long oars. It was reported in 1861 that 'an experienced man can easily tell whether Sprats or Herrings strike his oar from the increased resistance of the latter'. [94] Oars, of about 18 feet (5.49 m.) in length, were still in use on the Firth of Forth when West Coast fishermen operated there in the early 1930s, and some of the visiting crews themselves acquired oars. The oar served the additional purpose of locating mud-banks. [95]

The principle of feeling for herring was revived about 1929 with the adoption from Scandinavia of the feeling-wire, a coil of fine wire with a lead weight on its end, which, when trailed astern of a slow-moving boat, served the same purpose as the pole. Discussion of the feeling-wire must, however, wait until Chapter 11.

Chance Hauls

In the absence of natural appearances, 'chance hauls' or 'chancers' might be resorted to by the fishermen. These were most often tried in winter, when the weather was too loud for herring to be heard, and the 'burning' had gone from the water. In such conditions, an end of the net − seven or eight fathoms (12.80 or 14.63 m.), perhaps with a special 'bridle-stone' attached, both to sink the net and to retard the vessel's progress [96] − would be shot over the stern for 10 or 15 minutes, and checked periodically while the boat drifted. This was termed 'trying an end of the net' − in Gaelic, *Feuch a-mach ceann ach dé 'n coltas a th'ann* ('Try out an end of the net to see what the prospects are') [97] − or, in Tarbert and Carradale, a *bad*, [98] which is Gaelic for a 'piece'. Even if a single big herring was meshed in the wing − expressed as 'marking it' − the skipper would judge the distance his boat had drifted, cant to windward, and shoot his net. Good fishings were occasionally got on such a chance.[99] If a crew were seen getting up a net, heavy with meshed herring, the observant skipper might go to windward of the working boat and shoot away his own net, again on chance.[100]

The chance haul, at the extremity of the practice, was ventured without evidence at all of herring; but sound knowledge of the shallow coastal waters was a prerequisite to avoid damage to gear. The ideal ground for a 'chancer' was both 'clean' – that is, devoid of obstructions – and a known haunt of herring, and the most suitable times for shooting were evening, when the shoals began to move shorewards, and mornings, when they returned to deep waters. On a limited area of clean ground, to safeguard the net its length could be curtailed by 'hanging on' to the bulk of the second wing, and retaining it on board.[101] The ring of the net was therefore less likely to catch on rocks along the edge of the sand.

Fishermen were occasionally alerted to fishing prospects by oral reports of appearances of herring along the shores, and even of individual fish found stranded in tidal pools.[102]

Knowledge of Fishing Grounds

Knowledge of the ground, like the fishing skills necessary for its application, was largely acquired from older, experienced fishermen, but the sum of that knowledge was constantly being increased. When the Kintyre fishermen adopted clam-dredging, as a seasonal alternative to ring-netting, in the 1930s, a more complete understanding of the nature of the seabed along the shores of the Kilbrannan Sound and Lochfyne became possible. The iron dredges, being naturally less destructible than nets, could be towed across ground traditionally considered prohibitively rough. By taking bearings on the land, fishermen gradually learned where to expect sandy ground, in which clams were bedded, and where to expect rocky patches, on which their dredges would fill with stones. At the Brown Head, on the south-west shoulder of Arran, for instance, the accepted edge of the hard ground was in eight fathoms of water, inside of which the fishermen would not risk a haul. Clam-fishermen discovered, however, 'nice wee patches in three an' fower fadom', where ring-nets would 'run'.[103]

Knowledge of particular areas of ground could be remarkably accurate. One Dalintober fisherman, Henry Martin, who had fished extensively for clams about the Broon Heid, was asked one night, when he boarded the neighbour-boat to help with a haul there, 'Whoot laik's the gr'un' here?' (What like is the ground here?). 'Well,' he replied, 'this is the kinna gr'un' that ye'll get three or fower fadom o' a split. It's roon (round) stones below there, an' maybe barnacles an' things growin' on them jeest catches the net an' gies it a wee bit split.' To the surprise of the crew, the net, when hauled, had seven or eight short tears.[104]

Odd Tricks

Drift-net fishermen were, like their trawling counterparts, no more frustrated than when the herring shoals lay in deep water without rising. The trains of nets could be sunk to the seabed (p. 190), but foreign fishermen on Lochfyne were said to have stirred up the fish in a very deep hole of the loch by dragging weighted tree-trunks through it. The account − accurate or otherwise − is of special interest, because the informant was an elderly Ayrshire fisherman,[105] who had heard it from his father.

The early trawl-fishermen of Tarbert used similar means to rouse sluggish herring into the sweep of their nets. They attached branches of whin or broom to the sole-ropes, to 'tickle the herrin' lyin' on the bottom'.[106]

The use of light to attract herring to the surface was never seriously experimented with in ring-net fishing. The main reason is not difficult to isolate. The reaction from daylight of the herring was a phenomenon witnessed by fishermen on mornings innumerable throughout their working lives. Ronnie Balls noted, however, that the quality of light was an important factor, and cited the use of lamps in pilchard and other pelagic fisheries. 'The main point,' he wrote in *Fish Capture*, 'is that whereas daylight is general, the artificial light is local. It offers a light in the darkness. And in this case the light seems to win.'[107]

Twice in John McWhirter's career as a fisherman he witnessed the attraction of herring to light. On one of these occasions, he was helping to *redd* the net aft on a skiff, under sail, soon after a haul in The Loaden. As the crew worked away in the light of a flambeau, hung on the port side, suddenly the skipper, Archie McKay, called from the stern: 'My God − Look at this!' Looking astern, the fishermen saw clearly a 'ball' of herring following the boat. They 'shot back' and netted 15 baskets of herring.[108]

A similar experience off Ardrossan, about 1910, prompted a Campbeltown crew to experiment with light attraction in the herring fishery. When shooting the net, a wad of paraffin-soaked tow would be attached by wire to the cork-rope of the bag, and lit. 'The herrin' rose up tae that (and) wid play roon aboot it,' remembered one of the fishermen involved. Had the device been reliable, then it could have reduced the risk of herring sounding before the sole of the net had been closed, but such hitches as matches blowing out, in what had to be a quick operation, finally discouraged the fishermen.[109]

Electric lights were in general use by the early 1930s for locating herring ahead of a slow-moving boat. By training, intermittently, a searchlight over the bow, herring beneath the surface could be seen;[110] but there is no firm evidence that attraction was, generally, an intentional factor in the practice. A flashing electric light proved instrumental in keeping herring and mackerel in a net until the sole had been hauled, a practice relying on the principle of repulsion. By frequently switching on and off the outrigger light fitted to the wheel-house of the later decked boats, the fish could be driven repeatedly into the bag of the net. 'Ye could see them wi' the burnin', hittin' the back o' the net after the flash wis gone,' said J. T. McCrindle. 'But they wid come back again had ye no' continued wi' the light.'[111]

The Sleepy Season

This chapter would be incomplete without mention of that most frustrating of periods in the fisherman's year − failure of the fishery, in Gaelic *gort*[112] (scarcity).

The cause might be attributable to personal ill fortune – as expressed in the rueful proverb once current in Tarbert, *Nuair bhìos sgadan mu thuath, bidh Donnchadh Ruadh mu dheas* (When the herring's north, Red Duncan's south)[113] – but, at its worst, the effect was general. These periods occurred mainly during autumn, hence the expressions 'Harvest Slack'[114] and 'Blackberry Slack',[115] and would force the fishermen to remain ashore for a month or more. The following report of the Fishery Board for 1903 describes an especially bad season: 'July and August were the months in which fish were most plentiful, and from 5th September to 26th November no herrings were caught. The fishermen are of opinion that during that period, which they term the "sleepy season", herrings seek deeper water. The slack period varies in duration every year according to circumstances.'[116]

The Tarbert fishermen, as could be expected, had a word to explain concisely the absence of herring. The shoals were said to be *loching*, that is, keeping to the bottom of Lochfyne. On a September night the shoals would 'play the sea dry' and then vanish, a phenomenon referred to by Donald McIntosh (p. 164). Hugh McFarlane remembered: 'Workin' the drift-nets ye winna get what ye wid put in the fish-pan. Every night. Ye winna ken there wis a herrin' on the scene. Yet there wir plenty o' herrin'. When it wid rise agane, comin' on in the wintertim', there wis herrin' everywhere. An' ye wid wonder wherr they came oot o'. That's the impression it wid gie ye. But it wis a barren sea until they wid start tae move.'[117] In such bleak conditions, the policy of the drift-net fishermen – if at sea at all – would be, *Iomair mu thuath agus cuir mu dheas* (Row north and shoot south)[118] – the sooner to be home.

Fishermen were, from time to time, forced by indigence to take work ashore, usually locally at such short-term occupations as road-making, but they might leave the district to crew coastal steamers or to labour in Clydeside shipyards.[119] The drift-net fishermen of Campbeltown and Dalintober, however, annually passed the winter off-season working in distilleries,[120] and fishermen of the Silvercraigs district hired themselves out as agricultural labourers, chiefly at draining.[121] In March, 1930, the Ardrishaig fishermen were, most of them, gathering winkles, while seven or eight Tarbert crews were engaged in gravel-carrying,[122] a paying resort in lean seasons. The gravel was shovelled into rowing-boats in the bays at Skipness or Claonaig, and ferried out to the skiffs. When a full load had been secured, the crews would set off for Clydeside to discharge, usually at Queen's Dock, Glasgow.[123]

Weather Lore

Weather lore, like fishing lore, had to be understood in practice to be of value, and was of crucial importance to the generations of fishermen who lived before radio weather-forecasts. The tendency now is to dismiss it as so many old yarns, though the older fishermen continue to heed the more obvious signs, and comment on them over the radio wavelengths.

The strangest lore is associated with living creatures. On still, dark nights as the fishermen listened for herring, a thin whistling might be heard across the waters. This the Tarbert and Carradale fishermen called the 'cheepag' or 'cheepach' (probably from the English 'cheep' – the Campbeltown form of the word is 'cheeper'). Various theories exist as to the weird sound's origin, including 'foul air' – the convergence of wind currents[124] – electricity in the air,[125] and that it emanated from the little 'herring bird', as the storm-petrel was known to some fishermen.[126] The bird was said to sip the oil of the herring from the surface of the sea.[127]

The ornithologist W. R. P. Bourne, of the University of Aberdeen, suggested tentatively that the phenomenon may be attributable to the subdued, distant calls of various sea-birds – perhaps in particular kittiwakes – feeding, 'displaying', or moving ahead of the fronts associated with an atmospheric depression approaching from the west.[128] Whatever the explanation, the 'cheepag' was said to portend southerly gales, as was the great northern diver, known in Tarbert as the *murlach*[129] or *Ailean na Gaoth Deas* (Allan of the South Wind).[130] 'That's the fella that cries afore southerly win',' said Hugh McFarlane. 'Oh, ye may take on ye for the harbour. If ye wid hear the *murlach*, put on the sell (sail).'[131] Likewise, a whale leaping from the water while on a passage was considered a sign of approaching southerly wind,[132] though some fishermen were sceptical and suggested mere play of coincidence.[133] Ripples – or 'wee squalls' – within the troughs of waves when a breeze was 'on', were thought to signify a strong southerly wind's advance, and convinced fishermen would immediately seek an anchorage.[134]

The more familiar signs in the sky, such as red colouration in the morning, were respected, along with less common phenomena such as the 'dog', in Gaelic *fadadh cruaidh* ('hard kindling') or *an cuilean* ('the pup'), which may be a pun on the English name.[135] It is coloured as a rainbow, though a mere stump, and usually appears with a distant shower of rain. The fishermen did not like to see it, and pronounced its significance as 'broken weather',[136] or declared, 'There'll be a gale oot o' that.'[137]

The Tarbert fishermen called south-easterly wind the 'Weemen's Wind', because the harbour is completely sheltered from its air and the women of the village would unfailingly be astonished to hear their men returning long before dawn, though the sea might be foaming into Lochfyne.[138] A shifting wind, which is a presage of stormy weather, is alluded to in the Gaelic saying, *Tha a' ghaoth ag iarraidh nam port*, which the Tarbert men translated as, 'The wind is looking for harbours.'[139]

Severe winds occasionally compelled a crew to abandon its net on a lee shore. 'I have seen a gale of wind come on, when we have been obliged to leave the net and make clear of the shore altogether, although the net had a great quantity of herrings in it,' said John Bruce of Ardrishaig in 1864.[140] John McMillan of Tarbert remarked, in that same year, that herring were most abundant inshore when the weather was fine, but that, 'If a heavy gale comes on, the herrings rise and go to the deep water, and we will not see them either in deep or shoal water.'[141]

Moonlight

The ability of fishermen to catch herring was reduced during nights of bright moonlight. As John Weir remarked, 'They wir shiftin', an' they seemed tae see yer net, just lik' (in) daylight'. That 'wildness' in the herring's behaviour was frequently remarked on, and is expressed in the Tarbert saying: 'The herring likes the bonny moonlight, the mackerel likes the storm.'[142] The moon which succeeded the Harvest Moon was known to Tarbert fishermen as *Gealach Bhuidhe nam Broc* (The Yellow Moon of the Badgers).[143] 'That's the moon that catches winter,' said Hugh McFarlane. It was so named because 'the badger took home his bed wi' the moon – he wis foragin' wi' the moon'.[144] John Bruce remarked in 1864 that occasionally 'on moonlight nights we used our drift-nets, and carried our trawls back'.[145]

The difficulty of catching herring on bright nights was not much reduced with the advent of motor-power, deepened ring-nets, and electronic fish-detection devices, one of the few instances of the inefficacy of technology when waged against natural conditions in the herring fishery.

REFERENCES AND NOTES

1. Dr Archibald Campbell, remembered from boyhood in Ardrishaig. Letter to author, 8 July, 1977. Verified by D. MacVicar, 17 September, 1978 (H.D.S.G.). The word was also generally used by Tarbert fishermen, but latterly in a restricted context. If, when hauling the wings of a ring-net, no herring were to be seen hanging enmeshed, the remark might be, 'There's no' much o' a *coltas* (sign)', or, 'There's no' a *coltas* at all.' H. McFarlane, noted from, December, 1977 (quotations). Verified by W. McCaffer, 20 April, 1978 (H.D.S.G.).

2. 18 March, 1976. After the adoption of motor-power engines would be stopped temporarily to allow a period of silence necessary for successful listening. H. Martin, 3 May, 1974, and J. T. McCrindle, 3 May, 1976.

3. 28 April, 1974.

4. Respectively, H. McFarlane, 10 December, 1976; Dug. McFarlane, 14 March, 1978; W. McCaffer, 24 March, 1978 (all H.D.S.G.).

5. H. McFarlane, 10 December, 1976, and Dug. McFarlane, 14 March, 1978 (both H.D.S.G.).

6. D. MacVicar, 17 September, 1978 (H.D.S.G.).

7. J. Weir, 18 March, 1976.

8. 18 March, 1976.

9. D. McLean, 28 April, 1974.

10. 16 June, 1976, noted from (substantiated H.D.S.G.).

11. H.D.S.G. (Tarbert).

12. D. MacVicar, 18 January, 1979 (H.D.S.G.). The dialectal *an deoachainn* = *an seo* ('here').

13. H.D.S.G. (all Tarbert, but *toirm* also recorded in Carradale).

14. 27 April, 1974. H. Martin, 3 May, 1974: 'A heard the oulder fishermen sayin', when A wis young, that if ye heard a wile (wild) play o' herrin' on a good mornin' ye could say good-bye. That wis them on a passage. But, when ye wad hear wans jumpin' here an' there, jeest, that wis them kinna stationary.'

15. J. Conley, 5 December, 1974.

16. J. T. McCrindle, 3 May, 1976, and H. McFarlane, noted from.

17. *E.R.C.* (1864), 1175.

18. H.D.S.G. (Tarbert), and D. Mitchell, *Tarbert in Picture and Story*, Falkirk, 1908, 92. D. MacVicar, 17 September, 1978 (H.D.S.G.) gave *loisginn*.

19. Kintyre and Ayrshire. 'Singeing' was given, as an uncommon alternative, by D. McLean, 21 February, 1978 (H.D.S.G.).

20. Dr G. A. Boxshall, Crustacea Section, Dept. of Zoology, British Museum (Natural History). Letter to author, 30 January, 1978.

21. H.D.S.G. (Tarbert).

22. *Ib.* (Palatalised *l* becomes, in Kintyre, *j*, or Eng. *y* more or less.)

23. *Ib.*

24. *Ib.* One of the very few terms which survived, by adaptation, the elimination of the traditional sensory skills by electronic fish-detection equipment. In recent times (post-1945), the representation of a scant shoal of herring on the recording paper of an echo-sounder could be referred to as a 'thin *sràbhail*'. Angus Johnson and Robert MacDonald, Tarbert, noted.

25. D. MacVicar, 17 September, 1978 (H.D.S.G.).

26. H.D.S.G. (Campbeltown and Carradale).

27. 10 December, 1976 (H.D.S.G., substantiated).

28. 14 March, 1978 (H.D.S.G., term substantiated).

29. H.D.S.G.

30. *E.R.C.* (1864), 1182.

31. *Ib.*, 1176.

32. 8 December, 1975.

33. As above.

34. As above.

35. 5 February, 1976. An opponent of trawling, Archibald McEwan of Kames, Lochfyneside, complained in 1877: 'The trawlers make such a noise beating the gunwales of their skiffs that they frighten the fish, which generally escape seaward.' *E.R.C.*, 122.

36. J. T. McCrindle, 29 April, 1976, and A. Alexander, 2 May, 1976.

37. As above.

38. As above.

39. J. T. McCrindle, 3 May, 1976.

40. J. McWhirter, 26 April, 1975.

41. 5 February, 1976.

42. 27 April, 1974.

43. 29 April, 1974.

44. 2 February, 1975, recorded by W. Maclean.

45. H.D.S.G. (Tarbert), and D. MacVIcar, 17 September, 1978 (H.D.S.G.).

46. H. Martin, 4 June, 1974. Good fishings of herring taken nightly from Kilbrannan Sound during a week in June, *circa* 1912, all by shooting net on single gannets which had 'skited' into the water, for herring very close to the surface.

47. H.D.S.G. (Campbeltown and Carradale). The more emphatic 'skavin' lick' was current in Campbeltown. 'Skimpin'' was recorded in Tarbert.

48. 5 March and 26 April, 1975 (composite).

49. 29 April, 1976.

50. J. McWhirter, 5 March, 1975.

51. Dr A. D. Hawkins, Department of Agriculture and Fisheries for Scotland, Marine Laboratory, Aberdeen. Letter to author, 26 April, 1976.

52. H.D.S.G. (Tarbert).

53. D. MacVicar, 17 September, 1978 (H.D.S.G.).

54. H.D.S.G.

55. 8 December, 1975.

56. *E.R.C.* (1877), 129.

57. 1 May, 1974.

58. 28 September, 1975.

59. 1 May, 1974.

60. R. MacNab, 25 September, 1975, and H. McFarlane, 5 April, 1976.

61. 5 April, 1976.

62. 1 May, 1974.

63. 10 December, 1975.

64. 3 May, 1974.

65. 10 December, 1975 and 18 March, 1976 (composite).

66. 18 March, 1976.

67. 2 February, 1975, recorded by W. Maclean.

68. 10 December, 1975.

69. D. McIntosh, 27 April, 1974, and J. Conley, 5 December, 1974.

70. J. T. McCrindle, 29 April, 1976.

71. 2 February, 1975. (A strong odour of fish was often described, by Kintyre fishermen, as a 'ra' or 'raw' smell. G. Newlands, Campbeltown, 21 February, 1978, and C. Buchanan, Carradale, 20 May, 1978 (both H.D.S.G.).)

72. H.D.S.G.

73. J. A. Adams, D.A.F.S., Marine Laboratory, Aberdeen. Letter to author, 14 April, 1976.

74. H. McFarlane and J. Weir, both 5 April, 1976.

75. Denis McIntosh, 18 March, 1978 (H.D.S.G.).

76. H. McFarlane, 27 May, 1978 (H.D.S.G.). Gael. *bròg,* 'roe', with Eng. adjectival ending.

77. Charles Stewart, Campbeltown, noted Feb. 1978. Verified by D. McLean, 21 February, 1978 (H.D.S.G.).

78. H. McFarlane, 10 December, 1975.

79. J. T. McCrindle, 29 April, 1976, and H. McFarlane, 11 June, 1976.

80. J. McWhirter, 5 February, 1976, and (H.D.S.G.) H. Martin, 21 February, 1978, verifying 'seel doo' and 'hoppers'.

81. J. McCreath, 2 May, 1976. The phenomenon was attested to by Gilbert Clark, Port Charlotte, Islay, 21 April, 1978 (H.D.S.G.).

82. J. T. McCrindle, 29 April, 1976.

83. A. C. Hardy, *Fish and Fisheries,* London, 1959, 179.

84. J. T. McCrindle, 29 April, 1976.

85. H. Martin, 4 June, 1974.

86. J. Reid, 2 February, 1975, recorded by W. Maclean, and J. Weir, 10 December, 1975.

87. R. McGown, 8 December, 1975, and A. Martin, 4 May, 1976.

88. H.D.S.G.

89. H. McFarlane, 10 December, 1976 (H.D.S.G.).

90. D. McFarlane, 4 May, 1974.

91. H. McFarlane and J. Weir, both 28 November, 1975.

92. H. McFarlane, 28 November, 1975.

93. *Ib.*

94. Dr L. Playfair and Vice-Admiral H. Dundas, Report on the Sprat Fishery of the Firth of Forth, published with Draft Resolution by the Board of Fisheries and Statement upon the Herring Fisheries Act, 1860, 43. 1861. A.F. 37/143.

95. R. McGown, 29 April, 1974, and M. Sloan, 29 April, 1976.

96. H. McFarlane, 10 December, 1976 (H.D.S.G.).

97. D. MacVicar, 18 January, 1979 (H.D.S.G.).

98. H.D.S.G.

99. J. Conley, 1 March, 1975; R. Conley, 11 June, 1975; H. McFarlane, 26 April, 1976, etc.

100. D. McFarlane, 4 May, 1974.

101. H. Martin, 4 June, 1974 − example of 'half-haul', south end of Black Bay; D. McIntosh, 9 December, 1975 − example, below Dougie Morrison's (now ruined) house in Morrison's Bay, north of An Sruthlag; R. McGown, 8 December, 1975.

102. R. Conley, 11 June, 1975.

103. H. Martin, 3 May, 1974, and J. McWhirter, 5 March, 1975 (quoted).

104. H. Martin, 3 May, 1974.

105. T. McCrindle, 2 May, 1976. Mr McCrindle specified French fishermen.

106. D. McFarlane, 24 January, 1975.

107. London, 1961, 125.

108. J. McWhirter, 5 March, 1975.

109. D. McLean, 8 December, 1975.

110. J. McWhirter, 5 March, 1975; R. Conley, 11 June, 1975; J. Weir, 10 December, 1975.

111. 3 May, 1976.

112. H.D.S.G. (Carradale).

113. G. C. Hay, letter to author, 22 March, 1978.

114. D. McSporran, 30 April, 1974.

115. Charles Stewart, noted 20 February, 1978.

116. *F.B.R.* (1903), 242.

117. 10 December, 1976 (H.D.S.G.).

118. G. C. Hay, *op. cit.*

119. H. McFarlane, 20 April, 1974.

120. D. McSporran, 30 April, 1974.

121. Mrs J. MacBrayne, 22 February, 1977 (S.S.S.).

122. Inveraray Returns, 1 March, 1930. Campbeltown Fishery Office.

123. H. McFarlane, letter to author, 7 January, 1978.

124. As above, 5 April, 1976.

125. J. Weir, 18 March, 1976, and (H.D.S.G.) R. Ross, 7 April, 1978.

126. R. Conley, 27 April, 1976, and A. Martin, 14 May, 1976.

127. As above, and (H.D.S.G.) W. McCaffer, 20 April, 1978.

128. Letters to author, 26 August and 26 December, 1976.

129. H.D.S.G. Probably a corruption of *murabhlach* (*muir* + *balach,* i.e. 'sea-boy'), as noted in Tarbert by G. C. Hay. Letter to author, *op. cit.* Standard term is *muirbhuachaill,* i.e. 'sea-herd (boy)'.

130. Dug. McFarlane, 14 March, 1978, and H. McFarlane, 24 March, 1978 (both H.D.S.G.). Also G. Clark, Islay, 23 April, 1978 (H.D.S.G.).

131. 10 December, 1976 (H.D.S.G.).

132. R. Conley, 27 April, 1976.

133. J. McWhirter, 8 April, 1976.

134. Colin MacBrayne, Campbeltown, noted 14 January, 1978. Made known to him by the late Donald MacAnsh, Greenhill, Torrisdale.

135. G. C. Hay, *op. cit. Cuilean* recorded from Duncan McKeith, retired farmer, Saddell, Kintyre, 10 March, 1977 (S.S.S.), and Donald MacDonald, retired fishermen, Gigha, 11 November, 1978 (H.D.S.G.).

136. H. McFarlane, 5 April, 1976.

137. R. Conley, 27 April, 1976.

138. J. MacD. Hay, *Gillespie,* London, 1914, 304. Also, local tradition.

139. G. C. Hay, *op. cit.*

140. *E.R.C.* (1864), 1139.

141. *Ib.,* 1175.

142. J. Weir, 18 March, 1976.

143. J. Weir, as above, and H. McFarlane, 5 April, 1976.

144. H. McFarlane, as above.

145. *E.R.C.* (1864), 1139.

9
Seasonal Alternatives

RING-NETTING, though the main employment of Kintyre fishermen by 1900, and of Ayrshire fishermen by 1920, was not practised to the exclusion of all other fishing methods until the supersession of the skiffs by a larger and more seaworthy class of motor-boats (to be discussed in Chapter 11). Drift-netting, ground-netting, lines, and trawling proper were all methods of fishing periodically adopted by Clyde fishermen when conditions were unfavourable for ring-netting. The main period of abandonment of herring fishing was from the end of the Ballantrae Banks spawning fishery in March – by which time the fish were lean spents for which no profitable market existed – until May or June, when the herring begin to feed vigorously again and so build up their reserves of fat towards peak market condition in July and August.

Line Fishing

The first serious attempt to introduce long-line fishing to Kintyre on a profitable scale was probably undertaken by three Campbeltown merchants in the 1780s. They employed a fleet of wherries from northern Ireland, principally the port of Rush. Information on one of these vessels states its tonnage as 50, and its crew as numbering five men, including the master. The local fishermen were evidently instructed in the use of long-lines by the Irishmen, whose 'skill and management of baiting their hooks (and) shooting and hauling their lines with the wherries at a great distance from land, our people are entire strangers to'. The catches of these vessels were considerable. In 1787, for example, 27,390 dried ling and 280 dried cod were exported from Campbeltown to the Spanish ports of Lisbon, Malaga, and Bilbao. The final record of the catches of these boats appeared in 1794, when 1,714 dried ling were shipped to Dublin.[1]

Long-lining was evidently not again practised generally until some 40 years later, when two families from Morayshire, having settled in Campbeltown, worked the method so successfully that it was adopted by almost all the fishermen of the town, and by 1843 supported more than 500 families. Two sets of buoyed lines, each with from 1,000 to 1,500 hooks, were used, and 'whenever one which has been in the water eight or ten hours is drawn they shoot the other, and immediately proceed to land the fish'. The earlier method of white fish catching had been by handlines.[2]

That description almost certainly refers to 'small lines', which were set on the shallow banks for haddock, whiting, and flat-fish, whereas the lines employed by the men of Rush were undoubtedly set in deeper waters for cod and ling, and can be referred to as 'big lines'.

The principal bait for small lines was mussels, for which fishermen from Kintyre and Ayrshire went to Loch Riddon, Loch Striven, Loch Long, and the Clyde estuary. An open boat was towed astern of the skiff, which was anchored offshore until the fishermen had gathered by hand sufficient mussels to load the open boat. It was rowed off on the flood of the tide, and the mussels discharged into the skiff. Two tides were normally required to fully load a skiff.[3] The Tarbert fishermen gathered mussels also in Whitehouse Bay, West Loch Tarbert, by scraping them from the shore and into carts, with lap-nets. The carts were hired at a cost of several shillings each from local coal-merchants, and hauled by horse to and from the village.[4] The fishermen bedded the loads of mussels on an area of firm shore which they first enclosed with heavy stones to prevent the shellfish from being swept away by currents before they were able to fasten themselves to the ground. These enclosures were called *scaps* in Ayrshire and *scabs* in Kintyre.

The hour of gathering from the scaps was dependent on the state of the tide — work could not begin until the mussels were exposed. The women — usually a mother and her eldest daughter — draped in waterproof aprons and their fingers swathed in cloth for protection if the mussels were barnacle-encrusted,[5] managed the *shillin'* (shelling). A short-bladed knife — often a broken table-knife — sharpened to a point at the top edge of the blade, but shaped into a round towards the bottom edge, was used to scoop out the mussels into a dish, and care was taken to extract them whole.[6] Young boys who displayed an aptitude for shelling were often allocated that daily task, which, in the Maidens, Ayrshire, they fulfilled by rising at 4 a.m. and working until required to leave for school.[7] One mussel per hook normally sufficed, but if they were small or in poor condition, perhaps after spawning, as many as three or four meats would be needed to cover the 'haddock hook' which the Kintyre fishermen used;[8] but two meats generally sufficed on the smaller 'whiting hook' preferred by the Ayrshire fishermen.[9]

Each crew member owned a *baikie* (Ayrshire) or a *trough* (Kintyre),[10] which was a box square at one end and scooped at the other, into which the lines were baited. The main line, or 'back', was coiled into the hollow of the trough and the hooks arranged in rows on the scoop, beginning at the bottom, and each row slightly overlapping the preceding one. The hooks were set down carefully to prevent fouling.[11]

When the supply of scapped mussels had been exhausted, alternative forms of bait would be sought. The Ayrshire fishermen generally netted sand-eels in the clean bays close to their villages. The gear they used owed much to the design of the early 'trawls' or 'scringe-nets'. The main net was a length of old herring netting, complete with corks and lead rings, and into the centre of which a section of white gauze, 6 feet (1.83 m.) square, was laced. The sand-eels shoaled away from the enclosing wings of black netting and into the gauze, 'thinkin' they wir gettin' away'. In the last century, sweep-lines 100 fathoms (182. 88 m.) long might be attached to the ends of the small net, but within living memory the net alone was used, and could be shot and hauled to the shore by a single man or boy, obtaining sufficient fish for baiting.[12]

From Campbeltown, hauls were made with ring-nets – often cut down in size to, for example, 30 score of meshes deep and 80 fathoms (146.30 m.) long – usually in Kildalloig Bay; but the fishermen would, if necessary, search the Kintyre shore beyond Carradale for a 'baiting'. The drift-net fishermen's system of counting out herring in *casts* of three was employed in the allocation of bait. Thirty-three casts – or 99 herring – were considered sufficient to bait a trough. [13] The herring fillets would be cut on a board, incised in criss-cross pattern to prevent their slipping off it. The portions would be cut in a slant, rather than straight from one side of the fish to the other, the fishermen finding that by that method each fillet yielded more strips. If the fish were soft, they might be immersed in pickle for 10 minutes to stiffen them. [14]

The Tarbert fishermen also used herring as a substitute bait, but occasionally small dredges were used to rake up 'crechans' [15] (queen scallops) from the dense beds in Skipness Bay. The dredge – which consisted of a 3 feet (0.91 m.) -long 'spear' (a toothed iron bar) and a triangularly shaped poke of hand-knitted netting – was towed astern of a punt, rowed by two men, and four boxes of the dainty shellfish could be lifted in the course of an hour. [16]

Making and Working Lines

Each crew member possessed and maintained a line bearing 1,000 hooks spaced at intervals of a yard (0.91 m.). The main line, as already mentioned, was termed the 'back', and the attached hook-bearing lines *snids* (snoods). The spacing of the snids could be measured by stretching the back from the fingers of the outstretched right arm, across the chest to the point of the nose. [17] The snids were originally formed of two pieces. The shorter part was a line, lighter than the back. One end was knotted to the back, and to the other end was attached a 15- to 18-inch (0.38-0.46 m.) length of plaited horsehair, on the Scottish East Coast termed the *tippen*. [18]

A half-pound of horsehair would be purchased, usually from a ship-chandler's store, and washed in warm, soapy water. The Tarbert fishermen put the hair into an old stocking, which they tied about the top so that only the upper part of the bunch was exposed. Individual hairs would be plucked from the stocking, using the teeth, until the required number – from eight to 14 – had been extracted. These would be knotted together at one end and passed to a partner who, if skilful, would separate them into two lengths and 'twirl' them between his fingers until the snid was plaited immaculately along its length. The other end would then be knotted. Roset was customarily rubbed on to the working fingers to improve their grip on the hairs. More than a hundred snids might be produced at a single sitting, and some of these would be hung in bunches as spares. [19] In Campbeltown and Ayrshire mechanical spinners of a simple design were later adopted.

Shortly after the turn of the century, horsehair snids were dispensed with and the entire line was formed of twine. The twine snid would be attached by two hitches to the back, but leaving about 4 inches (0.10 m.) free. That length would be neatly twisted on to the snid proper, and knotted. The heavier upper part, so formed, helped prevent the snid's twisting up in a strong tide and fouling on the back. [20] To the lower end of the completed snid a hook would be 'tapped'[21] or 'beeten' (the Tarbert and Campbeltown terms respectively) with cotton thread, first rubbed over with roset to waterproof it and so prevent its rotting.[22] Moleskin thread, which could be heard 'singing' as it was tightened with the teeth, was traditionally used in Tarbert for tying on the hooks.[23]

The joined lines were set between small anchors. The ends of the line were marked by buoys called dhans or 'stooies'.[24] These were of varying design, but consisted basically of a cane, weighted at its base, buoyed above by strung corks, and with a flag attached at the top. From the submerged end of the dhan a heavy line, 28 or 30 fathoms (51.21 or 54.86 m.) long and termed the 'head-line', extended to the anchor. On to the ring of the anchor were attached two short beckets, the end of each spliced into an eye. To the eye of one was made fast the head-line, and to the eye of the other, the end of the back. A divided line was usually rigged by adding a middle anchor and buoy (or 'mid-heid'). If weather conditions worsened during the hauling of the lines, the latter half − safely anchored at both ends − would be abandoned, and the fishermen would run for harbour.[25]

Line-fishermen left harbour in darkness to arrive over the fishing ground as daylight was 'making'. By memorising the position of shooting and the progressive set of the lines, using landmarks, a good fishing ground could be returned to with accuracy. Sinkers were carried to *bicht* (bight) the lines. That is, when working over a good fishing bank of limited area, rather than set the lines straight along it, they would bicht them several times in zig-zag form, so increasing the concentration of hooks on the bank. Narrow stones were originally used for weighting the lines at the doubling points, but lead rings, which could more quickly be attached when shooting the lines, were later adopted.[26]

In shooting lines, the dhan was first dropped overboard, and, when the head-line had streamed out, the anchor was shot. The end of the first line would then begin to run out. The most experienced fisherman of the crew was invariably responsible for the shooting, and he would stand aft, holding the scoop of the trough to the side of the boat and casting away the coils of the back to lift the hooks clear and reduce fouling. The individual crew members' lines were joined by overhand knots as they were shot, usually two knots some inches apart, so that the joins were clearly visible, and the lines could easily be separated once hauled. (Before a line was baited into a trough, the bottom end would be kept clear by knotting it on the back handle of the trough, and before that line had been shot, the end would be untied and bent on to the upper end of the next line.) The fleet of lines was generally shot in less than two hours. As soon as the dhan marking the end was overboard, the crew would return to the start of the fleet and raise the anchor to begin hauling. One man hauled the back, standing on the port side of the boat, and when a hooked fish appeared he would catch the snid and swing the fish into a box. Another man unhooked the fish and coiled the back and snids into the appropriate trough. The other two men managed the oars.[27]

The marketing of the fish, in Campbeltown and in the villages of Ayrshire, was largely undertaken by 'hackers' (hawkers), both male and female. In Ayrshire, these sellers were equipped, of necessity, with horse and cart. The Maidens hackers sold fish in Maybole and from there proceeded to outlying farms.[28] In Campbeltown, barrows were adequate, as the fish were sold through the streets of the town. Occasionally, the last crews to reach Campbeltown would be unable to sell their catch in the market and would be compelled to trundle it from street to street on a net-barrow, after a morning's fishing, and with the next morning's trough of lines to prepare before sleep. Line-fishing did, however, provide an income which would otherwise have been lacking, and that income was regular, unlike the herring-fishing. A Dalintober saying expressed the fishermen's sentiments: 'When ye go tae the smaa' lin's ye can make the pan crack' (i.e. have bacon and eggs cooking in it).[29] The Tarbert fishermen, rather than sell a catch for an absurdly low price to the village buyers, might sell cod 'from Colintraive tae Skipness'. The fish were

34. Campbeltown fish-wives with barrow. Per W. Anderson, Campbeltown

weighed aboard the skiff and, when cut into half-stone portions, would be strung on the yard of the sail and carried through the villages to be sold at 1/6d per hunk.[30]

After a catch had been disposed of, the lines would have to be cleared in preparation for re-baiting. Ordinarily, they would be tipped from the trough on to the floor of the room in which baiting was to take place, and carefully *redd* into a basket or spare trough. While redding, each snid – or *sned/snedding*, the Campbeltown forms – would be caught some inches above the hook, the snid whirled several times around the hook, and a half-hitch passed over it, thus hanging it up on the snid. By so *sticking* the hooks, subsequent fouling was prevented, and the hooks would be less likely to go into the baiter's hand as he worked along the line. If the line was badly *fool* (foul), and the day was dry, redding could be done in the open. The coils of the line would be passed over the end of a pole, the pole jammed into a wall, and the coils then arranged along its length. Tangles could thus be examined easily and cleared with relative ease. The lines benefited additionally by receiving a drying.[31] The making of new lines was termed *minting*.[32]

The boats from which lines were originally worked were small and open, but a type distinct from trawl-skiffs. Popular with Campbeltown and Dalintober fishermen was the *Greencastle Skiff*, named after the Irish town from which the type was introduced, though by 1900 the main builder was in Portrush.[33] The 'Greenies', as they were commonly called, were from 22 to 25 feet (6.71-7.62 m.) overall length, between 6 and 7 feet (1.83 and 2.13 m.) of beam, and about 2½ feet (0.76 m.) of draught.[34] Bow and stern were said to be indistinguishable. Though safe when running before a sea, these boats, being so shallow-draughted, were utterly unsuitable for tacking, and for that reason the dipping lugsail was not habitually carried.[35] The sharp-ended 'Greenies' would be propelled to the fishing ground under four oars, two of which would be 'shipped' before the shooting or hauling of the lines began. In uncertain weather, a member of the crew would remain ashore, and, if the wind blew up, would proceed to sea in the big skiff and take the working crew aboard.[36]

In later years, ring-netting families motored to and from the fishing ground in the big skiff, with the 'Greenie' in tow, and shot and hauled the lines from the smaller boat, leaving one man in the skiff to follow them up. The fishermen soon discovered that the lines could be shot and hauled expediently from the line-skiff in tow, and, finally, the small boat was discarded and the entire operation was managed from the skiff, usually with a drag attached to reduce the 'way' of the boat and so ensure that the lines did not stream out too quickly.[37]

Big Lines

Big lines were of similar construction to small lines, but the components – back, snids, and hooks – were on a much larger scale. The principal distinction in the operation of the two methods was that the big lines were baited at sea as they were shot. There is evidence, however, that that distinction was not applicable until the present century, at least not in Tarbert. The big line fishermen there baited their lines at home into great troughs about 4 feet (1.22 m.) by 2 feet (0.61 m.) broad. Each man's line was formed of 30 'hanks' of heavy cord, and each hank was 30 fathoms (54.86 m.) long. These hanks, joined, constituted the back, to which the snids, 1 fathom (1.83 m.) or more in length, were attached, at 3-fathom (5.49 m.) intervals. By so spacing the big hooks, the lines could be shot under full sail and without undue risk of accident. Big lines, unlike small lines, were left out overnight, and if the weather was stormy might lie unattended for a week. They would be buoyed at five or six points along their length, to facilitate reclamation of broken lines, and additionally anchored with a stone in the middle. Skate, conger eel, and cod formed the main catch from the deep waters of Lochfyne. A stout gaff – in Gaelic, *clip* – was carried on the boats, and was especially useful for lifting on board large skate.[38]

35. The wretched end of a giant. A skate, taken by 'big line' in Kilbrannan Sound, c. 1935. The captors – crew of the skiff *Janet* of Campbeltown – photographed at Lochranza Pier, Arran, are *L.-R:* James Martin; and the brothers Peter, Donald and William Gilchrist. Per C. (Gilchrist) Munro, Drumlemble

Tarbert big line fishermen worked also in the Sound of Jura. To avoid the hazardous passage around the Mull of Kintyre, the fishermen – in accordance with tradition – carted their skiffs across the isthmus, which separates the east and west lochs by merely three-quarters of a mile. Two carts were required to transport a skiff. The vessel was first manhandled up the slip opposite the Tarbert Hotel, using ropes and rollers, and with some of the 20 or 30 men involved putting their backs to the hull. Once ashore, the bow of the boat would be raised, and the first cart pushed beneath her forefoot; then the stern would be lifted on to the other cart, the axles lashed together, and a horse hitched to one end. The fishermen would accompany the horse and carts across the isthmus, to 'gie the horse a help' on the braes. At the end of the drag the skiff would be launched in the shallow water of the upper loch. The lines were set in depths of 80 and 90 fathoms (146.30 and 164.59 m.) west of Ardpatrick, and streamed westwards towards MacArthur's Head on Islay, or north-westwards in the direction of the Small Isles. [39]

By the more usual method of working big lines, from Campbeltown and Ayrshire, the hooks were baited as the lines were shot. Preparatory to shooting, the lines would be coiled into a quarter-cran basket, with the hooks stuck in rotation around the rim of the basket, which was overlaid with cork. In shooting, one man removed the hooks from the rim and, repeatedly and alternately, passed one to a man on each side of him. Each man quickly stuck the hook into a whole herring or a half-herring, depending on the size of the fish, and threw it away from him. [40]

Some Campbeltown fishermen believed that cod caught with *buckie* (whelk) bait were firmer than those caught with herring bait, but the prolonged and tedious exercise of obtaining such bait would be resorted to only when herring were unobtainable. Ten or 15 creels, baited with rotting fish guts, would be set in Kilkerran Bay, Campbeltown Loch, for crabs; these creels, when lifted, would be re-baited with crushed crabs and set along Ru Stafnish, four miles south of the loch. As the creels were lifted, the buckies would be coped into the hold, and two men in the forecastle, 'oot the cold an' wet', would sit with mallets breaking the shells on a paving slab. [41]

When herring were unobtainable, the Tarbert big line fishermen might go out with a ring-net and fish for *gibearnach* (cuttle-fish). These they would bone before use. [42]

The decline and failure of line-fishing by 1935 was hastened by many factors. The installation of motor-power on the skiffs induced many fishermen to illegally trawl inshore waters for white fish; the advent of seine-netting on the Clyde, in the 1920s, coupled with the activities of steam-trawlers, depleted fish stocks and frequently resulted in the dragging away of fleets of lines; increasing numbers of fishermen were marrying women who were without knowledge of the skills customarily expected of them, and, in many cases, unwilling to learn them, because the work of womenfolk was, for the duration of the line-fishing season, as demanding and uncomfortable as that of the menfolk; finally, the larger class of boat, extensively introduced in the 1930s, enabled the fishermen confidently to extend the range of their ring-netting operations and so pursue a succession of distant fisheries throughout the year.

Dragging for Flat-fish

A novel method of trawling for flat-fish was introduced in the Firth of Clyde in the 1890s, evidently by Ayrshire fishermen. As, technically, it was not classifiable as otter-trawling, the fishermen could use it legally within the three-mile fishery limit. A standard ring-net, heavily weighted along the sole, was towed between two boats, or by a single boat, with one warp fast on the bow and one on the stern.[43]

The idea of catching white fish with surround nets was by no means novel on the West Coast of Scotland, but the earlier method had been to haul the nets to the shore. The fishermen of Oban had used 'scringe-nets' from the beginning of the nineteenth century, principally for saithe and flat-fish.[44] In Mull, too, 'scringing' was practised, for both herring and white fish.[45] By 1860 a 'trawl' was being used in Broad Bay, Lewis, for netting 'sillocks' (saithe), to the annoyance of the chamberlain of Lews Estate, who contended that its use was causing the destruction of salmon spawn 'at the mouth of the river' (*sic*).[46] And on Lochfyneside, 'splash nets' – narrow-meshed for the catching of 'cuddies' (young saithe) – were first specifically referred to in 1840.[47] Herring 'trawls', too, were employed for taking white fish. On a day in February, 1854, for instance, 300 boxes of whitings were shipped to Glasgow from Tarbert.[48]

By 1896, 42 boats from the Maidens, Dunure, and Girvan were engaged in white fish 'trawling',[49] and in the next year crews from Girvan introduced it to the Firth of Forth, where some of the local fishermen adopted it, while others reacted with hostility. Complaints had first been received in 1895 from Clyde line-fishermen, who claimed that the method 'approximated very closely to otter-trawling', and that the smallness of the mesh – in some cases only 1 inch (0.03 m.) from knot to knot – resulted in large quantities of immature and unsaleable fish being destroyed.

The latter objection was countered by the appearance of specially made nets of 2½ to 4 inches (0.07-0.10 m.) from knot to knot. By 1898 distinct pouches had been incorporated in the design of the nets, which increased their resemblance to otter-trawls. The Fishery Board became disturbed that year by the fear that the fishery would be 'detrimental . . . to the productiveness of the flat-fish grounds', and its perturbation was intensified by threats from steam-trawler owners that similar nets would be adopted by them for use on inshore grounds. These threats were fulfilled by Fleetwood trawlers operating in Broad Bay, Lewis, and a clash with local fishermen occurred. A prohibitive bye-law was quickly passed, and the method, which on the Firth of Clyde preceded seine-netting by almost 30 years, passed out of existence.[50]

Drift-nets

The year 1900, the beginning of a new century, marked – significantly – the virtual end of the tradition of drift-netting in Kintyre. The failure of the fishery on Lochfyne has been examined in Chapter 3, but in Tarbert, although ring-nets preponderated, some of the crews which had maintained both ring-nets and drift-nets allowed the latter to 'wear out', and did not replace them.[51] The failure of the Islay herring fishery by 1895 (p. 198) had caused the deterioration of numerous fleets of drift-nets in Campbeltown.[52] But in both Tarbert and Campbeltown, not a few families continued the tradition of visiting the Minches with drift-nets until the second decade of the century.

The design of drift-nets and the techniques of drift-netting changed little from the eighteenth to the nineteenth century, and survived, with few modifications, into the twentieth century. As a passive mode of fishing, drift-netting depended for its success upon the activity of the herring shoals. The nets were effective only when the fish struck into them and became enmeshed. That activity was termed 'working'. Said John McMillan of Tarbert, in 1864: 'Working is when they go into the nets. There are times when they work better than others.' The fishermen then believed that the movements of the herring were conditioned by the influence of the moon and by physical agitation, such as when a fresh body of herring mixed with the main shoal.[53]

The traditional measurement of the individual nets comprising a train was, in fishermen's language, a *barrel*, each length being about 90 yards (82.30 m.), or the quantity which a barrel would hold, although the measurement was more or less theoretical. In the narrow waters of Lochfyne the boats were equipped, in the mid-nineteenth century, with between eight and 15 barrels, but further south in the broader reaches of the Kilbrannan Sound as many as 22 barrels would be strung together.[54]

The individual barrels were themselves composed of smaller panels of netting, called *deepens* (deepenings). The length of each deepen was 12 yards (10.97 m.), and the depth 4 feet (1.22 m.). Five or six deepens would be joined to make a total depth of 20 or 24 feet (6.10 or 7.32 m.).[55] That assemblage was termed a 'net'. The bottom edge of the train, termed the *skonk* or *skunk*, remained free, but the upper edge was attached to a back-rope. The train was floated by buoys, fastened to the back-rope by strings 12 feet (3.66 m.) or more in length.[56] These strings were adjusted according to the depth of water the herring were known or supposed to be swimming at, but, as John Munro, an upper Lochfyneside fisherman, remarked in 1864: 'In some months of the year, they are at the very bottom, and we put our nets down to the bottom to take them . . . The top rope may be from 60 to 65 fathoms from the surface.'[57]

The net was shot at dusk, streaming out over the stern − if necessary, assisted overboard by one or two fishermen − as the boat moved slowly to leeward under sail or oars. The train, when completely shot, was secured to the bow of the vessel by a 20-fathom (36.58 m.)-long 'swing-rope'. In deep water, boat and nets together would be allowed to 'drive' to leeward throughout the night; but, if fishing close inshore, the windward end of the nets − that is, the far end − would be anchored.[58] As John Bruce of Ardrishaig explained in 1864: 'The tide would tear all their nets to pieces if they came into contact with the bottom while they were drifting.'[59]

If the herring were scarce, periodically − 'every half hour or oftener', according to the accounts of Greenock buss-masters in 1776 − the crew would check the nets to discover whether or not herring were meshing: 'This they do by warping along the back-rope, and here and there raising a piece of netting. By this means they not only find when they are upon good fishing ground but they see when the herrings

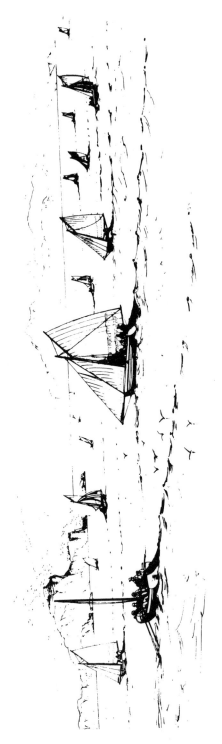

36. Shooting drift-nets in Lochfyne, an illustration of 1852

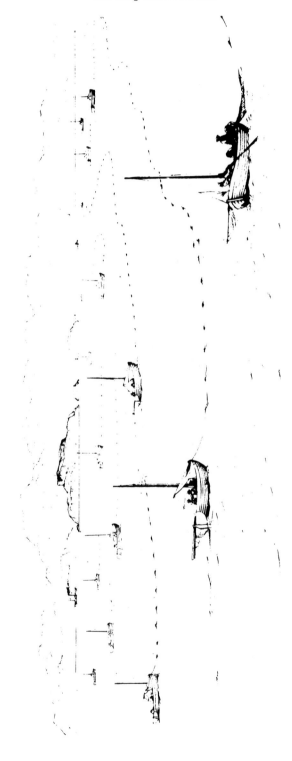

37. A fleet of Lochfyne drifters lying to their nets, 1852. In most of the boats a single man keeps watch at the stern, while his colleagues rest beneath a canopy spread over the bow. In the foreground, a recently arrived crew is shooting its net

swim high or low, and they raise or sink their nets accordingly by shortening or lengthening the ropes that are fastened to the buoys.'[60]

The 'privilege' of examining the nets of nearby crews was permitted, though not without anxiety that plundering would occur under the pretence of honest curiosity. That practice, which the fishermen called *preeing*, obviated needless drifting over unproductive ground. Although loathe to put themselves to the work of hauling and re-setting their nets, when fish were meshing in a neighbouring crew's nets, they would shift their own train to that vicinity. That shift was often the 'signal for the boats to haul in their nets and move somewhere else'.[61] A crew going preeing would detach its boat from the end of the train and tie on a large buoy. The practice was resorted to only on relatively clear nights, when they could be sure of recovering their nets.[62]

At dawn the nets would be hauled. One man aft pulled the back-rope, while another, standing on a broad beam extending across the middle of the hold, handled the netting. If the meshing was particularly heavy, he would be assisted by another man standing forward of him. A fourth man, in the fore end of the hold, coiled the back-rope and stacked the buoys in readiness for shooting again. The entire crew would be engaged in shaking herring from the nets and into the hold.[63]

The traditional type of drift-net lacked a sole-rope, and was strengthened along the bottom edge only by a *ra* (row) of heavy netting, four or five meshes deep, which helped the net withstand damage if it caught on the ground.[64] These nets the Ayrshire fishermen called 'sunk nets'. 'Surface nets', which were introduced to the Ayrshire coast in the earliest years of the twentieth century, were designed with a heavy sole-rope and hung like 'a dyke in the waater'.[65]

The surface nets were considered superior fishing instruments,[66] and were also less liable to be lost. The Ayrshire fishermen frequently set their nets along the busy shipping track between Arran/Ailsa Craig and the Ayrshire coast, and, as the trains of nets were hardly visible in darkness, steamers occasionally cut through a train, splitting back-rope and netting. The severed length of sunk nets would go adrift and might be lost. But although the back-rope of a surface net might be cut, the sole-rope, fathoms below, would escape damage and so hold the train together.[67]

About a quarter of the length of a 'barrel' of netting would be 'crooped' when setting it up to the light back-rope – 'The more croopin', the more herrin' ye wid get.'[68] On one end of the barrel, the back-rope was spliced into an eye, and on the other end the rope extended a couple of feet. When the individual barrels were to be linked, the rope extension on the end of one barrel would be 'tailed on' (attached by two hitches) to the eye on the opposite end of another barrel. Corks were strung on the back-rope at wide distances apart to keep the net hanging in fishing trim.[69]

The essential buoyancy of a fleet of drift-nets was provided by inflated bullock or pig bladders, which were preserved by Archangel tar, first heated to thin it, and then poured in through a small filler. An experienced man would manipulate the bladder vigorously to work the liquid into it, after which the neck of the bladder would be bound to an empty thread *pirn* (bobbin).[70] The finished buoy would be inflated through the ready-made hole, and then plugged. The Lochfyneside drift-net fishermen, if lacking a buoy, would instead tie on to the back-rope a string of perhaps 12 or 14 corks, which they called *peidirean àrcain*.[71]

REFERENCES AND NOTES

1. A. R. Bigwood, The Campbeltown Buss Fishery, 1750-1800 (unpublished M.Litt. thesis, University of Aberdeen, 1972), 120.

2. *N.S.A.*, Campbeltown, 457.

3. H. Martin, 3 May, 1974 and 10 June, 1976.

4. H. McFarlane, 11 June and 29 July, 1976.

5. H. Martin, 10 June, 1976.

6. H. McFarlane, 29 July, 1976.

7. J. 'Jake' McCrindle, 28 April, 1976.

8. H. Martin, 10 June, 1976.

9. J. 'Jake' McCrindle, 28 April, 1976.

10. H. Martin commented, 10 June, 1976: 'There wir East Coast men here, they came tae stop (stay) here, an' they wir at the lines, an' they dinna seem tae laik (like) tae hear ye sayin' "a trough". It wis aye a "baikie" they said.'

11. A. Martin, 14 July, 1976.

12. T. McCrindle, 2 May, 1976.

13. A. Martin, 14 July, 1976. These hauls were often 'chancers', usually at the Flet Rock in The Loaden (Kildalloig Bay) – H. Martin, 10 June, 1976.

14. H. Martin, 10 June, 1976.

15. Gael. *creachan* (H.D.S.G., Tarbert, Carradale, and Campbeltown). The dialect *creachal* was also current in Tarbert – G. C. Hay, noted 9 May, 1979.

16. H. McFarlane, 18 March, 1976.

17. A. Martin, 14 July, 1976.

18. Andrew Noble, Fraserburgh, and Mrs Doreen Shepherd, Forres, information supplied to author, 1977. H. McFarlane's unvaried pronunciation was 'hippin', but whether that represents the regular Tarbert form of the word cannot now be ascertained.

19. H. McFarlane, 11 June and 29 July, 1976. Tails could be got from distilleries in Campbeltown and Dalintober, or from farms in the district – H. Martin, noted 4 June, 1979.

20. A. Martin, 14 July, 1976.

21. From Gael. *tap*, 'thread a fishing hook'. H. McFarlane, 10 December, 1976 (H.D.S.G.), and in general recordings.

22. A. Martin, 14 July, 1976, and G. Newlands, noted 1 June, 1979.

23. H. McFarlane, 29 July, 1976.

24. H.D.S.G. (Tarbert, Carradale, and Campbeltown).

25. A. Martin, 14 July, 1976; 'mid-heid' – G. Newlands, noted 1 June, 1979.

26. H. Martin, 10 June, 1976, and A. Martin, 14 July, 1976.

27. A. Martin, 14 July, 1976.

28. Peggy Murray (Mrs J. T. McCrindle), 29 April, 1976.

29. J. McWhirter, 26 April, 1974, and H. Martin, 10 June, 1976 and (quotation) noted 4 June, 1979.

30. H. McFarlane, 30 October, 1974.

31. A. Martin, 14 July, 1976, and H. Martin, noted 4 June, 1979.

32. H. Martin, 10 June, 1976; H. McFarlane, 11 June, 1976; A. Martin, 14 July, 1976.

33. R.F.B., Campbeltown. Portrush builder's name, James Kelly. One Greencastle-based builder, McDonnell, referred to, in connection with the *Lizetta*, built in 1904.

34. *Ib.*

35. D. McLean, 3 June, 1974, and A. Martin, 14 July, 1976.

36. A. Martin, 14 July, 1976.

37. H. Martin, 4 June, 1974. Two types of drag used. 1. A conical canvas sea-anchor, with a rope to the point of the cone, by which the aperture there could be opened or closed to regulate the speed of the vessel. 2. A herring-box towed by two ropes, one on each handle. At reduced speed, the box would be towed square astern, but to increase the 'way' of the boat, one end would be slackened out, and the resistance of the box thus reduced.

38. H. McFarlane, 29 July, 1976.

39. As above.

40. A. Martin, 14 July, 1976.

41. J. McWhirter, 26 April, 1974.

42. H. McFarlane, 20 April, 1978 (H.D.S.G.).

43. *F.B.R.* (1898), x.

44. *R.R.C.* (1863), 17-18.

45. *Ib.*, 18.

46. W. Gillis, fishery officer at Broadford, Skye, 21 August, 1860, A.F. 37/19.

47. A. Sutherland, fishery officer at Ardrishaig, 12 February, 1840, A.F. 7/104, 220.

48. J. MacFie, fishery officer at Ardrishaig, 1 March 1854, A.F. 37/12.

49. *F.B.R.* (1896), 204.

50. *Ib.* (1898), x-xi.

51. *Ib.* (1900), 117.

52. *Ib.* (1899), 236.

53. *E.R.C.* (1864), 1180.

54. B. F. Primrose, *Paper* (1852), 1.

55. *Ib.*, and J. Stewart, commanding the *Princess Royal* cutter, 14 October, 1852, A.F. 37/5, in reply to a questionnaire on trawl- and drift-fishing compiled by B. F. Primrose, presumably as part of the research for his *Paper*.

56. J. Stewart, *op. cit.*, and J. Munro, fisherman, *E.R.C.* (1864), 1159.

57. *E.R.C.* (1864), 1159.

58. B. F. Primrose, *op cit.*, 2, and a Greenock buss-master, 1776, *Report on the Herring Fisheries*, 1798, 338.

59. *E.R.C.* (1864), 1139.

60. *Report* (1798), *op. cit.*, 338.

61. B. F. Primrose, *op. cit.*, 2.

62. *Report* (1798), *op. cit.*, 338.

63. H. McFarlane, 11 June, 1976.

64. As above, 26 April, 1976.

65. J. 'Jake' McCrindle, 28 April, 1976; T. Sloan, 29 April, 1976 (quoted); J. T. McCrindle, 29 April, 1976.

66. J. 'Jake' McCrindle, 28 April, 1976, and T. Sloan, 29 April, 1976.

67. J. 'Jake' McCrindle, 28 April, 1976.

68. H. McFarlane, 26 April, 1976.

69. As above, 11 June, 1976.

70. T. McCrindle, 2 May, 1976.

71. D. MacVicar, 16 September, 1978 (H.D.S.G.). *Peidirean* suggests *paidirean*, 'rosary', used more generally of a string of any beads − hence, presumably, a string of corks (*àrcain*).

10

All Points of the Compass

THE spread of trawling and ring-netting beyond Kintyre to the south and, later, north Minch, the west coast of Ireland, the East Coast of Scotland, and the Irish Sea, was accelerated by the introduction of the Lochfyne Skiff, a type of craft suited to fishing at a distance from home, having a forecastle, which provided the fishermen with a degree of comfort lacking in the open skiffs, and a spacious hold capable of carrying about 50 crans (or nine tons) of herring. The general adoption, in the 1930s, of a larger build of boat, fully decked and with much greater hold space, further increased the success of distant-water fisheries. But the first efforts were made with open boats.

Early Ventures North

It was claimed in 1856: 'The Highlanders cannot bear the trawl . . . Once it was introduced, they set upon it and tore it to pieces.'[1] That reference, factual or not, undoubtedly applies to northern fishermen. 'Scringing' − a simpler form of trawling, lacking long sweep-lines in its operation − had been practised by Oban fishermen since about 1800, and later became popular among Mull fishermen (p. 189). The earliest positive reference to trawling in the north Minch was heard by the Royal Commissioners of 1862, at Campbeltown. The informant had tried the method in the Sound of Scarp, off Skye.[2]

Kintyre trawlers seem to have first appeared in the North, to fish openly, in the late 1860s. They operated along the shores of Loch Hourn, a gloomy inlet of the Sound of Sleat,[3] and in Loch Eil, opposite Fort William.[4] That the Tarbert fishermen, schooled in the tricks of secret fishing, had been active in that area before then is a reasonable possibility. In 1861, for instance, the arrival at Fort William of a Tarbert smack, accompanied by a pair of skiffs, aroused the suspicions of local fishermen. The Tarbert crews, in their defence, reasoned that 'the name of the place they came from made the fishermen have suspicions of them', and accepted the fishery officer's warning of the 'danger they ran from the Native Fishermen at sea'.[5]

The fact that trawling had been legalised in 1867 did not deter local drift-net fishermen from violently demonstrating their opposition, and trawl-nets were frequently cut during disputes. The first serious disturbance was reported from Loch Hourn and nearby Loch Nevis in 1883, and H.M.S. *Jackal* was sent to restore order,[6] an ironic reversal of roles which likely did not escape the notice of the trawl-fishermen, whose legal rights that infamous ship had, in effect, arrived to protect. In 1884, the Lochfyne fishermen applied to the Fishery Board for protection from the native fishermen, and a cutter was stationed in that same area for the duration of the fishery, thus checking 'lawlessness'.[7] The trawl-fishermen were again at work, in Loch Eil, in the following year, and their presence was sufficient to provoke trouble. H.M. Cutter *Daisy* was ordered to the loch, and the trawl-fishermen were enabled to resume their work unmolested.[8]

The basic cause of the native resentment was the greater success of the trawl-fishermen, an aggravation almost as old as trawling itself. As one observer commented: 'The trawl fishermen were landing heavy takes of herrings while the native fishermen were getting scarcely any.'[9]

The Islay Fishery

In 1886, the interest of the Kintyre fishermen was directed to a fishing ground closer to home. In that year, in Lochindaal, Islay, a fishery unexpectedly began which would occupy the Kintyre fleets for six months annually, at both drift-netting and trawling, until its equally unexpected failure nine years later. Prior to 1886, all the herring taken in the loch had been for local needs and line bait exclusively,[10] and the native fishermen relied stubbornly on drift-nets.

Attempts had been made to introduce trawling to Lochindaal, but the few ambitious fishermen were violently opposed. One such man, Donald Anderson of Port Wemyss, paid £25 in 1873 for a trawl-net and gear at Campbeltown, and worked it unsuccessfully at Loch Hourn that year.[11] When necessity forced him reluctantly to use it in Lochindaal, he earned £100 for three catches of saithe.[12] But the local reaction was as he had feared it would be. A mob – 'ignorant, strong-headed and rebellious' – gathered at his door and 'threatened if they would get an opportunity they would murder myself and company, and should they get hold of the net they would tear it to pieces'. In November, 1874, Anderson appealed to the Board for a copy of the legalising Act of 1867, so that he could 'shut their Mouths'.[13] The outcome of Donald Anderson's bold attempt is not known, but the appearance of trawl-fishermen on a scale unimagined by the island opponents of the method was imminent.

By 1889, the Islay fishery, which then extended to Loch Gruinard on the north-west coast of the island, was attracting boats from Mull, Skye, Arran, Bute, Ayr, and the Firth of Forth, as well as from Kintyre.[14] In 1892, shoals were being located in the open sea between the Sound of Islay and Colonsay. The advantage then lay with the crews of the biggest class of boats. Port Askaig was the operational base of the fleet, and catches were taken there or discharged into buying-steamers in the Sound of Islay.[15]

The fishery reached peak productivity in 1893. By October, 309 boats, crewed by 1,082 men, had assembled, accompanied by nine carrying-steamers. The shoals shifted from the deep water between Islay and Colonsay into Loch Tarbert, which penetrates narrowly and irregularly the west coast of Jura, almost dividing the island. Although the fleet could work there in greater safety, the rocky entrance to the loch prevented the steamers from entering, until a channel through was discovered. In the first week of November more than 2,500 crans of herring were dumped in the loch because the fleet was prevented by stormy weather from reaching the steamers at anchor in the Sound of Islay.

P

The fleet's presence on that exposed Atlantic coast was hazardous. A gale on 17 and 18 November wrecked 16 boats and damaged nine others, at an estimated loss of £1,207 to the affected fishermen. The total destruction for the season amounted to 28 boats wrecked and an equal number damaged. By subscription in Campbeltown £480 was raised for the replacement or repair of local vessels.

The fishery produced 53,409 crans for the six-month season, but prices per cran dropped from a range of 16/- to 22/- in the early part of the season to 2/- to 8/- latterly. The importance of the Islay fishery to Kintyre crews by that year is shown in the yield of herring from the Kilbrannan Sound in 1893, 17,441 crans.

At the beginning of the season drift-nets only were worked, but trawls gradually came into use, with formidable results. It was not uncommon for one ring of a net to enclose a catch of herring which filled five, six, eight, and 10 skiffs, and 500 crans were secured once, realising £220.[16]

The Islay fishermen rejected trawling with as passionate a determination as the drift-net fishermen of Lochfyne had done before them, and their agitation resulted in the closure of Loch Gruinard to trawlers in 1892. Complaints against the trawlers continued, and in 1894 an official of the Fishery Board visited Islay to interview fishermen there.[17]

In that year, 50,681 crans were landed from the Islay and Jura fishing grounds,[18] but the total next season amounted to merely 12,000 crans.[19] A great fishery had ended.

The herring caught in these Atlantic waters were of an almost legendary size by the standards of the time. 'They had ribs on them lik' sheep,' one Campbeltown fisherman[20] was, with obvious exaggeration, assured. Robert McGown was told by his father that if the back-rope of the drift-nets was held taut in the hand, herring could be felt striking into the meshes.[21] That class of herring, which the fishermen called 'ocean herring', would be encountered again on the Irish west coast.

The low prices frequently received by the fishermen were attributed to their being 'at the steamers' mercy'.[22] One man who certainly profited by the Islay fishery was John Forgie of Tarbert, a block-maker to trade, who fitted out a dismasted sailing-ship, the *Banton Packet*, and had her towed to Port Askaig and moored. There he traded in 'everything fae the needle tae the anchor' – nets and other gear, oilskin clothing, seaboots, meat, and a variety of other foods. He even had a baker, Willie Hawthorne, making bread for the fishermen.[23]

Donegal

A less significant trawl-fishery – and, as such, but scantily documented – was opened up at Burtonport on the Donegal coast by Campbeltown crews in 1905. Conditions there were discovered to be favourable for trawling, and in November of that year several crews earned from £300 to £400, while the least successful 'more than covered expenses'.[24] The pioneering of the fishery was popularly credited to Hugh MacLean of the skiffs *Good Will* and *Good Hope*. He seems to have first operated there with drift-nets.[25]

The venture may not have been quite as exploratory as the Fishery Board report suggests. In 1883 six pairs of Campbeltown trawl-crews were engaged to fish on the Irish coast at £1 per cran and with a £20 bounty each crew.[26] Their specific destination was not reported, and may therefore have been one of the Irish Sea ports.

Much of the herring, 'swimming red' in the great Atlantic swell, was caught off the Island of Aran Mór. John McWhirter recalled his first trip to Burtonport, where catches were usually landed: 'There wis very little win', an' we kept aboot a mile off that island. Ye'd see the sea running away up the cliffs, a big bound, an' no wind, big bound an' flappin' o' the sail.' The fishermen waited until flood tide allowed them entry into Burtonport, and landed their fish. When the Irish heard that they had gone around Aran Mór with loaded boats they reproved them severely, and even the local priest arrived to add his voice.[27]

The native drift-net fishermen resented the activities of the Campbeltown trawl-fishermen. 'They thought we wir stealin' a' the herrin',' said John McWhirter. 'They dinna understan' a herrin' bein' rung at aal.' To appease local prejudice some skippers adopted the practice that had been successfully used to forestall trouble with Ayrshire fishermen on the Ballantrae Banks. They loaded the small local boats with trawled herring. Opposition afterwards quietened.[28]

The main mooring place was the Black Hole, a deep natural harbour west of Burtonport. Carrying-steamers and skiffs alike lay against the bare rock walls of the Hole. To amuse themselves the men fished for 'cuddins' (young saithe) with bent pins, or sported on the backs of young donkeys which ran wild on Aran Mór. They lived on the 'ocean herring' when they had them. So large and bony and strong-flavoured were these fish that the fishermen invariably halved them and cooked only the tail portion. On Saturday, a *junk* (large piece) of mutton would be bought at Burtonport for 2/6d, and that sufficed for several days.[29] The fishery continued into the late 1920s,[30] but was irregularly engaged in, and was as often unprofitable as profitable.

Ayrshire

Trawl-nets were first used by Argyll fishermen on the Ballantrae Banks in 1877,[31] despite the opposition of the Girvan, Maidens, and Dunure fishermen, whose ground-nets were frequently damaged by the operations of the trawlers.[32] Trawl-fishermen had worked in considerable numbers from Irvine and Saltcoats during the mid-nineteenth century. Twelve trawl-boats were operating from Irvine in 1859,[33] and in 1860 it was said of Saltcoats that, 'This appears to be the worst station on the Ayrshire coast for trawlers.'[34] The method had also been adopted on a casual basis by five crews from Ayr, and by some Dunure fishermen.[35] The repressive legislation of 1860 and 1861 seems, however, to have reduced the Ayrshire fleets to insignificance.

Crews of Girvan and Dunure fishermen were reported as having trawled in Whiting Bay, Arran, in the winter of 1895,[36] and eight years later an increase in the number of trawl-nets at Girvan was recorded. 'This,' the Fishery Board commented, 'is the more remarkable considering the Ayrshire fishermen's previous opposition to seine-trawling.'[37]

The North

The Minch fishermen's suspicious interest in the activities of visiting Kintyre fishermen was probably not without justification. Oral tradition suggests abundantly that small trawls were carried north along with drift-nets, but concealed in the stern sheets of the skiffs to be used secretly in secluded lochs and bays.[38] Tarbert fishermen, encountering the Blair family of Campbeltown in the North, would enquire knowingly: 'Hae ye the "ballycroobach" (a curtailed trawl)[39] wi' ye this year?'[40]

By 1910, trawling had already been engaged in, and discontinued, by small numbers of Skye and Loch Alsh fishermen. The method seems to have been introduced in the 1880s or 1890s,[41] possibly by fishermen who had seen it used in Loch Hourn and Loch Eil, or who had themselves used it as hired hands on the Carradale skiffs (p. 46). The reasons for its abandonment are more certain.

Widespread opposition may be accepted as the main reason. There is also evidence that the very practitioners of trawling were apprehensive of its effect on fish stocks. Gurnets had been trawled in Skye waters for several years, but had suddenly vanished from the coast. The failure of the fishery was generally attributed to the use of trawl-nets, and the older fishermen were convinced that the herring shoals would likewise be 'chased away'.[42]

In 1915, a complaint against ring-netting was registered by East Coast drift-net fishermen, the first officially lodged since the Loch Hourn fisheries in the early 1880s. They reported to the Oban fishery officer that Campbeltown boats were at work in Loch Bracadale, Skye, and were needlessly destroying large quantities of herring.[43] The available evidence indicates that two of the skippers involved were John McIntyre of the *Sweet Home* and Robert Robertson of the *Frigate Bird*.[44]

The Reid family of Lossiemouth, which had settled in Kyleakin, Skye, in 1901,[45] was also active in Loch Bracadale with a ring-net about that time. It was concealed in the hold of the *Isa Reid* because the fishermen mistakenly believed that they were violating the law.[46]

That same concern motivated James MacDonald of Skye to apply in 1915 to the Fishery Board for a certificate of proof that the use of the ring-net in the Minches was not illegal.[47] MacDonald contacted the Fishery Board again, in August, 1920, to 'hear what measure of protection your Board might be disposed to give in case of fishermen who disapprove carrying their opposition to an extreme point'.[48]

Other Skye fishermen were also apprehensive in 1920. One, Ewen Nicolson of Kyleakin, reported to the Board: 'Those who have no trawls object to us using trawls, and threatened to cut our nets, so will you please write us a letter to let those who object . . . see that it is quite legal to do so.'[49]

The unease of the ring-net fishermen may be explained by their minority status. On Skye and the adjacent mainland, in 1920, they numbered 70 men, with 15 boats, whereas 1,200 fishermen, manning 370 boats, worked drift-nets. Five hundred and forty-three fishermen put their signatures to a widely circulated petition to the Fishery Board, advocating the abolition of ring-netting north of Ardnamurchan.[50] The fishery officer at Kyle of Lochalsh commented: 'So far as I can see, it is not the whole-time fishermen who object to the seine-net so much as the crofter-fishermen, who are not wholly dependent on fishing for a livelihood.'[51]

The controversy abated for five years, during which period the method continued to spread. By 1925 a fleet of Mallaig ring-net boats was operating.[52] In that year, and the two successive years, the Board was petitioned repeatedly by drift-net fishermen on both sides of the Minch, and from the sources of these petitions can be deduced the expansion of the ring-net fishery to the Sound of Harris, Loch Eynort in South Uist, Gott Bay in Tiree, and the lochs of the Sutherland west coast.[53]

The most determined opposition was encountered on Scalpay, Harris. The fishermen there had the total support of the island-based herring-gutters and packers – most of them local women – who refused to handle ring-net herring.[54] Damage to the boats and gear of visiting Skye fishermen was threatened convincingly. As one Kyleakin fisherman commented: 'A lot of men can do damage who cannot pay for the same – the ringers left.'[55]

With the foundation in 1933 of *Comunn Iasgairean na Mara* (the Sea-Fishermen's League), the crofter-fishermen were provided with a medium for concerted protest. The League was concerned to restrict the operations within the Minches of both steam-trawlers and ring-netters, and active in its interest were John Lorne Campbell of Canna, the distinguished Gaelic folklorist and author, and the novelist, Compton Mackenzie.[56]

An investigation into the West Coast herring fisheries was conducted in 1933 by the Fishery Board. The main complaints of the crofter-fishermen were that ring-netting broke up and scattered the herring shoals, and that it was a wasteful method of fishing, great quantities of immature herring being surrounded and destroyed. These complaints, it may be noted, had been voiced by the nineteenth century drift-net fishermen of Lochfyneside, by then, significantly, a vanished population. Owners of ring-nets were encountered in various parts of the Outer Islands, including Lewis, Bernera-Harris, South Uist, Barra, and Eriskay, but virtually all had, however, ceased to use the method, under pressure of public opposition to it.[57]

The winter fishing off the Uists was carried on mainly in daylight, not least of all because the ring-net fishermen were unfamiliar with the coast and feared damage to boats and gear.[58] The herring were located swimming 'brown' or 'black' across the sands of the shallow bays.[59] East Coast drifters were employed to carry ring-net catches across the Minch to Mallaig. The arrangement suited the ring-net fishermen who were spared the often hazardous winter crossing, and whose fishing time was increased by their remaining constantly on the grounds. Payment to the drifter-crews ranged from a half to a third of the value of the catch.[60]

The standard craft used by Minch ring-net fishermen was the *Zulu*. Later Minch Zulus were specially built for ring-netting.[61] The fishermen of Avoch, on the Moray Firth, began visiting the Minches with ring-nets in the mid-1920s, in the *Skaffie* as well as the Zulu type.[62]

Until after the second World War, ring-net fishermen working Lewis waters dared not land their catches in Stornoway, but would instead cross the Minch to Ullapool or Mallaig.[63] In the post-war period, the fishermen of the Outer Islands adopted ring-netting. Scalpay-Harris, for two decades a centre of opposition to ring-netting, had become by 1970, ironically, one of the last strongholds of Minch ring-netting.

The East Coast

In January and February of 1930, 72 ring-netters from Kintyre, Lochfyneside, and the Ayrshire coast were at work in the Firth of Forth, at first from Burntisland to Bo'ness, and later off the coast of Fife.[64] It was the second season in which failure of the Clyde herring fishery had forced them through the Clyde-Forth Canal and into strange waters. Their nets, designed for West Coast waters, proved too deep for the estuary shoals of the Forth, and much damage was caused before the fishermen reduced their size.[65] The reaction of the local drift-net fishermen varied from community to community. The St Monans men are remembered as having been particularly resentful, and the ring-net fishermen were forced to leave the harbour, having been denied provisioning facilities. They instead based themselves at Pittenweem, where a liberal reception was enjoyed.[66]

Two deputations, protesting against the activities of the West Coast ring-net fishermen, were received in 1930 by the Fishery Board. It was argued that their ring-nets were on a larger scale than those used by local ring-net fishermen, operating principally from Newhaven, Fisherrow, and Cockenzie, and therefore more destructive to herring stocks.[67]

The winter herring fisheries of 1931 and 1932 were comparative failures, particularly in the upper reaches of the Forth, and no West Coast ring-netters were present. Complaints resumed, however, in 1933. In March, the East Fife branch of the Scottish Herring Producers' Association pressed for a delimitation of the fishing grounds, providing exclusive areas for ring-netting and for drift-netting.[68] In October, 902 members of the Association petitioned the Board with a recommendation that no ring-netting be permitted east of a demarcation line drawn from Rockhead Buoys, off Dysart, to Gullane Ness, Haddingtonshire. The main objections were of a familiar order. Ring-nets broke up the herring shoals and destroyed spawning beds. The 'impossibility' of the two methods being worked together on the same fishing grounds was cited. The constant movements of the ring-net boats when seeking herring damaged the fleets of drift-nets, and, to facilitate the actual operation of ring-netting, buoys were cut away from drift-nets.[69] (Ring-net fishermen themselves acknowledged the practice, but the reasonable among them would return buoys to their owners.)

The Board decided against action, pending the experience of another season, and impressed upon the drift-net fishermen the importance of reporting every case of damage to gear, and of providing evidence of the responsibility of ring-net fishermen for such damage.[70] Ninety-eight complaints were received during the period January to March, 1934. There were, allegedly, 69 drift-nets lost, 417 nets damaged, 74 buoys lost, and 26 buoys destroyed.[71]

The failure of the rival groups of fishermen to come to voluntary agreement decided the Board, in October, 1934, to draft a bye-law prohibiting ring-netting, for the season of 1935 only, within a defined area of the lower Firth. The Secretary of State for Scotland refused, however, to confirm the bye-law, and commissioned the sheriff of Inverness, Elgin, and Nairn, R. H. Maconochie, to enquire into the dispute. He reported an improvement of relations between the opposing sets of fishermen, and recommended that no official constraints be resorted to.[72]

The importance of the Firth of Forth fishery to West Coast ring-net fishermen diminished during the war years, and in the immediate post-war period the East Coast base of the ring-net fleets was Whitby on the Yorkshire coast.[73]

REFERENCES AND NOTES

1. B. F. Primrose, F.B.E., 587.
2. Captain Kerr, S.S. *Druid,* E.R.C. (1862), 17.
3. Patrick Campbell Ross, Plockton, *E.R.C.* (1877), 102: 'Seine trawling began in Loch Hourn seven or eight years ago . . . The trawl-fishermen are strangers, usually from Argyllshire.'
4. H. D. Ferguson, fisherman of Inveraray, *ib.,* 118: 'In 1869 trawlers gathered from all quarters in Loch Eil.'
5. J. McKiver, fishery officer, Fort William, 10 October, 1861, A.F. 26/9, 189-90.
6. *F.B.R.* (1883), xliii. H.M.S. *Jackal* joined H. M. Cutter *Daisy,* already there.
7. *Ib.* (1884), xlvi.
8. *Ib.* (1885), xliii.
9. *Ib.*
10. *Ib.* (1886), xlviii.
11. D. Anderson, letter to fishery officer at Campbeltown, 9 November, 1874, A.F. 22/6, 258.
12. A Levack, fishery officer at Campbeltown, rpt. on a visit of enquiry to Islay, 4 December, 1874, *ib.,* 263.
13. D. Anderson, *op. cit.*
14. *F.B.R.* (1889), 21.
15. *Ib.* (1892), xliii.
16. *Ib.* (1893), 191-2.
17. *Campbeltown Courier,* 18 August, 1894.
18. *F.B.R.* (1894), 187.
19. *Ib.* (1895), 201.
20. D. McSporran, 30 April, 1974.
21. 29 April, 1974.
22. D. McLean, 28 April, 1974.
23. H. McFarlane, 7 June, 1974 (main source); D. McIntosh, 27 April, 1974; R. McGown, 29 April, 1974.
24. *F.B.R.* (1905), 222.
25. D. McSporran, 30 April, 1974, and A. Stewart, 1 May, 1974.
26. *Campbeltown Courier,* 28 April, 1883.
27. J. McWhirter, 5 March, 1975.
28. As above.
29. As above, 26 April, 1975.
30. D. McSporran, 30 April, 1974. His last trip there was about the year 1929.
31. P. Wilson, fishery officer at Girvan, *F.B.R.* (1886), App. J.
32. W. Ware, fishery officer at Girvan, rpt. to Fish. Bd., 30 January, 1914, A.F. 62/355.
33. J. McKiver, fishery officer, Greenock, 28 November, 1859, A.F. 37/15.
34. J. McKiver, 13 November, 1860, Saltcoats, A.F. 37/22.
35. As above, 28 November, 1859, *op. cit.*
36. *F.B.R.* (1895), 207.
37. *Ib.* (1903), 249.
38. R. McGown, 29 April, 1974; D. McSporran, 30 April, 1974; D. McFarlane, 4 May, 1974; J. Weir, 27 June, 1975.
39. A ring-net reduced in size, either intentionally or of necessity, as when a badly torn part would be gathered quickly together at the back- or sole-rope, to allow the net to be worked for the remainder of the night. H. McFarlane, 24 February, 1978, and W. McCaffer, 24 March, 1978, both Tarbert; A. Black, 16 January, 1977, and D. Blair, 15 May, 1978, both Campbeltown (all H.D.S.G.). Complete meaning remains doubtful, but *crùbach* interpreted as 'lame'. D. MacVicar, 22 December, 1978 (H.D.S.G.), gave *baraille crùbach,* i.e. a lame barrel of drift-nets: 'A barrel o' nets that was cut down, maybe badly torn, and they whipped the end off it.' An alternative could be *ball (a) crùbach.*
40. D. Blair, 5 June, 1974.
41. K. MacRae, Portree, 19 February, 1974, recorded by W. Maclean.

42. D. Gillies, Kyle of Lochalsh, 7 April, 1975, recorded by W. Maclean.

43. Seine-Net Fishing for Herrings off the North-West Coast of Scotland, rpt. of Fish Bd., 13 October, 1928, A.F. 62 2163/4, 1.

44. R. Morans, 1 May, 1974; Duncan Blair, 9 May, 1974; N. Grant, Kyleakin, 18 February, 1974, recorded by W. Maclean. Mr Grant remarked: 'He (Robertson) carried drift-nets too up here, an' he would work in the lochs wi' drift-nets, but he had a ring-net, an' if he would get the chance he would sneak a shot.'

45. *Aberdeen Evening Express,* 14 May, 1931, obituary of John Reid.

46. N. Grant, 18 February, 1974.

47. Seine-Net Fishing . . . off the North-West Coast of Scotland, *op. cit.,* 1, and letter from J. MacDonald to Fish. Bd., 6 August, 1920, A.F. 62 2163/1.

48. Letter to Fish. Bd., *ib.*

49. Letter, 27 September, 1920, *ib.*

50. Seine-Net Fishing in the Waters North of Ardnamurchan, rpt. of Fish. Bd., 11 November, 1920, A.F. 62 2163/1. Signatures – 355, Loch Carron and Skye; 138, Harris; 70, Loch Broom and district.

51. J. Davidson, letter to Fish. Bd., 25 October, 1920, *ib.*

52. *F.B.R.* (1925), 27. Rptd. as having worked all summer in Harris waters.

53. Seine-Net Fishing . . . off the North-West Coast of Scotland, *op. cit.,* 3.

54. G. McGhee, fishery officer at Stornoway, 28 June, 1929, A.F. 62 2163/5.

55. A. D. Reid, letter to Secretary of State for Scotland, 2 September, 1929, *ib.*

56. F. Thompson, Compton Mackenzie and the Minch Fishermen, *West Highland Free Press,* 29 December, 1972.

57. West Coast Fishing Investigation, Fish Bd., 1933, A.F. 56/1030.

58. D. Gillies, 7 April, 1975, and J. McIntyre, 15 July, 1976.

59. As above.

60. K. MacRae, 19 February, 1974, and J. McIntyre, 8 May, 1974.

61. W. Maclean, information supplied to author.

62. G. Jack, Avoch, recorded by W. Maclean.

63. K. MacRae, 19 February, 1974.

64. *F.B.R.* (1930), 17.

65. *Ib.;* 'Dalintober', Firth of Clyde Boats on East Coast, *Campbeltown Courier,* 25 January, 1930; D. McIntosh, 24 April, 1974.

66. D. McIntosh, 24 April, 1974, and R. McGown, 29 April, 1974.

67. Narrative of Events Leading up to the Framing of Bye-law No. 42, 1, Fish. Bd. (1934), A.F. 56/12.

68. *Ib.,* 2.

69. Petition of the Scottish Herring Producers' Association Ltd., 4 October, 1933, A.F. 56/481.

70. Narrative, *op. cit.,* 2.

71. Drift-net Versus Ring-net Fishing for Herrings in the Firth of Forth, Fish. Bd. Rept., 26 December, 1934, A.F. 56/12.

72. Drift-net Versus Ring-net Fishing for Herrings in the Firth of Forth, Fish. Bd. rept., 22 June, 1936, A.F. 56/530.

73. Local tradition, Campbeltown.

11

The New Age

THE final development of trawl-fishing began in 1907 with the installation of a 7-9 h.p. *Kelvin* petrol-paraffin engine in the *Brothers*,[1] a Campbeltown skiff almost 35 feet (10.67 m.) in length and built in 1900 by John Thomson of Ardrossan.[2] The name of the skipper of that boat – the first fishing vessel on the West Coast of Scotland to receive motor-power – was Robert Robertson, and he can be described with justification as the pioneer of modern ring-netting.

By the end of 1908, 10 West Coast fishing vessels were motor-powered. Six Campbeltown boats – all about 35 feet (10.67 m.) long and 11 feet (3.35 m.) of beam, with draughts varying from 4 feet 7 inches to 6 feet 8 inches (1.40-2.03 m.) – were fitted with engines. Four of these were *Kelvins*,[3] manufactured in Glasgow by the Bergius Launch and Engineering Company, and installed at Hunter's Quay, near Dunoon. These engines cost £70, inclusive of a £20 installation charge, and were payable at £35 'down' and the remainder over the next year.[4] The other two models, an 8 h.p. *Thorneycroft* and an 11 h.p. *Fairbank*, cost £95 and £86 respectively. The latter produced, in calm weather, a speed of six knots, but the *Kelvin* model limited speed to between four and five knots. Only the *Fairbank* engine incorporated reverse-gear – 'It appears that with boats of such a small size the lack of reversing gear is scarcely felt' – but all were fitted with double-bladed propellers. These blades could be folded together, or 'feathered', to allow the boats free headway when under sail only.[5]

A *Kelvin* engine had also been installed on a skiff in the Inveraray district and one in the Ballantrae district, which included Ayr, Dunure, the Maidens, and Girvan. On the B.A.-registered vessel, two tanks were fitted, one for petrol – with which the engines were started – and one for paraffin, on which they ran.[6]

In the Rothesay district a skiff was specially built to receive a single-cylinder *Gardner* engine. The vessel, at 25 feet (7.62 m.) long, cost £75, while the engine cost £110. A half-gallon of paraffin, costing 3d, would drive her for an hour at 7½ h.p.[7]

The final engined boat, the *General de Wet* of Kyleakin, Skye, will be discussed in some detail at a later stage of this chapter.

Eighty-one boats were motor-powered on the West Coast of Scotland by the end of 1910.[8] Of that number, 40 were registered in Campbeltown and six in Rothesay. In that year 11 boats in the Ballantrae district were fitted with engines, and seven in Inveraray district.[9] By 1912 there were 72 motor-boats registered in Campbeltown alone, and the number of sailing-boats was reduced to 10. No pair of Campbeltown boats was entirely dependent on sail and oar, but there were 10 pairs which shared the benefits – and the costs – of a single engine.[10]

The arrangement was that the engined skiff towed the other to and from port, and from one part of the fishing ground to another, on windless days, and received at the end of each week a half of a share from that vessel's portion of the profit (p. 214).

38. Duncan 'Captain' Wilkinson's *Ellen* returning to Campbeltown, September 1908, after the installation of a 7-9 h.p. Kelvin engine at Hunter's Quay, Dunoon.
Per J. Short, Campbeltown

That transitional arrangement, which may have appealed initially to the fishermen whose imaginations were tradition-restricted, did not endure many years. The fishermen who paid out regularly for towage realised that they too could comfortably invest in engines on the basis of that half-share allocation, which was, in fact, exactly the portion of the profit claimed, for the engine, by those boat-owners who had paid it off.[11]

A table of comparative earnings compiled by the Fishery Board in 1912 revealed that the gross earnings of pairs of Campbeltown-registered motor-boats that year ranged from £840 to £2,100, whereas the pairs 'sharing' an engine earned substantially less, from £360 to £700. In the Inveraray district, where there were still pairs operating exclusively by sails and oars, the gross earnings of these pairs were even lower, at £60 to £300 per pair.[12] Even discounting the probability that the more advanced fishermen, who had quickly recognised the benefits of motor-power, were likely also to have been the more skilful and more industrious, such proof of the superior efficiency of motor-boats could not but have been generally acknowledged.

The report for the Ballantrae district in 1912 noted a 'tendency to go in for motors of greater power than hitherto'. Engines of 15-20 h.p. were being installed in newly built skiffs, and some boat-owners were removing the original 7-9 h.p. engines and replacing them with 15-20 h.p. engines.[13] The trend of increasing power would continue, with the West Coast fishermen retaining their faith in *Kelvin* engines, unlike their East Coast counterparts who generally preferred *Gardner* engines.

Detailed contemporaneous accounts of the effects of motor-power upon the ring-net fishermen and their craft are lacking, but informative reports on the Broadford (Skye)-registered drifter, the *General de Wet*, will suffice, though not strictly

39. Preparing to leave her moorings, the skiff *Fame* of Dalintober. The skipper-owner, Duncan Martin, stands by the tiller, while 'engineer' Donald 'Hairy' MacCallum looks up from the 'engine-room'. c. 1920

relevant. She was built in 1902 at Lossiemouth to the order of William Reid of Kyleakin, and was 56 feet 7 inches (17.25 m.) in length, 17 feet 8 inches (5.38 m.) of beam, and 7 feet 1 inch (2.16 m.) of draught. The engine fitted into her was a 34 h.p. two-cylinder *Gardner* model which was priced at £375, a high cost compared with the smaller *Kelvin* models. Thus powered, the *General de Wet* could attain a speed of almost eight knots, a decided advantage in reaching Kyle of Lochalsh in time for the first southern-bound train. After the motor had been installed, in 1908, Reid had a piece of the foremast cut off, and the lugsail shortened,[14] modifications which were applied to many of the ring-net skiffs when their dependence on sail was recognised to have diminished.

On ring-net skiffs the engines were mounted in the stern sheets and covered over by boards. The tiny compartment was illuminated by a hurricane lamp. Not all fishermen dispensed with ballast in relation to the additional weight of the engines, and some skiffs were consequently 'deadened' in the water.[15] The propeller shaft was led through the starboard quarter, and the hole strengthened by blocks of wood bolted to the hull. The stern-posts of many skiffs were insufficiently sturdy to support a shaft and propeller, and, in any case, the fishermen were happier, during the period when motor-power was something of an experiment, with the propeller as far clear of the net-hauling (port) side as possible.[16]

In 1913, engines incorporating reverse-gear – and consequently of heavier construction – were installed on ring-net vessels. Three boats built for Campbeltown owners were fitted with these improved engines, and one owner replaced his old engine with a new model.[17] Some fishermen were apprehensive of too great an encroachment upon hold space and declared, 'Ye'll hae nae room for herrin' if ye put reverse-gear in!'[18] The advantages were, however, indisputable. The original 'go ahead' engines, as the fishermen called them, complicated manoeuvres when approaching a quay or when bearing down, in darkness, on the 'winky' at the end of a net. The engine would be stopped, and exact judgement was necessary to ensure that the boat 'carried the way' sufficiently. Occasionally a second attempt would be called for.[19]

In 1913 paraffin had cost 7½d or 8d per gallon, an increase of almost 100 per cent on the price several years before.[20] A *Kelvin* engine of 26 h.p. was introduced on the market in 1919 at a cost of £351, compared with £125, the price which had been paid for the discarded engines of 15-20 h.p.[21]

The adoption of motor-power by the ring-net fleets of Kintyre had a swift and radical effect upon the character of the fishing industry. The break with sailing and rowing, and the associated skills, did not begin until the fishermen had affirmed their trust in motor-power by investing in the more powerful engines. Quite apart from considerations of tradition, the 7-9 h.p. engine was inadequate to drive a 15-ton skiff ahead in a strongly opposing wind or heavy sea. She would simply throw her head off course.[22] The earliest model of engine was therefore regarded merely as an auxiliary power, and of positive benefit only on windless days. That being so, the necessity of oars was removed, though the more cautious of fishermen retained a couple on board their skiffs until the middle of the second decade. Bowsprits had been used only in the making of long passages, and they too had become obsolete by that time.

Ringing in open waters, invariably an uncertain operation, became more practicable with the introduction of motor-power, and the result was the total abandonment, by 1914, of the once-preferred method of 'trawling' the net inshore to anchored skiffs. The fishermen, naturally, continued to work the shores, but there too employed the quicker and less laborious method of 'ringing'.

With the need to carry long ropes removed, sweep-line-boxes were discarded. Between 1912 and 1920, the foredecks of the skiffs were extended from the break of the deck to the forward beam (where an oarsman would formerly have sat). The bulkhead was not, however, shifted correspondingly, and the fishermen hung their oilskins under that cover to keep them dry.[23]

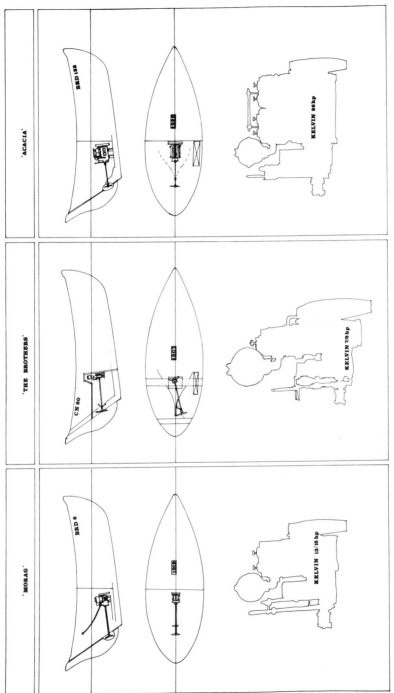

40. Boats and their engines – outlines. *L.-R*: The *Morag*, BRD. 6, 1919 (Kelvin 13-15 h.p.); The *Brothers*, CN. 97, 1907 (Kelvin 7-9 h.p.); The *Acacia*, BRD. 133, 1926 (Kelvin 26 h.p.)

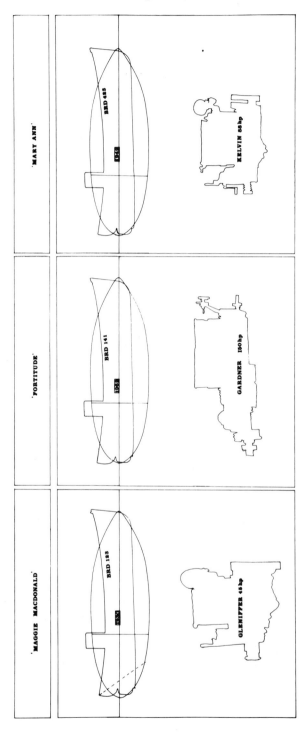

41. *L.-R*: The *Maggie MacDonald*, BRD. 123, 1930 (Gleniffer 45 h.p.); The *Fortitude*, BRD. 141, 1962 (Gardner 150 h.p.); *Mary Ann*, BRD. 423, 1949 (Kelvin 88 h.p.)

Finally, their dependence on the fleet of herring-carrying steamers was diminished, though the steamers would not disappear from the Clyde fishing grounds until about 1946.[24] As early as 1912 a Fishery Board report had referred to skiffs 'motoring' their catches to Fairlie railhead, in the absence of steamers.[25]

Post-War Developments

Those fishermen who returned, after the First World War, to the towns and villages of Kintyre and Ayrshire intent upon resumption of their occupation were confronted with a deplorable reality (multiplied, it is true, throughout the industries of that 'land fit for heroes') − not all would be able to obtain work. While the 'most successful and industrious' fishermen, of those who continued to fish during the war, had earned sufficient to 'keep themselves comfortable for a number of years',[26] numerous boats, left unattended during their owners' absence, had deteriorated.[27]

Soon after Armistice, appeals for assistance in resuming work were received by the Fishery Board from demobilised fishermen. The shortage of boats was such that many men had been unable to find berths on those which were still fit for sea, and an unprosperous season of 1919 − owing to poor quality of catches − deterred fishermen from investing in new vessels. The acute lack of fishing boats on the Clyde induced the Fishery Board to have specifications of a 'model Lochfyne motor skiff' produced, and tenders to build were invited from several firms experienced in the construction of that type of craft. The tenders received indicated that the cost of building would be about £1,100,[28] an estimate which may be compared with £140, the maximum cost of skiffs built in the Campbeltown and Inveraray districts just 15 years previously (p. 77).

But, as the Fishery Board's concern in 1919 was with the design of a vessel of the traditional type, as the basis of a renewed ring-netting fleet, in Campbeltown a fisherman of unrelenting ambition was preparing to set the development of the fishery on a radically altered course. He was Robert Robertson − then approaching his fortieth year − the man who, in 1907, had taken a gamble with motor-power. He was poised to gamble again, though undoubtedly he would not have admitted such a concept into the scheme of his thinking. He denied that luck, as an active factor in existence, had any validity. 'If it's got tae be,' he remarked towards the end of his life, 'that's it − it's got tae be.'[29] Luck or not − the speculation is unimportant − Robertson repeatedly impressed those who fished with him by his ability to know, as though by instinct, the right area in which to be, and the right time, for a herring fishery. 'It wis uncanny the things that happened, ye could put it that way,' said one who fished with him.[30] Robertson was a mere infant when, on 23 January, 1884, his 23-year-old father, Hugh Robertson, perished at sea with the Campbeltown schooner, the *Moy*, which foundered between Kintyre and the south end of Arran in a great storm.[31] The young widow later married Campbeltown fisherman John 'The Junk' McIntyre, and it was to his stepfather's occupation that the boy directed his aspirations.[32]

In 1921 Robertson − or 'The Hoodie', as he was better known − placed an order for a pair of boats with the St Monans, Fife, boatbuilders, J. Miller & Sons.[33]

Robertson's association with the yard, which lasted until his retirement from active fishing, began in 1921, when he travelled to St Monans to purchase two engines advertised by the yard. While there he showed William Miller the plan of a canoe-stern motor-boat, drawn by Glasgow naval architect W. G. McBride, and based on craft which Robertson had seen in Norway. He asked Miller if he would build to the plan, and fit the engines into the resulting vessels. Miller agreed.[34]

The first-launched of these craft was the *Falcon,* and she arrived in Campbeltown in April, 1922.[35] Her sister-ship, the *Frigate Bird,* arrived a month later. In design,

42. The innovatory canoe-stern ring-netter *Falcon* of Campbeltown after her launch in 1922. Per J. Miller & Sons, St Monans

these boats were startlingly unlike the Lochfyne Skiff, and as Robertson himself remembered two years later, with some satisfaction it may be supposed, general opinion was dismissive. 'Like all things new,' he wrote, 'these boats were looked upon as money lost.'[36]

At 50 feet 2 inches (15.29 m.) overall length, they were 10 feet (3.05 m.) longer than the largest of the skiffs, and their beam of 15 feet (4.57 m.) represented an increase of about 5 feet (1.52 m.) in that measurement. Decked over entirely, unlike the skiffs which were open but for a foredeck, they incorporated other features new to ring-net craft. A small wheel-house was fitted aft, providing shelter for the helmsman, who did not sit at a tiller – although an auxiliary tiller was provided – but stood at a steering-wheel, another innovation. The boats were each fitted with two *Gleniffer* paraffin engines of 18-22 h.p., which could be fuelled for six weeks, such was the space allowance for tanks. Acetylene (gas) lighting was installed throughout the vessels, which allowed for paraffin lamps to be dispensed with.[37] But the distinguishing feature which was most noted was each vessel's canoe-stern. The boats, which together represented an investment of £1,277 14/-, excluding costs of nets and other gear,[38] were, however, not entirely successful in their first year of operation, as Robertson himself later admitted.

Adjusting to the Innovations

The problems had not been so much consequences of defective design as difficulties in adjusting to the handling requirements of the boats. The replacement of tiller by wheel meant that 'steering . . . had all practically to be learned over again'.[39] As was explained to a later crew member, when herring were located the longer boats travelled deceptively far before being rounded on the shoal, and invariably the distance covered to get 'above' the shoal and surround it, on the return sweep, was too short. The ring of the net consequently missed its object.[40]

With experience, however, Robertson discovered that he could 'make a better ring with the wheel, as we can make any curve accurate all round'. The operation of ringing was less accurate with tiller-steering, the inclination of the helmsman being to 'pull or ease' the tiller unduly, with the result that 'the curve of the net is somewhat uneven'. The extent of the gamble which Robertson took with that design of boat is evidenced in his admission that in his experience a heavy build of vessel had been 'against the efficient working of the net'. In 1924, however, after almost two years of working the *Falcon* and the *Frigate Bird,* he could confidently state that the venture was tending to success. 'I believe before long,' he wrote, '(that) I will consider the weight and extra size all an advantage.'[41]

Whatever apprehensions he may have suffered during the tentative first year of operations, his faith in the boats was by 1924 absolute, and, characteristically, he was already planning to further over-reach the accepted limitations imposed upon ring-netting by the skiff design of craft. He intended that summer to instal motor winches and to 'try to seine herring with the top of the net below 18 fathoms, where the scientists tell us the daylight doesn't penetrate', which suggests a return to the traditional practice of sinking nets. He also intended fitting mizzen and fore-sails to the vessels to further reduce fuel expenses. He had, in 1923, saved £40 by buying paraffin in bulk and at cheaper cost at the railheads, which the increased fuel storage capacity of his boats enabled him to do. As he stated: 'The smaller boats cannot allow tank space.'

The 'extra comfort and seaworthiness' of the boats had, he wrote, persuaded his crews that they would 'leave the job if they had to go into the smaller boats again'. Their commitment to the new vessels had not been immediate, however, and that initial lack of confidence Robertson attributed to 'people, jealous I suppose, (telling) them that all earnings would go into expenses'.[42]

That objection would remain valid for most fishermen until the end of the 1930s, despite the obvious success of the *Falcon* and the *Frigate Bird,* and the vessels of that design which followed them. On the Clyde, where a share system prevailed, the portion of the total earnings of a pair of boats customarily allotted to each vessel, after deduction of food and fuel, etc., expenses, was one-and-a-half shares. The full share was 'for the net', and was intended to take care of repairs to and replacement of gear, and the half-share was allocated for the boat and engine (but if an engine was being paid up, then a full share would be allowed). Each experienced crew member received a share, and the young cook – if one was employed – received a half-share. The total number of shares between a pair of boats was therefore generally 12 or 13.[43] 'Fishermen,' the annual report of the Fishery Board in 1923 claimed, 'are not disposed to agree to any increase of this proportion, notwithstanding that such an increase may be warranted by the greater cost and superior efficiency of the boats of the latest type.'[44]

The resentment of the Campbeltown fishermen would not be concertedly vented until 1937. In April of that year, crews agreed unanimously that the allocation to each pair of canoe-stern boats should be 14 shares, from which each owner would receive two-and-a-half shares. From that allowance they would be expected to bear all expenses, except fuel, food provisions, and National Health Insurance. The owners rejected the proposals and presented a counter offer, which allowed them three shares each of the profits, their contention being that share-fishermen were not their employees, but 'co-partners in a venture in which the profits were shared'.

After several meetings of owners and crews, negotiations failed and the Transport and General Workers Union, of which almost all the fishermen had quickly become members, became actively involved. Towards the end of June a Board of Trade official arrived in Campbeltown to investigate the dispute, and a meeting of owners and two union officials was arranged to attempt settlement. The result of that discussion was that the owners effectively secured their case.

Their collective demand for three boat's shares was granted, and they conceded only that they would be responsible for paying for all engine breakages and overhauls. So ended the second strike in the history of ring-netting, and with it the fleeting involvement of fishermen in trades union membership. Ironically, the dispute saved the owners money, because the Mallaig fishery, in which the district fleet usually participated at that time of year, was a failure.[45]

To return, however, to the sequel of the account . . . The success of the *Falcon* and the *Frigate Bird* continued. In 1923, the second year of their operation, they had been 'by far the best fished pair in this district'.[46] But by 1924 the ring-net fishermen were still not 'quite satisfied as to the type and size of boat which will ultimately prove most profitable'. Robertson's example had, as yet, only influenced the owners of several of the larger skiffs to the extent of their having their vessels decked over. The cost of building remained high, but the Fishery Board warned that the renewal of the Clyde fishing fleets could not be postponed for many more years 'without reducing seriously the size of the fleet'.[47]

A winch which Robertson had had installed on board the *Falcon* in 1924 had not been successful. 'In fact,' he admitted in the following year, 'there is not yet any machine on the market capable of hauling this class of gear properly.' The ideal design of winch would enable a net to be hauled 'when the boat is in the trough of the sea, and then hold when the boat is thrown back, as the net would be hauled to pieces if the machine went ahead constantly'. He speculated that if a suitable winch were manufactured, one crew member could be dispensed with and production cheapened.[48] In that expectation he would be proved mistaken. Instead, mechanical hauling encouraged fishermen to increase the depth of ring-nets, and to weight and sink them, and the crews of the canoe-stern vessels were, in fact, supplemented.

Robertson's initiative in venturing to the south Minch when fishing was slack locally in 1924 was, he claimed, 'directly responsible for bringing in £4,500 to Campbeltown for one month's fishing with seven pairs . . . (which) would never have been got unless we had taken the lead'. His success was, he suggested, 'responsible for inducing the best men in the smaller boats to do work which the boats were really not capable of, and which might have had disastrous consequences'. The *Falcon* and the *Frigate Bird* had landed four catches at Castlebay, Barra, when the other boats could not venture the trip, although some skippers were willing to risk the Minch crossing. 'We were just afraid they might lose their lives,' he wrote. His boats enabled him to 'wait longer for fishing' and to 'go to the best markets with reasonable weather'. The principal disadvantage of which he was conscious was that the ring-net could not be worked in rough weather. His initial anxiety on the size of the boats had been dispelled, and, indeed, replaced by a belief that 60-foot (18.29 m.) vessels would have been 'even better'.[49]

Robertson's next annual report, for 1925, revealed even greater optimism, verging on ebullience: 'When these boats were introduced, I ventured to prophecy that in ten years all the herring boats would be 50 feet in length. I think that my prophecy will be realised before that time.'[50] His statement is remarkable considering that no new vessels had been built for Campbeltown, excepting his own pair. William Duthie, the fishery officer there, was quite as confident: 'Any new boats which are likely to be added to the fleet will very probably be near the size of the two large motor boats'. That no such development had begun, Duthie attributed to the conservatism of the local fishermen, who did not care for 'going out of the Firth of Clyde to get boats built but prefer to try to bring their present boats into line with the larger boats'.[51] Robertson, too, was critical of the trend to improve rather than replace boats. 'In the meantime,' he wrote, 'they are adding to their present boats and to my mind foolishly spending money on larger motor engines. They are also in some cases increasing the carrying capacity of their present ones, which may have a tendency of making them difficult of navigation when they have any quantity of fish on board.'[52] He returned to the early problem of the manoeuvrability of the *Falcon* and the *Frigate Bird*, to affirm that he had 'mastered this difficulty'. He wrote: 'I can now do much more than can be done with (small) boats . . . when fishing. I may say this took a little time as the difficulty of turning inside a small area after being accustomed to the smaller boats had got to be got over by experience.'[53]

The continuing conservatism of the Kintyre fishermen was embodied in three new boats, the *Thalassa*, the *Paragon*, and the *Maid of Honour*, which arrived in

Carradale in 1926. All were built by Noble of Fraserburgh, and though completely decked and provided with wheel-houses, each incorporated the traditional raked skiff-stern. More significantly, they were just over 43 feet (13.11 m.) long. [54]

The next new Campbeltown-registered boat was built by J. Miller & Sons for Robertson and a new partner, John Short. Named the *Crimson Arrow*, she was constructed on lines similar to the innovatory boats of 1922, but was reduced in length to 41 feet 9 inches (12.73 m.), with a beam of 12 feet 11 inches (3.94 m.), and a draught of 4 feet 11 inches (1.50 m.). [55]

The sole reason for the scaling-down – there were skiffs of traditional design almost her equal in length – was the hard work involved in basketing herring into the *Falcon* and the *Frigate Bird*. The larger build of boat had, of necessity, a higher freeboard, which meant a heavier lift for the men operating the baskets. The return

43. The *Nil Desperandum* leaving St Monans harbour, 1928. The 'Dandy' winch is visible. Per J. Miller & Sons

to a bigger build of boat would be postponed until 1928, when suitable winches became available, and experiments with power-discharging of herring could begin. [56]

In the two succeeding years, three boats were built for Campbeltown ring-net fishermen, the *Faustina* in 1927, at the yard of Walter Reekie, St Monans, for the McGown family, and the *Bengullion* in 1928, also at St Monans, for the Blair family. The former was 42 feet 5 inches (12.93 m.) in length, and the latter 45 feet 1 inch (13.75 m.). [57] Both were of canoe-stern design, completely decked, and with a wheel-house.

In August 1928, the *Nil Desperandum* – 46 feet (14.02 m.) in length and 14 feet 6 inches (4.42 m.) of beam – was launched from Miller's yard for Robertson and John Short. Her design incorporated an experimental feature. The wheel-house was positioned forward, at the break of the deck, thereby freeing the 'useful' part of the deck from obstructions.[58] That modification resulted from the recognition that mechanical hauling of ring-nets was no longer the testing problem it had been. The implications of the advance were discussed by Robertson and Tom Miller, and they agreed that to operate the ring-net by mechanical power, with maximum efficiency, a clear after deck would be essential. Tom Miller therefore designed the *Nil Desperandum* in accordance with that concept.[59]

Mounted aft of the hold hatch, and on the port side of the vessel, was a winch, re-designed for ring-netting by William Miller, and marketed as the *Dandy*. The body

44. The *Acacia* of Kyleakin, forward wheel-house visible, with the *Star of the Sea*, a 46-foot Zulu ring-netter of conventional design, alongside. c. 1930.
Per N. Grant, Kyleakin

of the machine could be swivelled in any direction and operated at either of two speeds, one suited to hauling the net and the other to discharging herring. A 'protected' clutch, invented by the builders, enabled the winch to be set to automatically stop hauling when the strain became excessive, and to resume hauling when the strain slackened, requirements specified by Robertson four years previously (p. 215). The engine was a 30 h.p. *Kelvin-Ricardo*.[60]

Despite the success of the *Nil Desperandum*, belief in the value of a forward wheel-house proved mistaken, and a year later it was re-positioned aft.[61] Two factors had operated against the design from the start. In the forward wheel-house the

45A-B. Canoe-stern craft, the loveliest form in the evolution of true ring-netters. The forerunners of the Ayrshire fleet.

A. The *Mary Sturgeon,* owned by the Gemmell brothers of Dunure.

B. The *Golden West,* owned by the McCrindle brothers of the Maidens. Both vessels were launched in 1926 at the Cockenzie yard of Weatherhead and Blackie.
Per W. Weatherhead, Dunbar

skipper was unable to simultaneously and conveniently steer the boat and observe the shooting of the net; and, as the fore end of a boat going through a heavy swell lifts and falls to a greater degree than the after end, the strain on the legs of the steersman was severe. A later factor decided the issue. The derrick necessary for discharging herring from the net by *brailer* (pp. 231-4), rather than by basket, could be supported more securely by an after wheel-house.[62]

The *Nil Desperandum's* forward wheel-house had not been a novel idea in ring-netting. Alexander Reid of Kyleakin had, in 1926, designed a 40-foot (12.19 m.) Zulu ring-netter, the *Acacia*, incorporating just such a feature. The *Acacia's* design had a curious origin. Her shape was based on the lines of a herring gull, a bird judged by Reid to be well adapted for riding the seas, having a broad forward 'beam'.[63]

The first canoe-sterned ring-net boat owned in Ayrshire was the *Unitas*. Built in 1912, with a raked transom-stern, she was purchased from her Saltcoats owner by John, William, and James Munro of Dunure, who had a canoe-stern added to her, but retaining the transom in position. Her length was thus increased from 39 feet 11 inches (12.17 m.) to 41 feet 6 inches (12.65 m.). She was registered B.A.21 on 18 May, 1925.[64] By the end of 1928 eight canoe-sterned ring-netters had been built for Ayrshire fishermen.[65] The length of these boats ranged from 44 feet (13.41 m.) to 50 feet (15.24 m.), and all were built at the Cockenzie yard of Weatherhead and Blackie, excepting one, which was a Miller boat.[66]

In January, 1930, the *Falcon* and the *Frigate Bird* became the property of Dan Conley, a former partner in the company, and Robertson travelled to St Monans to order another pair of vessels.[67] These, the *Kestrel* and the *Kittiwake*, were launched in June of that year and registered under the joint ownership of Robert Robertson, John Short, and John Wareham. The first-launched vessel, the *Kestrel*, was fitted with the first *Gleniffer* diesel engine, of 60-80 h.p., and the *Kittiwake* was identically powered. Both boats were 52 feet 2 inches (15.90 m.) in length, exactly 2 feet (0.61 m.) longer than the *Falcon* and the *Frigate Bird*, and were 15 feet 7 inches (4.75 m.) of beam and 6 feet 6 inches (1.98 m.) of draught.[68] In 1932/33 three boats were built for Robertson and company, the *Nulli Secundus*, the *King Fisher*, and the *King Bird*. The *Nulli Secundus* was 48 feet (14.63 m.) in length, and the other two fractionally exceeded that measurement.[69]

Robertson, his eyesight failing, would retire from active fishing several years later, his technological vision at last recognised by the majority of fishermen. In his lifetime he had, in the study of fishing methods, travelled extensively in the world, to Norway, to the U.S.A., and to Australia, where he settled temporarily. He died, in his 59th year, on 20 January, 1940, three days before the fifty-sixth anniversary of his father's drowning, an event which, ironically, had given him to the fishing industry. He was pre-eminent among ring-net fishermen and, if tradition proves just, will be so remembered.

Thirteen new boats, built on the East Coast of Scotland, of canoe-stern design, were launched in 1933 for Campbeltown and Carradale owners. All were about 48 feet (14.63 m.) long, and were fitted with diesel engines, electric lighting, and winches. Some of these boats could attain a speed of nine knots per hour, and in the

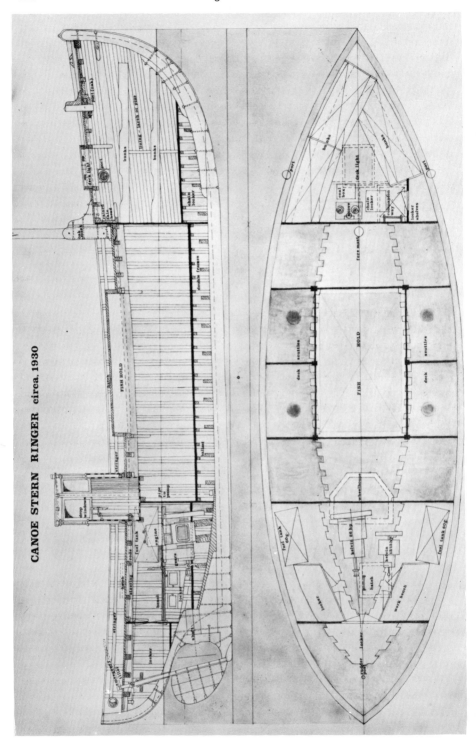

CANOE STERN RINGER circa, 1930

course of a night could encircle Arran in search of herring. The skiffs, which retained paraffin-driven engines of lesser horse power, were too slow to follow, or else their skippers were deterred by the prohibitiveness of the cost. The Kintyre herring market could be relieved by the larger boats' conveying their catches to the railheads, such as at Ayr. On at least one occasion during that year a buying-steamer refused a catch at sea and proceeded to Fairlie; the fishermen immediately set off for Ayr and consigned their catch to Glasgow, where it was sold before the arrival of the carrier's load, which had to be disposed of at a loss, the market already having been supplied.[70]

That expansion was parallelled on the Ayrshire coast. In 1933, 14 canoe-sterned vessels were launched, nine for Dunure owners, four for Girvan owners, and one for Ballantrae owners. The Cockenzie yard of Weatherhead and Blackie continued to rank, among Ayrshire fishermen, as the most popular builder, with eight of these vessels to its credit. Four Dunure boats, the *Morning Star II*, the *Marigold*, the *Nancy*, and the *Quest* were, however, 'home-built' at the Govan yard of Maclean.[71]

Winches

The first motor-driven winch fitted on a West Coast fishing vessel was evidently designed and assembled by William Reid of Kyleakin in 1921. He was then engineer on the 20-year-old Zulu, the *Star of the Sea*, and his intention was to dispense with manual discharging of herring at the quayside. To the flywheel shaft of the engine – a 13-15 h.p. *Kelvin* – he fitted a 'V'-pulley, connected by a belt to a second 'V'-pulley, formed by bolting an inverted plummer-block to the underside of the deck above. From that latter pulley, a shaft was led along the port side of the fish-hold hatch to a third pulley at the fore end of the hatch. A half-turn on that pulley with a bight of the discharging tackle completed the mechanics of the device.[72]

The year in which the ideal winch became available is uncertain. The evidence, incomplete as it certainly is, suggests 1928. The first reference to 'capstans' in Fishery Board reports appeared in 1926, but it was meagrely detailed, revealing nothing of the design of the model or models.[73]

The first attempt to design a winch specifically for ring-netting seems to have been undertaken in 1925 by Thomas Reid & Sons of Paisley, at the instigation of Hugh Anderson, skipper-owner of the *Hawk* of Dunure. The resultant winch, which was belt-driven by bevel-gearing from the main shaft of the *Hawk's* engine, operated successfully. Complaints were, however, made to Anderson of the noise caused by the winch on the fishing grounds, and the model was re-designed by Reids to operate by worm-gearing, which satisfactorily reduced noise. Both of these models were single-barrelled,[74] and were thus capable of hauling only a single sole-rope, the after one.

The next model to appear incorporated two barrels, mounted one above the other on a perpendicular stock, enabling both ends of the sole-rope to be simultaneously hauled. But the sole-rope itself could not be hauled around the barrels of the winch. The rigging arrangement by which drift-nets were pulled in by steam capstan was

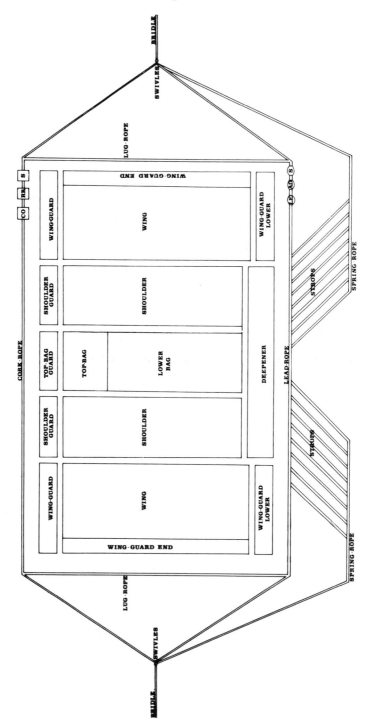

47. Ring-net rigged for mechanical hauling

adopted, but where and by whom the first ring-net was so rigged cannot be established with certainty. That Ayrshire fishermen were responsible is the most considerable possibility.

In rigging a ring-net for mechanical hauling, to the sole-rope was attached, by 'stops' or 'stoppers', a length of rope termed the 'spring'. These stoppers, which in 1929 numbered 18, were 2 fathoms (3.66 m.) in length, excepting the centre two which were half-a-fathom (0.91 m.), and were secured equidistantly along the entire length of the sole.[75] In practice, as soon as each end of the net had been raised to the boat by winching up the bridle-ropes, the fore and after ends of the spring-rope were passed around the respective barrels, and hauling began. As the first stopper, on either side, was brought up, the sole-rope between it and the end of the net lost its tension, and a man at the rail heaved that slack rope in and coiled it at his feet. When the first stopper was raised to the barrel, the turns of the after spring-rope around the upper barrel were 'thrown off' until only a half-turn remained bearing the strain; then the spring was heaved on again, and the stopper-end guided around the barrel and inboard; two or three turns were again taken, and the hauling of the spring continued until, stopper by stopper, the entire sole was raised inboard.

That model of winch did not permit of the stoppers on the fore end of the net being spliced to the spring, like those on the after end. The barrels being mounted one above the other, the stoppers on the fore spring-rope, which were winched on the lower barrel, could not have been passed around that barrel without completely throwing off the after spring-rope on the upper barrel, which would have resulted in much of the sole-rope and netting, already gained by hauling, being carried overboard by the weight of gear yet to be hauled. The expedient adopted was that the stoppers, though spliced to the sole-rope, were simply tied to the spring-rope by a rolling hitch on a loop, which, as the stopper emerged from the water, could be slipped quickly by the man engaged in raising the slack of the sole, and so pass freely around the barrel.[76] These were re-tied when *redding* the fore end of the net,[77] or when shooting it again.[78]

Both the bridle-ropes and the spring-ropes were led to the winch barrels through a moveable metal 'fair-lead', incorporating two rollers angled from a stock, which fitted into a metal-plated hole in the rail opposite the winch. That fair-lead would be removed after the sole of the net was entirely raised, and before the manual hauling of the shoulders and bag began.

Sunk Nets

Fishermen, in common with workers in other industries of a traditional character, are generally cautious of an innovation until its value and reliability have been indisputably established. In the case of the winch, however, all caution was abandoned, in 1928 and 1929, in the face of the gruelling labour in which they were then engaged.

The fishery, at the end of 1928, was going on in Loch Striven, a deep and narrow loch penetrating eight miles into mountainous Argyll. A fleet of about 300 ring-net and drift-net boats, based at Rothesay, assembled in October, but the herring, though known to be abundant, sounded to depths of 60 and 70 fathoms and held the bottom. The fishermen were forced to modify the rigging of their nets and sink them to the shoals. 'For the working of the net in these conditions,' the Fishery Board reported, 'winches were necessary, and nearly all the vessels which were equipped with winches or succeeded in obtaining them shared in a remarkably heavy fishing.'[79]

The majority of the Kintyre boats initially lacked winches and experienced a poor season. An incident, remembered by Duncan McSporran, illustrates the mood of exultancy in which those fishermen whose boats were fitted with winches could indulge. An Ayrshire crew called to a Campbeltown crew, one night in Loch Striven: 'Get oot the road an' let men catch herrin' that *can* catch them!'[80]

The greater success of these Ayrshire boats was apparent to the majority of fishermen still hauling nets by hand, and the demand for winches was instantaneous and quite general. By 1929 a perfected model had appeared on the market. It was a compact double-barrelled machine, and its design has survived, basically unaltered, until the present day. The barrels were mounted separately and side by side, and thus the fore stoppers could also be spliced to both the sole-rope and the spring-rope and hauled around the barrel without complication. Slip-knots were afterwards dispensed with by most fishermen. The model manufactured by McBain of Berwick sold at the remarkably low price of about £28, and quickly dominated the market.[81]

The buoyancy of ring-nets rigged for sinking could, according to a report of 1929, be reduced by chopping off 1½ stones of corks along the back-rope, and by adding to the sole-rope 1½ to 3 hundredweights of lead rings or lengths of chain. The three buoys on the bag of the net were strung to float at from 6 to 11 fathoms (10.97 or 20.12 m.) above the sunk net. Once the net had been shot in a semi-circle and had settled on the bottom, each boat towed its end around until they met, and then the regular hauling procedure was adopted. The towing of the net lasted 10 or 15 minutes,[82] which, ironically, alarmed many of the Tarbert district fishermen, who claimed that it was 'practically trawling'.[83] Some crews deepened their nets – ordinarily 50 or 52 scores of meshes – by an additional 20 or 30 scores,[84] a significant precedent. The best-fished pair of Tarbert boats in that fishery operated an 80-score net.[85]

Sweep-lines of greatly increased length were, of course, necessary. Both the sinking of ring-nets and the use of long ropes were old devices, not three decades abandoned, but there was an obvious difference between the fishing method of the nineteenth century and that of the twentieth. Whereas the original nets had been hauled across the sea bed on the ends of the sweep-lines, in the late 1920s the haul was an upward one, and that was made possible by winch power. Half-a-coil of rope – 60 fathoms (109.73 m.) – was generally sufficient, but the use of full coils was not extraordinary.[86] The fishermen tried to accurately judge the depth of water before shooting, believing that excessive rope caused mudding of the net.[87]

The Ayrshire fishermen not only extended the strings attaching the buoys to the back-rope, but also attached an additional buoy, on an even longer string, to the middle. Termed the *flotter* (floater), that buoy was a means of raising the back-rope

48. On the *Johan* of Loch Carron, Wester Ross, backs are bent to handle a heavy meshing of herring in the bag. Loch Duich, c. 1950. Per A. Finlayson, Auchtertyre, Ross-shire

49. Hauling a ring-net on the *Summer Rose* of Dunure, 1951. *L.-R:* John 'Iain' Gemmell, taking in the cork-rope, John Main, skipper-owner John Gemmell, Tommy Lawrence, Jimmy Weir, and William 'Elkie' Munro. Photograph courtesy of the *Glasgow Herald* and *Evening Times.* Per J. Gemmell, Ayr

of the bag as soon as the net had begun to lift. Great quantities of herring had been known to spill over the back before the net had risen completely to the surface. [88]

Weight was not necessary along the centre of the sole, but was distributed from the shoulders out to the ends. Bunches of lead rings, eight or nine strung together at intervals along the sole-rope, or, alternatively, lengths of chain, were originally used as weight, but the rings in particular tended to cause damage to the netting if they worked in through the 'keys' of the setting up, and certain fishermen used, instead, lead piping which they cut into lengths of about 1 foot (0.31 m.) and tied to the sole-rope with a light line. [89] Lead rings, if used, could be run on to the stoppers and tied

50. Drying herring in the bag, 1951. The boats *Summer Rose* and *Stormdrift* of Dunure are 'squared'. On the bow, John 'Iain' Gemmell is fending off effortlessly, the boat-hook resting on his shoulder. The other fishermen, *L.-R:* are: William Munro, Tommy Lawrence, John Grieve, and John Main. Photograph courtesy of the *Glasgow Herald* and *Evening Times*. Per J. Gemmell, Ayr

next to the sole-rope, similarly obviating the risk of damage, but additionally advantageous in that the increased weight was borne by the winch and not by the man hauling the slack of the sole-rope. [90] A 56-lb. weight could be attached by a rolling hitch to the end of each bridle-rope, as the sweep-line was being tied or clipped to the eye. When hauling was in progress, and the bridle-rope was approaching the fair-lead, the weight was quickly swung inboard; as the winch-man paused, the knot was slipped, and hauling continued. [91]

Expert fishermen were careful not to reduce excessively the buoyancy of their nets. The consequence of chopping off too many corks was that when the hauling of the net began, the back-rope lacked sufficient buoyancy to rise independently, and extra men would be required to handle it. 'Ideally,' said Matthew Sloan of the Maidens, 'when ye started tae heave in the soles, the corks had sufficient floatin' power, an' buoys an' everythin' attached wid come up of their own accord, so that what ye wir catchin' in yer net wisna escapin' oot over the corks.'[92]

Wide Wings

The rigorous working conditions of that sunk net fishery in 1928/29 were productive also of an improvement in the design of ring-nets which proved of permanent value. The wings of the net, which at 34 or 36 rows to the yard were uniform with the mesh size in the shoulders and lower bag, were removed and replaced with netting of a much wider mesh.

One of the first wide-winged nets was probably that of neighbouring Maidens skippers 'Uncle' Tommy Sloan and 'Jeely' Johnny McCrindle, of the *Twin Sisters* and the *Silver Spray*. That net was of peculiar design, being technically without shoulders. On to each end of a 100-yard (91.44 m.) bag of narrow netting was laced a 100-yard section of netting 24 rows to the yard. It has been claimed that that net caught more herring than any other net of its class.[93]

The experience of the few Campbeltown fishermen who experimented with the shoulderless net was less inspiring, and suspected loss of quantities of marketable herring out through the wings before the bag was reached induced many of them to re-assemble their nets, incorporating shoulders to the reduction of the extent of wide netting.[94] Similar doubts of loss of catch affected these Maidens skippers, and that net's replacement was of more regular design. The narrow netting in the middle was increased to 140 yards (128.02 m.), in effect re-creating the shoulders, and the wings were decreased accordingly.[95]

Technicalities aside, the benefits of wide wings were generally recognised. Meshings in the wings invariably increased the fishermen's labour and delayed the raising of the bag. The introduction of wide wings enabled the fishermen to gain the shoulders, and then the bag, much more quickly, though many, perhaps initially cautious, adopted wings of 29 rows, which proved, when working among good-sized herring, to be as unmanageable as 34- and 36-row wings had been among mixed herring.[96] The development would later be towards increasingly wider wings, with the lessening of the fishermen's suspicion that wide netting allowed too great a part of the catch to escape. The most valid argument against that suspicion was that wide netting, having less material constitution than standard netting, offered less resistance when hauled, and thus, as Matthew Sloan put it, 'was easily moved through the water an' scooped up a lot o' herrin' that a narrower net wis no' capable o' catchin''.[97]

That the idea of wide wings was not fully realised until the late 1920s is the more surprising considering the periodic net modifications which lay in that direction. In 1900, for instance, the Fishery Board had noted in Campbeltown an 'improvement . . . by some of the most enterprising fishermen (who are) increasing the size of the mesh so as to allow the small immature fish to escape when encircled'.[98] About five years later the Cook family of Campbeltown assembled drift-nets, which had lain unused in their net-loft after the failure of the Islay fishery, and created wings with them. These wings were 28 rows to the yard, but the motive was one of economic expedience rather than of technical innovation. [99]

The Feeling-Wire

The great depths in which the herring shoals in Loch Striven and the Kyles of Bute were lying increased the uncertainty of their being located by natural means. Herring which had been 'playing' on the surface were invariably disturbed by the sound of activated engines and quickly sounded, and the sole remaining evidence of their former presence might merely be a cluster of gulls. [100] In the fourth year of the sunk net fishery an important, if obvious, fish detection instrument was adopted on a general scale.

The use of the *feeling-wire* was first acknowledged by the Fishery Board in its report for 1931, but the belief in Kintyre is that the device had been secretly and successfully employed by a pair of Maidens crews in the few preceding years. [101] That belief is, arguably, supported by the Fishery Board's claim to have introduced the instrument to select Kintyre and Ayrshire crews in 1925, though after a trial their unanimous judgement, according to the Board, was that 'the idea was impracticable'. [102]

The presence on the Clyde fishing grounds of Swedish hake-seining boats in the latter half of the decade had, in any case, afforded the ring-net fishermen familiarity with the feeling-wire. Robert McGown, of the *Faustina* of Campbeltown, was introduced to it by several Swedes who accompanied him to sea on a summer evening in 1928 to observe the working of the ring-net. They took with them a feeling-wire, which was standard equipment on their vessels. 'Our means of fishin'' then was,' said Mr McGown, 'if a gannet struck, ye went an' circled round that spot. So, they were puttin' out this wire, trailin' it astern o' the boat, an', och, this wis too slow for us, because ye'll never get over any ground. An' then somebody'd be sayin', "There's a gannet struck! Let's go an' get round that gannet!" They'd tae haul in the wire. Och, we said, that's only a silly way, workin' about slow wi' a thing lik' that.' The Swedes left the wire on board the *Faustina*, but it lay unused by the incurious fishermen until 1930 or 1931. [103]

The feeling-wire was operated on the same principle as had been the pole which the earliest trawl-fishermen employed along the shallow bays of Lochfyne for 'feeling' herring (p. 172), but the fishermen of the twentieth century were so far removed from that practice that, with few exceptions, they rejected outright its possibilities. By 1932, however, the wire was said to have 'come into common use', and to its employment the fishermen credited as much as 75 per cent of that year's catch. [104]

How, then, was the instrument assembled? – Considering its effectiveness, with startling simplicity. To a length of snare-wire or fine-ply twine, 50 to 100 fathoms (91.44-182.88 m.) in length, was attached an ordinary lead ring; the side of a herring-box, complete with handle, served as a bearer for the twine. [105] Twine was not considered sufficiently 'keen', and was generally replaced with snare-wire which, in turn, was superseded by piano-wire.[106]

When lowered over the stern of a slow-moving boat, the wire registered the impact of fish striking it, and with experience a fisherman could estimate the depth and density of a shoal by the vibrations transmitted along the wire to his forefinger. The slowness of the operation and the frequency with which leads were lost from the end of the wire induced some fishermen to spin three-ply wire, using an ordinary speed-brace with a hooked bit,[107] and to increase the weight to 5 lbs.,[108] the mould for which was, in some cases, based on the form of a gannet egg.[109] The stronger wire and the increased weight at its end allowed the boats greater headway, and with quicker towing the tension of the wire was increased, thus increasing its sensitivity.[110]

In the operation of 'wiring about', a man stood aft of the net and, if he 'felt them thick', motioned or called to his skipper and began to wind in the wire, a frustrating task if the herring were being sought at great depths. In later years, he might clamber across the net and take up position on the port shoulder of the boat, continuing to use the wire as the net was being set, either to 'feel' the ground or to satisfy himself that the shoal was still present as the net was shot. The wire would be discarded during the 'burning fishing', and a watch maintained on the bow. In any case, the wire shed 'fire' astern of the boat, which tended to scare fish.[111]

In the midst of a dense fleet of boats wiring in the same direction, discretion might be practised when notifying a skipper of the presence of fish, to prevent an unscrupulous crew astern noticing a sudden indicative movement or hearing a cry, and shooting ahead of the crew technically and morally entitled to take the ring.[112]

Echo-Sounders

Ironically, in the initial years of the feeling-wire's adoption, an electronic device which would ultimately supersede all forms of natural fish detection, and the feeling-wire itself, had begun to be marketed. Echo-sounders, operating on the principle that, by measurement of the time-lag between the transmission and reception of sound-waves (which travel through sea-water at a constant speed of approximately 4,800 feet per second), the depth and nature of sea bottoms could be determined, were made available by both the Marconi and Kelvin Hughes companies about 1930, but were originally conceived as navigational aids. Marconi's *North Sea* model, which had been developed for fishing craft, was of the optical type, that is, it did not graphically record information. A thin beam of light was focused on a mirror which reflected it on to the scale of the sounder. The mirror tilted on transmission, deflecting the light upwards into a peak, and tilted again on reception to a point of the scale representing the depth.[113]

The first fisherman to fully appreciate the fish detection potential of echo-sounding was drift-net skipper Ronnie Balls of the *Violet and Rose* of Yarmouth. He discovered that intermediate peaks appeared on the 'spotline' of the instrument, between those marking surface and sea bed, and that these irregular peaks coincided with the presence of fish. From 1933 to 1937 he meticulously recorded the

R

association of 'spotline' peaks with subsequent catches, and his discoveries, among others, encouraged manufacturers to experiment in improving echo-sounders as fishing aids.[114]

As early as 1932 – merely a year after the feeling-wire achieved general acceptance – an echo-sounder was installed on board a Campbeltown ring-netter,

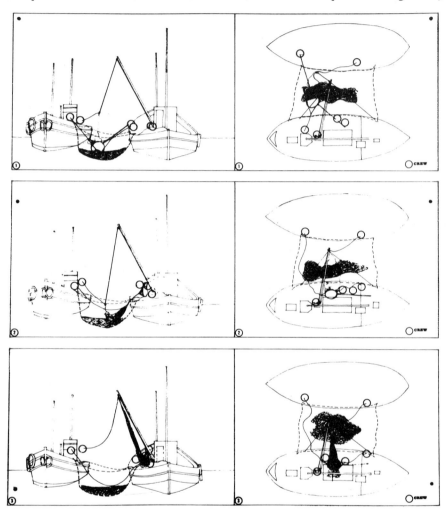

51. Boating a catch using tail-brailer

the *Nobles*, built that year in Fraserburgh. Predictably, perhaps, it was regarded sceptically by the fishermen and was evidently discarded after less than two years' trial. The skipper, James Mackay, wrote tactfully in 1933 to the manufacturers, the Marconi Company, to state that the machine had 'proved very successful so far . . . With the new outboard projector the machine can "see" herring at all speeds up to our maximum of eight or nine knots . . . In fact, we find if we do not "see" the fish upon the machine it is useless to put out the net, which is, of course, a great saving of time, labour and gear.'[115]

The Campbeltown fishery officer's report on the initial year of the experiment was brief and unmistakably dismissive of the fishermen's efforts: 'One of the best fished pairs this year carries an echo-sounding device. The crew, however, did not appear to persevere with it for spotting herrings, and do not attribute their success to its use.'[116] Like the feeling-wire, the echo-sounder was but briefly tested and, having been condemned, disappeared until proof of its value could not be ignored. As it happened, the idea of electronic fish detection quietly disappeared on the West Coast of Scotland for more than a decade.

52. Brailing herring. Hoisting the brailer on the *Ribhinn Bhàn* of Skye, 1973. Photograph by W. J. Maclean

Brailers

The five-year period, 1928 to 1933, was one of radical change in the ring-net fishery. An ideally designed winch had been manufactured, after experimentation, and quickly adopted; the value of the feeling-wire had been recognised; wide wings became a standard feature of the ring-net; and the problems of capturing herring in deep waters had been eradicated by the successful rigging of nets for sinking. By the beginning of that period, the initiative to develop the industry had shifted from Robert Robertson in Campbeltown, to Ayrshire. The young men of the Maidens, Girvan, and Dunure had returned from the Great War determined that drift-netting on that coast should be abandoned.[117] It took them only a decade to master ring-netting, and with fresh intelligence and impatient vigour to hasten the introduction of additional change.

In January, 1933, there returned to the village of Dunure from North America a fisherman, Jock Simson, not yet 32 years of age. A native of Dunure, he had gone to the herring fishing as a boy of 14 years, but, in 1924, at the age of 23, the death of his mother decided him to emigrate and to work on his brother's cattle ranch in Colorado. His discontent with ranch work was such that in the spring of 1926 he abandoned the land and travelled to Seattle, where he joined the crew of a purse-seiner preparing for the Alaskan herring fishery. Until 1933, each spring and summer was spent fishing off Alaska, and his winters he passed as a shipyard

53. Brailing herring. Emptying the brailer on the *Ribhinn Bhàn*.
Photograph by W. J. Maclean

worker in Seattle or sardine- and tuna-fishing from California and Oregon. But in January, 1933, he was back in Dunure, and before long realised that the existence of motor-winches on the ring-net boats would enable him to introduce ideas derived from his experience of the Alaskan purse-seine fishery. His main intention was to rig a brailer capable of discharging herring from the net quickly and in quantities hitherto impossible.[118]

Designs of brailers had already been experimented with, notably by Robert Robertson. In 1931 he had an aluminium-framed dipping-basket constructed, but the disadvantages of that kind of brailer were three-fold: four men were required to operate it, one at the winch and three tipping it; on breezy nights it swung dangerously; and, finally, its capacity of about a cran was inadequate.[119]

Jock Simson believed that he had the solution to the problem, and approached Dunure skippers for support. The reaction was derisive until he spoke with John 'Sharpy' Gemmell of the *Mary Sturgeon*, who was convinced of the practicability of Simson's proposals and immediately ordered the necessary materials. That original brailer consisted of a conical bag of narrow-meshed, heavy netting attached to a galvanised iron tube, bent into hoop shape, of 4 or 5 feet (1.22 or 1.52 m.) diameter, and with the projecting ends bound to an 8-foot (2.44 m.) wooden pole. A line attached to a swivel (to prevent its twisting) on the tail of the bag passed through blocks, at the head and foot of the brailing-pole, to the winch.[120]

One man dipped the hoop and mouth of the bag into the mass of fish and, when a sufficient quantity of herring was judged to have been scooped into the brailer, the

54. Brailing herring. Emptying the brailer, as photographed from the starboard side of the *Summer Rose* of Dunure, 1951. Photograph by courtesy of the *Glasgow Herald* and *Evening Times*. Per J. Gemmell, Ayr

hoop was hauled up over the rail of the boat by an attached guy. The tail of the bag was then hoisted up by winch and the fish poured out in a rush. The brailing-pole was supported by a wire back-stay and two manila side-stays. When not in use it was lashed to the lowered mainmast, but when required could be quickly swung outboard. That design of brailer was not superseded, though minor modifications were introduced.

Jock Simson's brailer was quickly adopted by Clyde and other ring-net fishermen with suitably large and decked boats. A minority of fishermen, largely of the older generation, resisted the innovation, as an anecdote from Donald McIntosh of Carradale illustrates. On the first night that his brailer was put to use, an over-ambitious fisherman from Torrisdale filled the bag entirely with herring, so that the winch could not budge it. An elderly man on the neighbouring boat, who had been sceptically observing the operation, could withhold his opinion no longer and bellowed out in his nasal voice: 'Get the basket! Get the basket! That's no use!' [121] The stick-basket continued to be carried on many boats for some years after the brailer's introduction, being more convenient for the emptying of small catches of herring from the net.

55. Discharging a catch of herring from the hold of the *Summer Rose* of Dunure, 1951. The fishermen are filling a basket and carefully picking out any mackerel mixed through the herring. The after hold 'pocket' is in the foreground. As the level of the herring is reduced, the pocket-boards will be removed one after the other until the fish are being scooped from the platform. On the left is Tommy Lawrence and on the right is William Munro. Photograph courtesy of the *Glasgow Herald* and *Evening Times*.
Per J. Gemmell, Ayr

Pursing-Rings

A bold, but not entirely successful innovation in the rigging of the net was also introduced by Jock Simson. Like the brailer, it derived from his experience of purse-seining. The device, if wholly practicable, would have enabled the sole of the ring-net to be closed rapidly, thus reducing the possibility of loss of catch. It was based on the most distinctive feature of the purse seine-net, the purse-string, a continuous line rove through a series of rings attached to the sole of the net, and which, when winched, draws the lower walls of netting together into a purse. Applied to ring-netting, the experiment, as such, failed, because the sole of the net was drawn to the side of the boat in an unmanageable lump, and so quickly that the back-rope and 'flow' netting could not be hauled at the same rate. The operation of boating the net had, therefore, to be suspended until the sole had been raised.[122]

It was suggested that in deep-water fishing the innovation would have been a constant 'winner',[123] but as ring-netting was as much an inshore method, the incidence of tearing was increased. The bunched net, when it contacted hard ground, tore more extensively,[124] and in a *thrawn* (awkward) tide the netting, when swept into the rings, caught and ripped.[125] Jock Simson himself referred also to great loss of herring out through the wide wings when the sole was winched up by

56. Discharging herring from the *Misty Isle* of Kyleakin. At Uig Pier, Skye, 1973. Photograph by W. J. Maclean

the continuous purse-string, again a consequence of the rapidity of the operation.[126]

A modified design, incorporating from two[127] to six[128] rings spliced at intervals of three or more fathoms from the centre of the sole outwards on each side ensured that the bulk of the net could be hauled inboard concurrently with the sole, and that the final stage of the operation was completed quickly and effectively. By Jock Simson's original design, each metal ring was suspended on a short bight of rope, spliced into an eye at each end, and these eyes lashed with twine to the sole-rope. Through a succession of rings the line was led.[129]

The Post-War Period

Pre-war ring-net craft had been characterised by canoe-sterns and lean quarters, but the introduction of winches allowed for the handling of deeper nets, and thus a demand was created for increased deck space. That demand, coupled with the need for greater buoyancy aft to support more powerful and, consequently, heavier engines, resulted in the stern design's undergoing a transition from canoe to cruiser. It can be stated, as a general truth, that pre-war ring-netters were of canoe-stern design, and post-war ring-netters of cruiser-stern design.[130] (A singular exception to that rule was the transom-sterned ringer-trawler, the *Prospector*, built for William Anderson of Dunure in 1973.[131])

Motor road-transport had, by the late 1940s, transformed the marketing of herring in the Clyde fishery. The 'herring-screws' had by then entirely disappeared, and ring-net fishermen working the Kilbrannan Sound and Lochfyne could land their catches at Campbeltown or Tarbert, to be sold, boxed, and transported by lorries to the city markets. On one morning in 1948, during a heavy fishery in Lochfyne, almost 100 lorries were crammed into Tarbert awaiting the arrival of the fleet.[132]

57. Boxing herring on the quayside, John Gemmell Snr. and John 'Iain' Gemmell Jnr. of the *Summer Rose* of Dunure, 1951. Photograph by courtesy of the *Glasgow Herald* and *Evening Times*. Per J. Gemmell, Ayr

In the post-war years began the decline of ring-netting in the Firth of Clyde, the Minch herring fisheries assuming greater importance to Kintyre and Ayrshire fishermen. Seine-netting for white fish had increasingly been adopted, in the 1950s, and in that period bottom-trawling for 'prawns' (nephrops) was introduced. A number of fishermen concentrated on these methods, which led to a reduction in the size of their vessels to less than 40 feet (12.19 m.), the maximum permitted for seining within the three-mile fishery limit. For most fishermen, however, until the 1960s, these alternative methods were adopted only during spring and early

58. The *Seafarer* of Campbeltown setting off for 'The North' after her annual overhaul, spring 1947. On deck, *L.-R:* Dugald Blair Jnr., Duncan 'Young Captain' Wilkinson Snr., Charlie Kelly, Edward Lafferty, James Gillies, and Duncan Wilkinson Jnr. In wheelhouse, Dugald Blair Snr. Per D. Blair, Machrihanish, Kintyre

59. The launch of the cruiser-stern ringer *Mary McLean* at St Monans, 1950, for the McArthur family of Campbeltown. The *Mary McLean,* still owned by that family, is one of the few fishing boats of her age on the West Coast of Scotland which has, in the old style, retained a varnished hull. Photograph by G. M. Cowie, St Andrews.
Per D. McArthur, Campbeltown

summer – when earlier generations would have gone to sea with lines or simply 'tied up' – and experience demonstrated that the 58 to 61 foot (17.68-18.59 m.) -long range of craft was ideal, being sufficiently large to work seine and trawl nets with greater convenience than the 40-foot boats, yet not so large that the manoeuvrability necessary for ring-netting in enclosed waters was restricted.[133]

These dual-purpose craft assumed many features which distinguished them from the original decked boats used in ring-net fishing. Motor-power increased beyond the 80 to 100 h.p. range of the immediate post-war years, to exceed 200 h.p.; from the mid-1950s, an additional winch, mounted on the foredeck for seining and trawling, was carried; the size of wheel-houses was increased to accommodate an expanding range of electronic navigational and fish-finding equipment; and, since the adoption of mid-water trawling, hydraulic 'power-blocks' – for the mechanical hauling of these nets – have been fitted to the after side of wheel-houses.[134]

By the early 1970s, mid-water trawling had overtaken ring-netting as the most productive method of herring fishing in the Firth of Clyde and Minches, but its productivity has been of brief duration. While in deep, offshore waters a more efficient method than ring-netting, it is also more indiscriminately destructive, as was at once apparent. Towed at speed between two boats, all fish within its sweep – large and small, marketable and unmarketable – are taken, and, by compression in the narrow-meshed cod-end, there perish or are injured beyond recovery. In Carradale particularly, of the Clyde fishing communities, a vigorous but futile campaign for its prohibition was conducted, the arguments advanced bearing remarkable similarity to those which had accompanied the advance of the ring-net method. And so change succeeds change; efficiency begets greater efficiency, until progressive ideals, by their very fulfilment, are all undone.

REFERENCES AND NOTES

1. *On and On Forever* (pamphlet), Kelvin Diesels Ltd., reprinted from *Motor Boat and Yachting,* 4 August, 1972. Also *F.B.R.* (1907), Campbeltown district report (vessel unnamed), and local tradition.
   A fishing boat, the aptly-named *Pioneer,* had been fitted experimentally with motor-power in 1905. D.J.T., rpt. to Fish. Bd., 8 January, 1907, A.D. 56/1467.
2. R.F.B., Campbeltown II.
3. *F.B.R.* (1908), Motor-Power Applied to Fishing Boats, xlvii.
4. A. Campbell, service manager, Kelvin Diesels Ltd., Glasgow, letter to author, 2 August, 1976.
5. *F.B.R.* (1908), *op. cit.,* xlvii.
6. *Ib.,* xlvii.
7. *Ib.,* xlviii.
8. *Ib.* (1910), xi.
9. *Ib.,* 232-7.
10. *Ib.* (1912), xiv and xvi.
11. D. McFarlane, 4 May, 1974.
12. *F.B.R.* (1912), xv.
13. *Ib.,* 231.
14. A. Munro, fishery officer at Broadford, rpts. to Fish. Bd., 15, 19, and 24 October, 1908, A.F. 56/1470.
15. H. Martin, 4 June, 1974.

16. H. Martin, 3 May, 1974; A. Martin, 10 February, 1975; J. Reid, Kyleakin, 9 February, 1974, recorded by W. Maclean.

17. *F.B.R.* (1913), 231.

18. H. Martin, 3 May, 1974.

19. As above.

20. *F.B.R.* (1913), xxi.

21. *Ib.* (1919), 61.

22. J. McWhirter, 5 February, 1976, and J. Weir, 18 March, 1976.

23. R. McGown, 8 December, 1975, and J. Weir, 18 March, 1976. The addition of a wash-board, 3 or 4 inches (0.08 or 0.10 m.) high, across the after end of the extended foredeck may also be mentioned. It served to shed water clear of the hold, and also ensured that any catch of herring remained in good condition. D. McLean, 5 December, 1974, and G. Newlands, 1 February, 1976.

24. J. M. Steven, article on the history of the Lochfyne fishing industry, 1968 (unpublished).

25. *F.B.R.* (1912), 227.

26. *Ib.* (1918), 38.

27. *Ib.* (1919), xvi.

28. *Ib.* xvii.

29. J. McIntyre, 15 July, 1976.

30. As above.

31. *Campbeltown Courier*, 2 February, 1884. All hands lost: Captain John Walker, Donald MacInnes, Hugh Robertson, Robert McCracken, and a then-unknown man shipped in Belfast.

32. Mrs Marion (McIntyre) Lyon, R. Robertson's step-sister, noted 31 January, 1979.

33. *F.B.R.* (1921), 87.

34. W. P. Miller, letter to author, 19 December, 1977.

35. R. F. B., Campbeltown II, 2. Partner in ownership of boats was Daniel Conley, who would skipper the *Frigate Bird*.

36. Review of Fishing with the *Falcon* and the *Frigate Bird* from June, 1922, dated 18 March, 1924. Elicited by Fish. Bd. Letters File, December, 1923 – March, 1926, 40. Campbeltown Fishery Office.

37. *Campbeltown Courier*, 29 April, 1922.

38. J. Miller & Sons Ltd., St Monans. The *Falcon* cost £614 14/-, and the *Frigate Bird* £663.

39. R. Robertson, Review, *op. cit.*, 39.

40. J. MacIntyre, 15 July, 1976.

41. Review, *op. cit.*, 39.

42. *Ib.*, 39-40.

43. *F.B.R.* (1918), 28.

44. *Ib.* (1923), 12.

45. Campbeltown fishery officer's report for 1937. Reports File, March, 1926 – December, 1940, Campbeltown Fishery Office.

46. G. McGhee, fishery officer at Campbeltown, rpt. appended to R. Robertson's review, *op. cit.*, 41.

47. *F.B.R.* (1924), 14.

48. Review of 1924, dated 9 February, 1925, *op. cit.*, 153.

49. *Ib.*, 152-3.

50. Review of 1925, dated 12 March, 1926, *op. cit.*, 242.

51. Notes appended to above review, *ib.*, 244.

52. Review of 1925, *op cit.*, 242.

53. *Ib.*

54. R.F.B., Campbeltown B, 42-4.

55. *Ib.*, 60.

56. W. P. Miller, letter to author, 5 January, 1978.

57. R.F.B., Campbeltown B, 67, and A, 3.

58. *Fishing News*, 11 August, 1928.

59. W. P. Miller, letter to author, 19 December, 1977.

60. *Fishing News*, *op. cit.*

61. W. P. Miller, letter to author, 5 January, 1978.

62. As above, letter 19 December, 1977.

63. W. Maclean, in communication with author.

64. Mungo Munro, Dunure, letter to author, 7 January, 1978, and R.F.B., Ayr III, 133.

65. The first was the *Mary Sturgeon,* launched in May of 1926 for the Gemmell family of Dunure. She was followed two months later by the *Golden West,* built for the McCrindle brothers of the Maidens. In 1927 were built the *Betty* (George Forbes and sons, Girvan), the *Mary Munro III* (John Grieve and sons, Dunure), the *Teresa* (Andrew Sloan and Hugh McCrindle, Girvan), and the *Margarita* (John and James McCrindle, the Maidens). In 1928 the *Britannia* (Gibson family, Dunure) and the *Nighthawk* (James, William, and John Edgar, Dunure) were launched. R.F.B., Ayr III.

66. *Ib.*

67. Campbeltown fishery officer, rpt. 18 January, 1930. Reports File, *op. cit.*

68. R.F.B., Campbeltown A.

69. *Ib.*

70. Campbeltown fishery officer's report for 1933, Reports File, *op. cit.*

71. R.F.B., Ayr IV.

72. W. Maclean, in communication with author.

73. *F.B.R.* (1926), 12.

74. T. Reid, chairman, Thomas Reid & Sons Ltd., letter to author, 9 July, 1976.

75. W. Duthie, fishery officer at Campbeltown, rpt. to Fish. Bd., 29 January, 1929, A.F. 62/2162.

76. J. T. McCrindle, 29 April and 3 May, 1976.

77. As above, 29 April, 1976.

78. J. 'Jake' McCrindle, 28 April, 1976.

79. *F.B.R.* (1928), 22.

80. 30 April, 1974.

81. W. P. Miller, letter to author, 19 December, 1977.

82. W. Duthie, *op. cit.*

83. Inveraray Returns, 29 December, 1928. Campbeltown Fishery Office.

84. W. Duthie, *op. cit.*

85. Inveraray Returns, 5 January, 1929, *op cit.*

86. J. T. McCrindle, 29 April, 1976, and T. McCrindle, 2 May, 1976.

87. J. T. McCrindle, 29 April, 1976.

88. T. Sloan and J. T. McCrindle, 29 April, 1976; J. McCreath and T. McCrindle, 2 May, 1976.

89. M. Sloan, 29 April, 1976.

90. A. Alexander, 2 May, 1976.

91. M. Sloan, 29 April, 1976; T. McCrindle and A. Alexander, 2 May, 1976.

92. 29 April, 1976.

93. T. and M. Sloan, 29 April, 1976.

94. J. McWhirter, 10 December, 1974.

95. M. Sloan, 29 April, 1976.

96. D. McLean, 5 December, 1974.

97. 29 April, 1976.

98. *F.B.R.* (1900), 269.

99. R. Morans, 26 September, 1975. Practice also referred to by J. McWhirter, 10 December, 1974. He did not, however, specify the mesh-size of the drift-nets so used.

100. T. Sloan, 29 April, 1976.

101. D. McIntosh, 27 April, 1974; R. McGown, 29 April, 1974; R. Conley, 11 June, 1975.

102. *F.B.R.* (1931), 23.

103. R. McGown, 29 April, 1974.

104. *F.B.R.* (1932), 12.

105. Duncan Blair, 1 May, 1974; J. T. McCrindle, 29 April, 1976; A. Alexander, 2 May, 1976.

106. J. 'Jake' McCrindle, 28 April, 1976, and J. T. McCrindle, 29 April, 1976.

107. J. McCreath, 3 May, 1976.

108. Duncan Blair, 1 May, 1974; J. T. McCrindle, 29 April, 1976.

109. A. Alexander, 2 May, 1976.

110. J. T. McCrindle, 29 April, 1976.

111. A. Alexander, 2 May, 1976.

112. J. McIntyre, 8 May, 1974, and K. MacRae, Portree, 19 February, 1974, recorded by W. Maclean.

113. It All Began with the *Violet and Rose* (undated pamphlet), Marconi International Marine Company Ltd.

114. *Ib.*

115. A.F. 56/1042.

116. *F.B.R.* (1932), 12.

117. J. McCreath, 2 May, 1976.

118. J. Simson, letter to author, 17 May, 1976.

119. J. McIntyre, 15 July, 1976.

120. J. Simson, letters to author, 17 May and 5 June, 1976.

121. D. McIntosh, 27 April, 1974.

122. J. 'Jake' McCrindle, 28 April, 1976.

123. D. McSporran, 30 April, 1974.

124. D. McSporran, 30 April, 1974, and J. McCreath, 2 May, 1976.

125. A. Alexander, 2 May, 1976, and J. T. McCrindle, 3 May, 1976.

126. Letter to author, 8 July, 1976.

127. J. McCreath, 2 May, 1976.

128. J. T. McCrindle, 3 May, 1976.

129. J. Simson, letter to author, 8 July, 1976.

130. W. P. Miller, letter to author, 19 December, 1977.

131. *Fishing News*, 30 November, 1973.

132. J. M. Steven, *op. cit.*

133. *Fishing News*, Noble's West Coast Ring-Netters, 31 March, 1972.

134. *Ib.*

12

Conclusion

THE journey from Glasgow to my home in Campbeltown, along the road that loops the bays of the western shore of Lochfyne, unfailingly saddens me, as though I were travelling through a country peopled by ghosts. Abandoned cottages have succumbed to ruination or, more ignominiously, surrendered their history to summer occupants; fishermen's stores overflow with weeds and litter; the stumps of net-poles stand rotting in the ebb; the croft-lands, for centuries tilled and harvested, have reverted to wildness; and with all that unhuman dissolution has receded the Gaelic people and their language, an unreturning tide. As much was foreseen, indeed foretold, by the final generations of Lochfyne drift-net fishermen, whose condemnation of 'trawling' was revived – point by point, with significant correspondence – by the Hebridean crofter-fishermen.

A terrible reality is forcing itself upon the consciousness of fishermen, that there is a natural limit to what the sea can give of its resources. This century has been one of unprecedented technological advance, of development so rapid and commanding that its effects will prove comprehensible only in bitter retrospect.

'Industrialisation' invaded the lives of fishermen relatively late. The factory-production of netting and ropes aside, the fishermen's self-sufficiency remained intact throughout the upheavals which, in the eighteenth and nineteenth centuries, swept away numerous traditional crafts and ushered in the horrors of capitalism.

The economics of fishing being governed by a compound of natural factors, varying unpredictably year by year, ensured the industry's security from external interference with the 'means of production'. And so it was left to fishermen themselves to modify or replace the design of their fishing craft, to improve fishing gear and methods, and to experiment with labour-saving devices. Remote and independent technologists did not enter significantly into the affairs of fishermen until motor-power began to be installed in fishing craft in the early years of the century. Since then the dependence of fishermen upon technological aids has increased, generation by generation, so that, in the present time, the development of the industry is prefigured in the glossy pages of technical journals.

The natural checks upon over-fishing have been steadily assaulted and broken down, fisherman and scientist advancing together on a course of unrelenting destruction. Traditional conservation principles – the trawling of immature and spawning herring especially agonised drift-net fishermen – cannot now temper that advance, because perhaps the greatest transformation of all is that which has overtaken fishermen themselves. I can perhaps do no better than quote an old Carradale fisherman, who remarked to me, with what degree of perspicacity I could not divine: 'The fishermen used to work *with* nature; now they work *against* nature.'

The preparation of this book began with a single, simple object: the documentation of the history of ring-net fishing and of the social conditions in which the generations of fishermen lived and worked. Some two years ago, however, a current of unease began to run counter to that contented plan. I realised that the final product of that wonderful flourishing of skill and knowledge has been the eradication not only of the communities of drift-net fishermen from Lochfyne to Loch Ryan, and from the Butt of Lewis to Barra, but, finally, of the ring-net fishermen themselves. The introduction of mid-water trawling to Clyde waters was but the final touch of the unrestrained hand of 'progress'.

At the time of writing (1979), the Minch herring fishery is in its second year of suspension. This season's catch quota, on the Clyde, is set at 2,000 tonnes, exactly half of the previous season's quota. The nightly allowance per fisherman alternates between one-and-a-half and two 'units' of 100 kilogrammes (220 lbs.), but many of the crews – mid-water trawlermen exclusively – have been failing consistently to realise even these comparatively low maxima. The indications are that the Clyde will have to be designated, with all other Scottish herring grounds, a 'no go' area.

This book will end, as it began, with personal comment. The most profound and influential personal lesson of these five years of questioning and gathering has been this: that Western society, by its criminal contempt for the fellow-creatures which share its corner of the planet, has brought itself to the edge of an ecological and moral crisis from which, without the exercise of immediate and unswerving restraint, there can be no withdrawal. The 'good old days' have gone, and how suddenly.

Appendix
Fishermen's Place-Names

THE examination of all place-names contained in Figs. 60, 61, and 62 has not been considered either necessary or practicable. Difficult and self-explanatory place-names have been disregarded, and analyses of major names – towns, villages, and prominent land features – already exist in published forms, wholly reliable or otherwise. Instead, emphasis has been placed on hitherto unrecorded – and, in many cases, recent, i.e. nineteenth and twentieth century – fishermen's place-names. Derivations are from Gaelic unless otherwise stated.

Main sources of information in the compilation of maps and notes: –
Tarbert: Hugh McFarlane; David McFarlane; William McDougall; Robert Ross; George Campbell Hay; F. S. Mackenna.
Carradale: Donald McIntosh; Dugald Campbell; Matthew 'Rob' McDougall; Donald 'Fergie' Paterson; Archibald Paterson; Graham McKinlay.
Campbeltown: Duncan McSporran; Henry Martin.

All source material is lodged in the archive of the School of Scottish Studies, University of Edinburgh.

The maps were made by the author.

Notes on Fig. 60, Map of Fishermen's Place-names along the Tarbert West Shore.
Airigh Fhuar: Cold Shieling. A township on hill, said to have been deserted by emigration to Canada sometime during the nineteenth century. (D. McF.) The form *Airigh Ùr*, New Shieling, given by D. McF.
Allt Beithe: Birch Burn, close to which stood a township, now ruinous.
Bàgh mu Dheas: South Bay. Also *Lùb mu Dheas.*
Bàgh mu Thuath: North Bay. Also *Lùb mu Thuath.*
Bàrr Mór: Big Ridge.
Battle Isle, The: This English name, which now prevails utterly, is a translation of *Eilean na Còmhraig.* For *còmhrag* (battle), H. McF. maintained *corrag* (finger) – i.e. *Eilean na Corraig* – offering this traditional account: – During a contest for possession of the island, a combatant sliced off one of his fingers and threw it ashore, thus securing his party's claim. The stress on the first syllable of *còmhrag* indicates, however, the correctness of that derivation.
Beebul, The: From *Am Bìobull*, The Bible, in which English form it is now generally known. A large square rock on the shore, which has, from seaward, the appearance of a closed book. (H. McF.).
Bight Lucky: The bay at Morrison's Mill, celebrating its excellence as a herring haunt. (H. McF.) Though English, the name's construction – i.e. noun preceding adjective – is Gaelic.

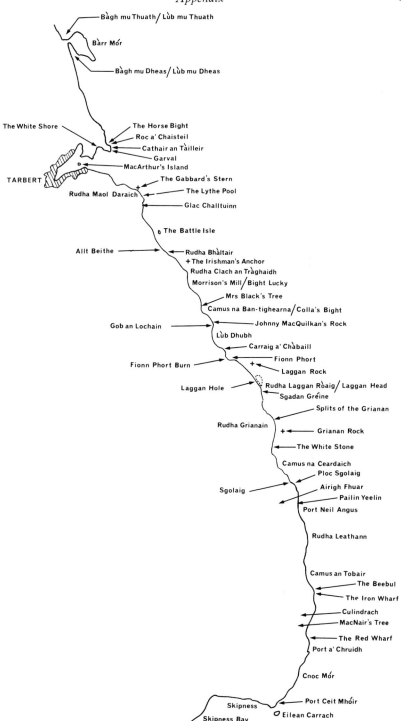

Bàgh mu Thuath / Lùb mu Thuath
Bàrr Mór
Bàgh mu Dheas / Lùb mu Dheas
The White Shore
The Horse Bight
Roc a' Chaisteil
Cathair an Tàilleir
Garval
MacArthur's Island
TARBERT
The Gabbard's Stern
The Lythe Pool
Rudha Maol Daraich
Glac Challtuinn
The Battle Isle
Allt Beithe
Rudha Bhàltair
The Irishman's Anchor
Rudha Clach an Tràghaidh
Morrison's Mill / Bight Lucky
Mrs Black's Tree
Camus na Ban-tighearna / Colla's Bight
Gob an Lochain
Johnny MacQuilkan's Rock
Lùb Dhubh
Carraig a' Chàbaill
Fionn Phort Burn
Fionn Phort
Laggan Rock
Laggan Hole
Rudha Laggan Ròaig / Laggan Head
Sgadan Gréine
Splits of the Grianan
Rudha Grianain
Grianan Rock
The White Stone
Camus na Ceardaich
Ploc Sgolaig
Airigh Fhuar
Sgolaig
Pailin Yeelin
Port Neil Angus
Rudha Leathann
Camus an Tobair
The Beebul
The Iron Wharf
Culindrach
MacNair's Tree
The Red Wharf
Port a' Chruidh
Cnoc Mór
Port Ceit Mhóir
Skipness
Eilean Carrach
Skipness Bay

60. Map of fishermen's place-names along the Tarbert West Shore

Camus na Ban-tighearna: The Lady's Bay, 'lady' = 'gentlewoman'. Also *Colla's Bight.*

Camus na Ceardaich: Bay of the Tinker/Smith. (D. McF.)

Camus an Tobair: Bay of the Well. See Index.

Carraig a Chàbaill: The Rock of the Cable. A streak of white stone in rock face, suggestive of a cable.

Cathair an Tàilleir: The Tailor's Seat, as it is now generally known. A chair-shaped rock on the shore. (H. McF.)

Cnoc Mór: Big Knoll. 'A high bit o' land.' (H. McF.)

Colla's Bight: Colla, Gaelic personal name. Undoubtedly a later fishermen's place-name. Recorded only from H. McF., but on several occasions. Also *Camus na Ban-tighearna.*

Eilean Carrach: 'The ragged island, rock shelves . . .' (H. McF.)

Fionn Phort: White Port, being a sandy bay. (H. McF.) A watering place of skiff fishermen. On what used to be a fine green there, fishermen would amuse themselves by playing 'rings', a game of trying to throw rings over a wooden peg hammered into the ground. (W. McF.) In the spoken form of the name, an intrusive 'a' occurs between the words.

Gabbard's Stern, The: A large tidal rock, shaped like the stern of a gabbard or coastal sailing-vessel. The sides fall square and then angle inward to form a keel. (H. McF.)

Glac Challtuinn: Hazel Hollow. (D. McF.)

Gob an Lochain: The Point of the Wee Loch. *The Lochan* – in itself a distinct place-name – is a small rain-water pool on the point. (H. McF.)

Irishman's Anchor, The: A submerged rock south of The Battle Isle, though the landmark for it also bears the name. (H. McF.) Gaelic form now redundant.

Iron Wharf, The: Has been dismantled for its scrap value, leaving only the cement bases in which the uprights were lodged. (W. McD.)

Johnny MacQuilkan's Rock: In Gaelic, *Creag Eonaidh* (Johnny's Rock) or *Creag Mhic Cuilgein* (MacQuilkan's Rock). See Index.

Laggan Hole: A depression, some 12 fathoms (21.95 m.) deep, in the seabed off Laggan Head, and running north for approximately 200 yards (182.88 m.). (W. McD.) As with Laggan Rock, takes its name from the once-cultivated hollow which extends from the back of the Head south along the bay. (H. McF.)

Laggan Rock: Evidently not a rock in the regular sense, but a cluster of boulders some 50 yards (45.72 m.) offshore. (W. McD.)

Lùb Dhubh: Black Bay. 'It wis dark in below the hills, it wis black for the herrin', an' the herrin' went in close.' (H. McF.)

Lùb mu Dheas: South Bay. (G.C.H.) Also *Bàgh mu Dheas.*

Lùb mu Thuath: North Bay. (G.C.H.) Also *Bàgh mu Thuath.*

Lythe Pool, The: Lythe are young pollack. Deep water off the face of Rudha Maol Daraich. (W. McD.)

MacNair's Tree: Stood close to the farm-house at Culindrach, once farmed by MacNairs. The tree was a landmark in darkness; its silhouette could be seen when the house itself was blotted out. (H. McF.)

Morrison's Mill: A burn. After heavy rain the spate foams off the hill as though descending steps, like water coming off a mill-wheel. (H. McF.)

Mrs Black's Tree: No longer stands, having been cut down many years ago to furnish knees for a punt. (D. McF., W. McD.) See Index.

Pailin Yeelin: An iron fence. Corrupt, but may have been hybrid *Pailin* (Scots, a stake-fence) *Iaruinn* (Gaelic, iron).

Ploc Sgolaig: Ploc, a round mass. Applied to a rock a short distance offshore at Sgolaig.

Port Ceit Mhóir: Big Kate's Port.

Port a' Chruidh: The Cattle Port, where cattle were shipped. (H. McF.)

Red Wharf, The: Derives from the red sandstone blocks − traditionally believed to have been quarried in Arran − with which the jetty was built. Now virtually derelict. (W. McD.)

Roc a' Chaisteil: The Castle Rock, referring to the landmark which directed hand-line fishermen to the submerged rock. When Tarbert Castle was opened on Garval Point, then the rock lay directly below. (H. McF.)

Rudha Bhàltair: Walter's Point (the regular Gaelic form of 'Walter' is *Bhàtair*). *Uamh Bhàltair* − Walter's Cave − said to have been in hill above point, but G.C.H. could not locate it.

Rudha Clach an Tràghaidh: The Ebbing Stone Point. The ebbing stone − a bare rock − lies several yards to seaward of the point. (H. McF.)

Rudha Grianain: Sunny Point. (D. McF.)

Rudha Leathann: Broad Point. (D. McF.)

Rudha Maol Daraich: Point of the Oak Headland. (D. McF.)

Sgadan Gréine: Herring of the Sun, a rock. See Index.

Sgolaig: Norse, perhaps from *skali-víkn,* 'shieling-bay'. Boats were drawn ashore and tents or huts pitched there in more recent times. (H. McF.)

Splits of the Grianan: Vertical splits in rock. A shooting mark. A ring-net would not 'run' there, and hauls would be taken south of the Splits, where the ground was clean. (H. McF.)

White Stone, The: A narrow streak of white rock some 10 feet (3.05 m.) in length. (W. McF.)

Notes on Fig. 61, Map of Fishermen's Place-names along the Shores North and South of Carradale.

Allt Creeve: Allt Craoibhe, The Tree Stream.

Ard Craig: Àrd (high) + *creag* (rock).

Balfadyen: Probably *Baile Phàidein,* Pàidean's Farm. Cf. *MacPhàidein* (MacFadyen).

Barmollach Rock: An offshore rock, the tangles on the top of which are visible with a big ebb. (D. McI.) Takes its name from the farm directly onshore.

Bennian/Bunnian Buie: A submerged offshore rock, and a ring-net 'fast'. May be *Am Binnean Buidhe,* The Little Yellow Peak, borrowed from the crest of onshore *Creag a' Chreamha* (The Wild Leek Rock), which is so named. (G. McK.)

Champion Port, The: The *Champion* was a smack wrecked there during the First World War. (D. McI.)

Crooban, The: Crùban (crab). A submerged rock, buoyed.

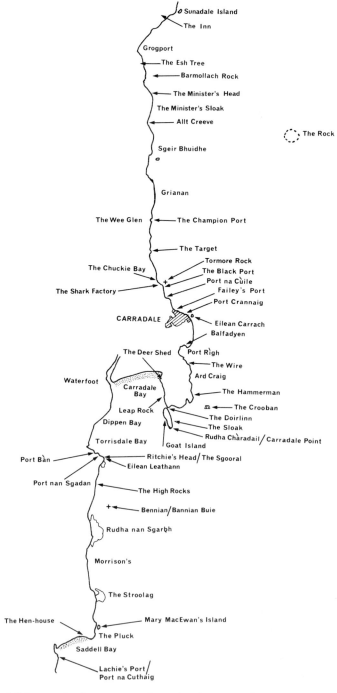

61. Map of fishermen's place-names along the shores north and south of Carradale

Doirlinn, The: The ebbing shoreline which, at low tide, connects Goat Island to the mainland.

Eilean Leathann: Broad Island. Actually a tidal rock close inshore. (D.C.)

Esh Tree, The: Esh = ash.

Failey's Port: 'Failey' was the bye-name of a crofter named Paterson who lived above the port until his death c. 1890.

Hammerman, The: Hammerman = smith. A distinctive shore rock.

Hen-house, The: Dugald Morrison's house on the shore of Saddell Bay. Said to have been introduced by Dugald himself, alluding to the smallness of the building. (D.C.)

Lachie's Port: Where lived Lachlan Galbraith, Saddell ferryman to the Glasgow steamers in the early part of the present century. Also *Port na Cuthaig.*

Leap Rock, The: When the wind was strong easterly and Carradale skiffs had to be moored in the shelter of the eastern shore of Carradale Bay, at certain stages of the tide fishermen landing from the small ferry-boats had to leap ashore from the rock. (D.P.)

Mary MacEwan's Island: Said to commemorate an unfortunate girl who, asleep after bathing in the sea, was fastened to the rock by limpets which gathered and clamped on her long hair. She drowned in the flood tide. (G. McK.)

Morrison's: The bay on the shore of which stood the now-ruined house of a family of crofter-fishermen of that name.

Port Bàn: White Bay.

Port na Cùile: Bay of the Nook.

Port na Cuthaig: The Cuckoo Port. Also *Lachie's Port.*

Port Rìgh: King's Bay. Said to commemorate a landing there by Robert Bruce. (M. McD.)

Port nan Sgadan: The Herring Port.

Ritchie's Head: The surname of a ferryman at Torrisdale who conveyed passengers and goods to the off-lying Glasgow steamers. Predominantly a Campbeltown fishermen's name. Also *The Sgooral.*

Rock, The: A submerged peak. A haunt of herring, but risky fishing ground. Appears on Admiralty charts as 'Erins Bank'.

Rudha nan Sgarbh: Point of the cormorants or shags.

Sgeir Bhuidhe: Yellow Reef. It ebbs dry and is connected at low tide to the shore.

Sgooral, The: Ordnance Survey gives *Sgorshuil.* Also *Ritchie's Head.*

Shark Factory, The: A shed in which the liver-oil of basking-sharks was processed in the immediate post-Second World War years.

Sloak, The: Sloc (hole, pit, etc.). A ring-net haul along the eastern shelf of Goat Island.

Stroolag, The: An Sruthlag, The Wee Burn. Fishermen watered there at a spring rising from the shore below high water-mark. See Index.

Target, The: The site of a rifle-range of the Carradale 'Volunteers'. (D. McI.)

Tormore Rock: Tòrr Mór, Big Hill. A submerged rock, its name borrowed from an onshore feature. (A.P.).

Wire, The: A fence.

S

Bunlarie

Allt na Beiste

Henry's

Sandy's Point

Bobby's Rest

The Hole of the ▮

Isla Ross

The Black Bay

The Carry

The Gull Rock

Ardnacross Bay

The Geelet

The Stackie Rock

The Stackie

The Smerbies

Isla Muller

Macreenyan's Point | Kilcousland

Black Rock Bay   The Light

DALINTOBER   The Ootir Buie

The Gauger's Rock

Trench Point   The Studdie   The Riddlings

The Isle Slip

CAMPBELTOWN   Pork Bay

Maggie   Donal Stott's   The Cuilleam

Baan's Hole   Broo   Davaar   Port a' Chappal

The Doirlinn   Island

The Muckle Rocks   MacVoorie's Rock   The White Ley

The Yellow Point

The Long Point

Bella's Bay   Mackinnon's Point   THE LOADEN

The Flat Rock

The Rocky Burn   Rae's

The Guns

The Bicht of the Lady

Achinhoan Heid

The Otter   The Bloody Bay

The First Water

62. Map of fishermen's place-names along the shores north and south of Campbeltown

Notes on Fig 62, Map of Fishermen's Place-names along the Shores North and South of Campbeltown.

Allt na Beiste: Stream of the Beast.

Bella's Bay: Named after Bella Bannatyne, who kept a public house in Campbeltown and whose home was by the shore of the bay.

Bicht of the Lady, The: Bicht = bight or bay.

Black Rock Bay: See Index.

Bobby's Rest: A tombstone, now sunken and overgrown, erected in memory of a favourite dog by Captain John MacGregor who lived in retirement at Ballochgair almost a century ago.

Carry, The: From *carraig* (rock). Applied to point.

Cuilleam, The: Etymology unknown. Gaelic form given solely on the ground of the awkwardness of an English spelling, which would be 'Coolyam'. A rock face.

Dalintober: Dail an Tobair, Field of the Well.

Doirlinn, The: Tidal isthmus.

Donal' Stott's Broo: Broo = brow, or slope, but nothing is now known of the person named.

First Water, The: A stream. *The Second Water,* further south, flows through Balnabraid Glen.

Flat Rock, The: See Index.

Gauger's Rock, The: A gauger, or exciseman, serving in Campbeltown during the last century, customarily bathed at that rock.

Geelet, The: The sheltered channel opposite Peninver village. Possibly English 'gullet'.

Gull Rock, The: See Index.

Henry's: A ring-net haul, the name generally considered to derive from a farmer at High Ugadale, Henry Semple.

Isla Muller: Miller's Island. A small promontory at the mouth of Smerby Burn. It is believed that the house of the miller who worked Smerby corn-mill was situated there.

Isla Ross: Eilean an Rois, The Island of the Promontory. *The Hole of the Isla Ross* was an anchorage of fishing boats.

Isle Slip, The: Stone-built jetty.

Light, The: Davaar Lighthouse.

Loaden, The: The fishing ground between Davaar Island and Achinhoan Heid. Possibly diminutive of *lod* (pool).

Mackinnon's Point: Named after a person who lived in Point House. Used principally by line-fishermen, for whom it was a fishing mark.

MacVoorie's Rock: Presumably *MacMhuirich,* in Kintyre now rendered exclusively 'MacMurchy'.

Maggie Baan's Hole: A tidal pool off St Clair Terrace, Dalintober, where, in a period remote from living memory, boats discharged cargoes.

Muckle Rocks, The: Scots muckle = great. Offshore rocks where the N.A.T.O. jetty now stands. (D. McS.)

Ootir Buie, The: An ebbing ledge of wrack-covered rock below The Light.

Port a' Chappal: According to D. McS. meaning is 'The Mare's Port', which would give Gaelic *Port a' Chapuill.* This would seem to correspond with *The Horse Bay,* so named because—according to Edward 'Teddy' Lafferty—a dead horse was mysteriously washed ashore here.

Rae's: The cottage formerly occupied by a salmon-fisherman of that name.

Riddlings, The: Scree on the face of Davaar Island. Fishermen listening for herring on still nights might hear, occasionally, stones dislodging and falling. (H.M.)

Sandy's Point: Personal name, of unknown origin. Also known as *Mecky's Point,* 'Mecky' being a south Kintyre diminutive of 'Malcolm'.

Smerbies, The: Offshore rocks, named from Smerby Farm.

Stackie, The: Rock mass on shore. From *staca.*

Stackie Rock, The: Shore rock, about 30 feet (9.14 m.) high. (D. McS.)

Studdie, The: Scots for 'anvil', which, seen from a boat entering or leaving Campbeltown Loch, the rock distinctly resembles.

White Ley, The: Ley = lay, a fishing landmark or position. The White Ley extends from mid-height on the south-facing slope of Davaar Island to the summit and was a light-coloured strip of grass. Principally a line-fishermen's mark. (D. McS.)

Index

Notes: Continually recurring references have been omitted from this index, particularly when of a general character, e.g. 'drift-net fishermen', 'drift-net boats'. Such omissions are indicated in the index, under the relevant references. Personal names unaccompanied by occupational description are generally those of fishermen.